Lecture Notes in Mathematics

Edited by A. Dold, F. Takens and B. Teissier

Editorial Policy
for the publication of monographs

1. Lecture Notes aim to report new developments in all areas of mathematics – quickly, informally and at a high level. Monograph manuscripts should be reasonably self-contained and rounded off. Thus they may, and often will, present not only results of the author but also related work by other people. They may be based on specialized lecture courses. Furthermore, the manuscripts should provide sufficient motivation, examples and applications. This clearly distinguishes Lecture Notes from journal articles or technical reports which normally are very concise. Articles intended for a journal but too long to be accepted by most journals, usually do not have this "lecture notes" character. For similar reasons it is unusual for doctoral theses to be accepted for the Lecture Notes series.

2. Manuscripts should be submitted (preferably in duplicate) either to one of the series editors or to Springer-Verlag, Heidelberg. In general, manuscripts will be sent out to 2 external referees for evaluation. If a decision cannot yet be reached on the basis of the first 2 reports, further referees may be contacted: the author will be informed of this. A final decision to publish can be made only on the basis of the complete manuscript, however a refereeing process leading to a preliminary decision can be based on a pre-final or incomplete manuscript. The strict minimum amount of material that will be considered should include a detailed outline describing the planned contents of each chapter, a bibliography and several sample chapters.
Authors should be aware that incomplete or insufficiently close to final manuscripts almost always result in longer refereeing times and nevertheless unclear referees' recommendations, making further refereeing of a final draft necessary.
Authors should also be aware that parallel submission of their manuscript to another publisher while under consideration for LNM will in general lead to immediate rejection.

3. Manuscripts should in general be submitted in English.
Final manuscripts should contain at least 100 pages of mathematical text and should include
– a table of contents;
– an informative introduction, with adequate motivation and perhaps some historical remarks: it should be accessible to a reader not intimately familiar with the topic treated;
– a subject index: as a rule this is genuinely helpful for the reader.

Lecture Notes in Mathematics 1715

Editors:
A. Dold, Heidelberg
F. Takens, Groningen
B. Teissier, Paris

Subseries: Fondazione C. I. M. E., Firenze
Adviser: Roberto Conti

Springer
Berlin
Heidelberg
New York
Barcelona
Hong Kong
London
Milan
Paris
Singapore
Tokyo

N.V. Krylov   M. Röckner   J. Zabczyk

Stochastic PDE's and Kolmogorov Equations in Infinite Dimensions

Lectures given at the 2nd Session of the
Centro Internazionale Matematico Estivo
(C.I.M.E.) held in Cetraro, Italy,
August 24 – September 1, 1998

Editor: G. Da Prato

Fondazione
C.I.M.E.

Springer

Authors

Nikolai A. Krylov
Department of Mathematics,
Computer Science & Statistics
University of Illinois at Chicago
M/C 249, 851 Morgan Street
Chicago, IL 60607, USA

Jerzy Zabczyk
Instytut Matematyczny
Polskiej Akademii Nauk
ul. Sniadeckich 8
00-950 Warszawa, Poland

Michael Röckner
Department of Mathematics
University of Bielefeld
Universitätsstrasse 25
33615 Bielefeld, Germany

Editor

Giueppe Da Prato
Scuola Normale Superiore
Piazza Cavalieri, 7
56126 Pisa, Italy

Cataloging-in-Publication Data applied for

Die Deutsche Bibliothek - CIP-Einheitsaufnahme

Stochastic PDE's and Kolmogorov equations in infinite dimensions : held in Cetraro, Italy, August 24 - September 1, 1998 / N. V. Krylov ... Ed.: G. Da Prato. - Berlin ; Heidelberg ; New York ; Barcelona ; Hong Kong ; London ; Milan ; Paris ; Singapore ; Tokyo : Springer, 1999
  (Lectures given at the ... session of the Centro Internazionale Matematico Estivo (CIME) ... ; 1998,2) (Lecture notes in mathematics ; Vol. 1715 : Subseries: Fondazione CIME)
  ISBN 3-540-66545-5

Mathematics Subject Classification (1991): 60H10, 60H15, 60G15, 31C25, 60J60

ISSN 0075-8434
ISBN 3-540-66545-5 Springer-Verlag Berlin Heidelberg New York

The use of general descriptive names, registered names, trademarks, etc. in this publication does not imply, even in the absence of a specific statement, that such names are exempt from the relevant protective laws and regulations and therefore free for general use.

Typesetting: Camera-ready TₑX output by the authors/editors
SPIN: 10700262   41/3143-543210 - Printed on acid-free paper

Preface

Kolmogorov equations are second order parabolic equations with a finite or an infinite number of variables. They are deeply connected with stochastic differential equations in finite or infinite dimensional spaces. They arise in many fields as Mathematical Physics, Chemistry and Mathematical Finance.

These equations can be studied both by probabilistic and by analytic methods, using such tools as Gaussian measures, Dirichlet Forms, and stochastic calculus.

Three courses, of eight hours each, have been delivered. N. V. Krylov of the University of di Minnesota presented Kolmogorov equations coming from finite-dimensional equations, giving existence, uniqueness and regularity results.

M. Röckner has presented an approach to Kolmogorov equations in infinite dimensions which is based on an L^p–analysis of the corresponding diffusion operators with respect to suitably chosen measures, extending the classical Dirichlet form approach.

Finally, J. Zabczyk, started from classical results of L. Gross, on the heat equation in infinite dimension, and discussed some recent results, including applications to control theory and mathematical finance.

Afternoon sessions have been devoted to research seminars delivered by the participants.

I wish to thank the lecturers and all the participants for their contribution to the success of the School. I thank the CIME scientific committee for giving me the opportunity to organize the Meeting and the CIME staff for their continuous help.

Scuola Normale Superiore (Pisa)　　　　　　　　Giuseppe Da Prato
March 1999

Table of Contents

On Kolmogorov's equations for finite dimensional diffusions

N.V. Krylov

School of Mathematics, University of Minnesota, Minneapolis, MN, 55455

The purpose of these lectures is to present a more or less self contained exposition of Kolmogorov's equations for finite dimensional diffusion processes which are treated as solutions of Itô's equations. Kolmogorov's equations provide probabilistic solutions for elliptic and parabolic second order partial differential equations. On the one hand, this allows one to prove the solvability of such equations and, on the other hand, quite often one can get a substantial information about probabilistic quantities by solving these equations.

We start with proving the solvability of Itô's stochastic equations with random coefficients. This is done by using Euler's method, which turns out to be very powerful in many situations. Then we mainly concentrate on equations with nonrandom coefficients and prove the Markov property of solutions. Next step is deriving Kolmogorov's equations under the assumption that the probabilistic solutions are smooth functions of initial data. After this we give sufficient conditions for the smoothness in the case of equations in the whole space. Final sections are devoted to Kolmogorov's equations in domains when the probabilistic solutions are not necessarily continuous let alone differentiable. In this case Kolmogorov's equations are understood in the sense of generalized functions.

All material of these lectures, perhaps apart from the last sections, is classical and can be found in many articles and books. For this reason we only cite the original article [KO] by Kolmogorov himself which opened up the new chapter in probability theory we will be talking about.

It is a great pleasure to thank Director of CIME R. Conti and Scientific Director of the CIME Session on Stochastic PDE's and Kolmogorov Equations in Infinite Dimensions G. Da Prato for inviting me to deliver these lectures and for excellent organization of the session in such an exquisite place as Cetraro (Cosenza).

1. Solvability of Itô's stochastic equations

Let $(\Omega, \mathcal{F}, P)$ be a complete probability space and let $(w_t, \mathcal{F}_t)$ be a d_1-dimensional Wiener process on this space defined for $t \in [0, \infty)$, with σ-algebras $\mathcal{F}_t$ being complete with respect to $\mathcal{F}, P$. Assume that, for any $\omega \in \Omega$,

1991 *Mathematics Subject Classification.* 58G32,60J60,60J65
Key words and phrases. Stochastic equations, Kolmogorov's equations
The work was partially supported by NSF Grant DMS–9625483

$t \geq 0$, and $x \in \mathbb{R}^d$, we are given a $d \times d_1$-dimensional matrix $\sigma(t, x)$ and a d-vector $b(t, x)$. We assume that σ and b are continuous in x for any ω, t, measurable in (ω, t) for any x, and $\mathcal{F}_t$-measurable in ω for any t and x. We also assume that, for any finite T and R and $\omega \in \Omega$, we have

$$\int_0^T \sup_{|x| \leq R} \{\|\sigma(t, x)\|^2 + |b(t, x)|\} \, dt < \infty. \tag{1.1}$$

Notice that, for a matrix σ, by $\|\sigma\|^2$ we mean $\sum_{ij} |\sigma^{ij}|^2$.

Remark 1.1. It is shown in [KR1] that, under the monotonicity condition (1.2) and the coercivity condition (1.3) below, condition (1.1) is satisfied if

$$\int_0^T \{\|\sigma(t, x)\|^2 + |b(t, x)|\} \, dt < \infty \quad \forall x \in \mathbb{R}^d.$$

Theorem 1.2. *In addition to the above assumptions, for all R, $t \in [0, \infty)$, $|x|, |y| \leq R$, and ω, let*

$$2(x - y, b(t, x) - b(t, y)) + \|\sigma(t, x) - \sigma(t, y)\|^2 \leq K_t(R)|x - y|^2, \tag{1.2}$$

$$2(x, b(t, x)) + \|\sigma(t, x)\|^2 \leq K_t(1)(1 + |x|^2), \tag{1.3}$$

where $K_t(R)$ are certain $\mathcal{F}_t$-adapted nonnegative processes satisfying

$$\alpha_T(R) := \int_0^T K_t(R) \, dt < \infty$$

for all $R, T \in [0, \infty)$, and ω. Also let x_0 be an $\mathcal{F}_0$-measurable d-dimensional vector. Then Itô's equation

$$dx_t = \sigma(t, x_t) \, dw_t + b(t, x_t) \, dt, \quad t \geq 0 \tag{1.4}$$

with initial condition x_0 has a solution, which is, moreover, unique up to indistinguishability.

Remark 1.3. It follows from inequalities like $|b(t, x)| \leq |b(t, 0)| + |b(t, x) - b(t, 0)|$ that assumptions (1.1), (1.2), and (1.3) are satisfied if

$$\int_0^T \left[\|\sigma(t, 0)\|^2 + |b(t, 0)|\right] dt < \infty$$

and the Lipschitz condition

$$\|\sigma(t, x) - \sigma(t, y)\| + |b(t, x) - b(t, y)| \leq K|x - y|$$

holds for all $t \in [0, T]$ and $x, y \in \mathbb{R}^d$, where K is a constant. However (1.2) and (1.3) are also satisfied if, for instance, $d = 1$, $\sigma \equiv 0$, and $b(t, x)$ is *any* decreasing function of x equal zero at $x = 0$.

Before proving the theorem we deduce two lemmas.

Lemma 1.4. *Let y_t be a continuous nonnegative $\mathcal{F}_t$-adapted process, γ a finite stopping time, N a constant, and assume $Ey_\tau \leq N$ for any stopping time $\tau \leq \gamma$. Then*

$$P\{\sup_{t\leq\gamma} y_t \geq \varepsilon\} \leq N/\varepsilon$$

for any $\varepsilon > 0$.

For the proof, it suffices to take $\tau = \gamma \wedge \inf\{t \geq 0 : y_t \geq \varepsilon\}$, to notice that $\tau \leq \gamma$ and

$$\{\omega : \sup_{t\leq\gamma} y_t \geq \varepsilon\} = \{\omega : y_\tau \geq \varepsilon\},$$

and to use Chebyshev's inequality.

Lemma 1.5. *Suppose that, for $n = 1, 2, ...$, we are given continuous d-dimensional $\mathcal{F}_t$-adapted processes x_t^n on $\Omega \times [0, \infty)$ such that $x_0^n = x_0$ and*

$$dx_t^n = \sigma(t, x_t^n + p_t^n)\, dw_t + b(t, x_t^n + p_t^n)\, dt \quad t \geq 0$$

for some $\mathcal{F}_t$-adapted processes p_t^n which are measurable in (ω, t). For $n = 1, 2, ...$ and $R \geq 0$, let $\tau^n(R)$ be stopping times such that
(i) $|x_t^n| + |p_t^n| \leq R$ if $0 < t \leq \tau^n(R)$,
(ii)

$$\lim_{n\to\infty} E \int_0^{T\wedge\tau^n(R)} |p_t^n|\, dt = 0 \quad \forall R, T \in [0, \infty), \tag{1.5}$$

(iii) there exists a nonrandom function $r(R)$ such that $r(R) \to \infty$ as $R \to \infty$ and

$$\lim_{R\to\infty} \overline{\lim_{n\to\infty}} \, P\{\tau^n(R) \leq T, \sup_{t\leq\tau^n(R)} |x_t^n| \leq r(R)\} = 0 \quad \forall T \in [0, \infty). \tag{1.6}$$

Then, for any $T \in [0, \infty)$, we have

$$\sup_{t\leq T} |x_t^n - x_t^m| \overset{P}{\to} 0 \tag{1.7}$$

as $n, m \to \infty$.

Proof. Without loss of generality (see (1.1)), we assume that $|b(t, x)| \leq K_t(R)$ for $|x| \leq R$. Furthermore, it is easy to find stopping times $\tau(R)$ such that $\tau(R) \overset{P}{\to} \infty$ as $R \to \infty$ and $\alpha_{t\wedge\tau(R)}(R)$ is a bounded process for any $R > 0$. For instance, it suffices to take

$$\tau(R, u) := \inf\{t \geq 0 : \alpha_t(R) \geq u\},$$

to observe that, obviously, $P\{\tau(R, u) \leq R\} \to 0$ as $u \to \infty$, to find $u(R)$ so that

$$P\{\tau(R, u(R)) \leq R\} \leq 1/R,$$

and to put $\tau(R) = \tau(R, u(R))$. For such $\tau(R)$, the transition from $\tau^n(R)$ to $\tau^n(R) \wedge \tau(R)$ preserves all the assumptions of the lemma and, therefore, without loss of generality, we may and will assume that $\alpha_t(R) \leq u(R)$ for $t \leq \tau^n(R)$, where $u(R)$ are some finite constants.

Observe that now

$$\lim_{n \to \infty} E \int_0^{T \wedge \tau^n(R)} |p_t^n| K_t(R)\, dt = 0 \quad \forall R, T \in [0, \infty). \qquad (1.8)$$

Indeed, to see this, we merely split the integrand in (1.8) into a sum using

$$1 = I_{K_t(R) \geq i} + I_{K_t(R) < i}$$

and then, using (1.5) and the inequalities $\alpha_t(R) \leq u(R)$ and $|p_t^n| \leq R$ for $t \leq \tau^n(R)$, we let first $n \to \infty$ and then $i \to \infty$, relying upon the dominated convergence theorem in the second passage to the limit.

Next, fix n, m, and R temporarily and put

$$x_t = x_t^n, \quad y_t = x_t^m, \quad p_t = p_t^n, \quad q_t = p_t^m, \quad \psi_t = \psi_t(R) = \exp(-2\alpha_t(R) - |x_0|).$$

According to Itô's formula,

$$d(|x_t - y_t|^2 \psi_t) = \psi_t \left[2(x_t - y_t, b(t, x_t + p_t) - b(t, y_t + q_t)) \right.$$
$$\left. + \|\sigma(t, x_t + p_t) - \sigma(t, y_t + q_t)\|^2 - 2K_t(R)|x_t - y_t|^2 \right] dt + d\beta_t,$$

where $\beta_t = \beta_t^{nm}$ is certain local martingale with $\beta_0 = 0$. To transform the right-hand side, we represent $x_t - y_t$ as $(x_t + p_t) - (y_t + q_t) - p_t + q_t$. Then we use the monotonicity condition (1.2) and the inequalities

$$2|x_t - y_t|^2 \geq |(x_t + p_t) - (y_t + q_t)|^2 - 2|p_t - q_t|^2, \quad |p_t - q_t|^2 \leq 2R(|p_t| + |q_t|),$$

the last of which holds at least for $0 < t \leq \gamma^{nm}(R) := \tau^n(R) \wedge \tau^m(R)$. Finally, we also use $|b(t, x)| \leq K_t(R)$ for $|x| \leq R$ and find that

$$|x_t^n - x_t^m|^2 \psi_t(R) \leq \lambda_t^n(R) + \lambda_t^m(R) + \beta_t^{nm}(R) \quad t \leq \gamma^{nm}(R), \qquad (1.9)$$

where

$$\lambda_t^k(R) := 4(R + 1) \int_0^t |p_s^k| K_s(R)\, ds.$$

From (1.9) it follows at once that, for any $R, T \in [0, \infty)$ and any stopping time $\tau \leq T \wedge \gamma^{nm}(R)$,

$$E|x_\tau^n - x_\tau^m|^2 \psi_\tau(R) \leq E\lambda_{T \wedge \tau^n(R)}^n(R) + E\lambda_{T \wedge \tau^m(R)}^m(R). \qquad (1.10)$$

Owing to (1.8), the right-hand side in (1.10) tends to zero, and since τ is arbitrary, by Lemma 1.4 we get

$$\sup_{t \leq \gamma^{nm}(R) \wedge T} |x_t^n - x_t^m|^2 \psi_t(R) \xrightarrow{P} 0$$

as $n, m \to \infty$. The factor $\psi_t(R)$ can be dropped since it is positive and independent of n, m. We now see that, to prove (1.7), it only remains to show that

$$\lim_{R \to \infty} \overline{\lim_{n \to \infty}} \, P\{\tau^n(R) \le T\} = 0 \quad \forall T \in [0, \infty). \tag{1.11}$$

So far we have not used the growth condition (1.3). By using this condition and assuming $K_t(R) \ge K_t(1)$ for $R \ge 1$, which does not restrict generality, similarly to (1.9) we get that, for $t \le \tau^n(R)$,

$$(1 + |x_t^n|^2)\psi_t(1) \le (1 + |x_0|^2)e^{-|x_0|} + \lambda_t^n(R) + \beta_t^n,$$

where β_t is a local martingale starting at zero. Then similarly to (1.10) we obtain that, for any stopping time $\tau \le T \wedge \tau^n(R)$,

$$E|x_\tau^n|^2 \psi_\tau(1) \le N + E\lambda_{T \wedge \tau^n(R)}^n(R),$$

where N is the least upper bound of $(1 + |x|^2)e^{-|x|}$. From the above stated properties of $\lambda_t^n(R)$ and from Lemma 1.4 we further obtain that

$$\lim_{c \to \infty} \sup_{R \ge 0} \overline{\lim_{n \to \infty}} \, P\{ \sup_{t \le T \wedge \tau^n(R)} |x_t^n|^2 \psi_t(1) \ge c \} = 0.$$

Again since $\psi_t(1)$ is independent of n and R, this term can be dropped. It follows that, as $R \to \infty$,

$$\overline{\lim_{n \to \infty}} \, P\{ \sup_{t \le \tau^n(R)} |x_t^n| \ge r(R), \tau^n(R) \le T \} = 0.$$

Combining this with the assumption (1.6) we get (1.11). The lemma is proved.

Remark 1.6. In the above proof we did not use the continuity of σ and b in x.

Proof of Theorem 1.2. We are going to use Euler's method. Define the processes x_t^n so that $x_0^n = x_0$ and

$$dx_t^n = \sigma(t, x_{k/n}^n) \, dw_t + b(t, x_{k/n}^n) \, dt$$

for $t \in [k/n, (k+1)/n]$. Clearly x_t^n satisfy

$$dx_t^n = \sigma(t, x_{\kappa(n,t)}^n) \, dw_t + b(t, x_{\kappa(n,t)}^n) \, dt, \tag{1.12}$$

$$dx_t^n = \sigma(t, x_t^n + p_t^n) \, dw_t + b(t, x_t^n + p_t^n) \, dt,$$

where $\kappa(n, t) = [tn]/n$, $p_t^n := x_{\kappa(n,t)}^n - x_t^n$. Also define $\tau^n(R)$ as the first exit time of x_t^n from the ball $B_{R/3} = \{x \in \mathbb{R}^d : |x| < R/3\}$, and let $r(R) := R/4$. Then, for $0 < t \le \tau^n(R)$, we have $|p_t^n| \le 2R/3$, $|x_t^n| \le R/3$. In particular, the event entering (1.6) is empty and this condition is satisfied. Moreover,

$$-p_t^n = \int_{\kappa(n,t)}^t \sigma(s, x_{\kappa(n,t)}^n) \, dw_s + \int_{\kappa(n,t)}^t b(s, x_{\kappa(n,t)}^n) \, ds.$$

Hence according to well known inequalities for stochastic integrals, for any $\varepsilon, \delta > 0$, we have

$$P\{|p_t^n| \geq 2\varepsilon, t \leq \tau^n(R)\} \leq P\{\int_{\kappa(n,t)}^t \sup_{|x| \leq R} |b(s,x)| \, ds \geq \varepsilon\}$$

$$+ P\{\int_{\kappa(n,t)}^t \sup_{|x| \leq R} ||\sigma(s,x)||^2 \, ds \geq \delta\} + \delta/\varepsilon^2. \quad (1.13)$$

In view of (1.1), this shows that the left-hand side goes to zero as $n \to \infty$. In other words, the product of $|p_t^n|$ and the indicator of $\{t \leq \tau^n(R)\}$ tends to zero in probability. Since this product is less than $2R/3$, its expectation is bounded in t, n and tends to zero as $n \to \infty$. An application of Fubini's theorem and the dominated convergence theorem proves that (1.5) holds.

Now by Lemma 1.5 and the completeness of the space of processes with respect to the uniform convergence in probability, we get that there exists a continuous process x_t such that

$$\sup_{t \leq T} |x_t^n - x_t| \xrightarrow{P} 0 \quad \forall T \in [0, \infty).$$

Here one may replace t in x_t^n and x_t with $\kappa(n,t)$. Moreover, one can do this only for x_t^n since x_t is continuous in t. Then, using the continuity of σ and b in x and passing to the limit in (1.12), we see that x_t is indeed a solution of equation (1.4) with the initial condition x_0. Its uniqueness and even continuous dependence on the initial data we prove in the following theorem. The theorem is proved.

Theorem 1.7. *Let the assumptions of Theorem 1.2 apart from assumption (1.3) be satisfied. Let x_0, x_0^n be $\mathbb{R}^d$-valued $\mathcal{F}_0$-measurable vectors such that $P\text{-}\lim_{n \to \infty} x_0^n = x_0$. Assume that equation (1.4) has solutions x_t and x_t^n with initial data x_0 and x_0^n, respectively. Then*

$$P\text{-} \lim_{n \to \infty} \sup_{t \leq T} |x_t^n - x_t| = 0 \quad \forall T \in [0, \infty). \quad (1.14)$$

Proof. Without loss of generality, we may and will assume that $x_0^n \to x_0$ almost surely. Then

$$\phi_t(R) := \exp(-\alpha_t(R) - \sup_n |x_0^n|)$$

satisfies $\phi_t > 0$. Next, for

$$\gamma^n(R) = \inf\{t \geq 0 : |x_t^n| + |x_t| \geq R\} \wedge T$$

as in the proof of Lemma 1.5 we find that

$$|x_{t \wedge \gamma^n(R)}^n - x_{t \wedge \gamma^n(R)}|^2 \phi_{t \wedge \gamma^n(R)}(R) \leq |x_0^n - x_0|^2 \phi_0(R) + m_t^n(R),$$

where $m_t^n(R)$ is a local martingale starting at zero. Since $|x_0^n - x_0| \to 0$ in probability, we have $E|x_0^n - x_0|^2 \phi_0(R) \to 0$, and Lemma 1.4 implies that

$$P\text{-}\lim_{n\to\infty} \sup_{t\leq T} |x_{t\wedge\gamma^n(R)}^n - x_{t\wedge\gamma^n(R)}|^2 \exp(-\phi_{t\wedge\gamma^n(R)}(R)) = 0,$$

$$P\text{-}\lim_{n\to\infty} \sup_{t\leq T} |x_{t\wedge\gamma^n(R)}^n - x_{t\wedge\gamma^n(R)}| = 0 \qquad (1.15)$$

for any finite T, R. Add to this that

$$P\{\gamma^n(R) < T\} \leq P\{\sup_{t\leq T}(|x_{t\wedge\gamma^n(R)}^n| + |x_{t\wedge\gamma^n(R)}|) \geq R\}$$

$$\leq P\{\sup_{t\leq T} |x_{t\wedge\gamma^n(R)}^n - x_{t\wedge\gamma^n(R)}| \geq 1\} + P\{2\sup_{t\leq T} |x_t| \geq R - 1\}.$$

Then from (1.15) we get

$$\lim_{R\to\infty} \overline{\lim_{n\to\infty}} P\{\gamma^n(R) < T\} = 0,$$

which along with (1.15) implies (1.14). The theorem is proved.

Above we have only used Lemma 1.5 when the probability in (1.6) is zero. Later on we will need the following result in the proof of which the verification of (1.6) is not that trivial.

Theorem 1.8. *Let the assumptions of Theorem 1.2 be satisfied. Let $\sigma^n(t,x)$ and $b^n(t,x)$ be some functions satisfying the conditions imposed on σ and b before Remark 1.1 and let processes y_t^n be such that $y_0^n \to x_0$ in probability and*

$$dy_t^n = \sigma^n(t, y_t^n)\, dw_t + b^n(t, y_t^n)\, dt \quad t \geq 0. \qquad (1.16)$$

Define

$$\gamma^n(R) = \inf\{t \geq 0 : |y_t^n| \geq R\}$$

and assume that, for all $R, T \in [0, \infty)$, we have

$$\int_0^{T\wedge\gamma^n(R)} [\|\sigma^n - \sigma\|^2 + |b^n - b|] (t, y_t^n)\, dt \xrightarrow{P} 0. \qquad (1.17)$$

Then $y_t^n \to x_t$ uniformly on $[0, T]$ in probability for any $T \in [0, \infty)$, where x_t is the process from Theorem 1.2.

Proof. Let $x_t^{2n} = y_t^n - p_t^{2n}$, where

$$p_t^{2n} = \int_0^t (\sigma^n - \sigma)(s, y_s^n)\, dw_s + \int_0^t (b^n - b)(s, y_s^n)\, ds - y_0^n + x_0,$$

and let $x_t^{2n+1} = x_t$, $p_t^{2n+1} = 0$. Also let $\tau^n(R)$ be the first exit time of $|x_t^n| + |p_t^n|$ from $(-1, R)$ and $r(R) = R/2$. Clearly $\tau^{2n}(R) \leq \gamma^n(R)$ and $|p_t^n|, |y_t^n| \leq R$ for $0 < t \leq \tau^{2n}(R)$. From this and (1.17) similarly to (1.13), we obtain that, as $n \to \infty$,

$$\mu^n(T,R) := \sup_{t \leq T} |p_t^n| I_{0 \leq t \leq \tau^n(R)} \xrightarrow{P} 0.$$

This certainly yields (1.5) and since the probability in (1.6) is less than $P\{\mu^n(T,R) \geq R/2\}$, condition (1.6) is fulfilled as well. Other conditions of Lemma 1.5 are also satisfied and we conclude that $x_t^{2n} = y_t^n - p_t^{2n} \to x_t$ uniformly on $[0,T]$ in probability for any $T \in [0,\infty)$. It only remains to notice that, for $R, \varepsilon > 0$, we have

$$P\{\sup_{t \leq T} |p_t^{2n}| \geq \varepsilon\} \leq P\{\mu^{2n}(T,R) \geq \varepsilon\} + P\{\tau^{2n}(R) \leq T\}$$

$$\leq 2P\{\mu^{2n}(T,R) \geq \varepsilon\} + P\{\sup_{t \leq T} |x_t^{2n}| \geq R - \varepsilon\},$$

which after letting first $n \to \infty$ and then $R \to \infty$ implies that $p_t^{2n} \to 0$ uniformly on $[0,T]$ in probability for any $T \in [0,\infty)$. The theorem is proved.

Exercise 1.9. Prove that for $d = 1$, the equation $x_t = -\int_0^t x_s^3 \, ds + w_t$ has a unique solution defined for all times.

Exercise 1.10. Prove that for $d = 1$, the equation $x_t = \int_0^t x_s^3 \, ds + w_t$ has a unique solution defined before it blows up to ∞ or $-\infty$.

2. Markov property of solutions

We saw that under some conditions equation (1.4) with random coefficients can be solved by using Euler's method. It turns out that it is useful to formalize this property of equations.

In this section we assume that the coefficients σ and b of our equations are nonrandom functions satisfying the conditions stated before Remark 1.1.

2.1 Regular equations

We will be dealing with the equation

$$x_t = x + \int_0^t \sigma(s+r, x_r) \, dw_r + \int_0^t b(s+r, x_r) \, dr \quad t \geq 0, \qquad (2.1)$$

where s and x are some nonrandom data.

To introduce general Euler's approximations, let us call $I = \{t(k)\}_{k=0}^\infty$ a *partition* of $[0,\infty)$ if $t(0) = 0$, $t(k) \uparrow \infty$ and $\Delta(I) := \sup_{k \geq 0}(t(k+1) - t(k)) < \infty$. For every partition $I = \{t(k)\}$ of $[0,\infty)$ we define *Euler's approximation* for (2.1) recurrently as a process x_t, denoted by $x_t^{s,x}(I)$, such that $x_0 = x$,

$$x_t = x_{t(k)} + \int_{t(k)}^t \sigma(s+r, x_{t(k)}) \, dw_r + \int_{t(k)}^t b(s+r, x_{t(k)}) \, dr \qquad (2.2)$$

if $t \in [t(k), t(k+1)]$, $k = 0, 1, 2, \ldots$. It is useful to note that $x_t^{s,x}(I)$ is well defined under the above hypotheses and satisfies the equation

$$x_t = x + \int_0^t \sigma(s + r, x_{\kappa(r)})\, dw_r + \int_0^t b(s + r, x_{\kappa(r)})\, dr, \qquad (2.3)$$

where $\kappa(r) = t(k)$ for $r \in [t(k), t(k+1))$, $k = 0, 1, 2, \ldots$. Denote $C = C([0, \infty), \mathbb{R}^d)$ the space of all continuous $\mathbb{R}^d$-valued functions with metric

$$|x_\cdot - y_\cdot|_C = \sum_{n=1}^{\infty} 2^{-n} \arctan \sup_{t \le n} |x_t - y_t|$$

under which C is a complete separable metric space. Observe that $x_\cdot^n \to x_\cdot$ in C if and only if $\sup_{t \le T} |x_t^n - x_t| \to 0$ for every $T < \infty$.

Definition 2.1. Suppose that for every $s \ge 0$ and $x \in \mathbb{R}^d$ equation (2.1) has a unique solution on $[0, T]$ for every $T \in (0, \infty)$. Denote this solution by $x_t^{s,x}$. Moreover, suppose that for every $R \in [0, \infty)$ as $\Delta(I) \to 0$, we have $x_\cdot^{s,x}(I) \to x_\cdot^{s,x}$ (convergence in C) in probability, uniformly with respect to $s \in [0, T]$ and $x \in B_R$, i.e. for every $R, T \in [0, \infty)$, and $\varepsilon > 0$

$$\sup_{s \in [0,T]} \sup_{|x| \le R} P\{ \sup_{t \in [0,T]} |x_t^{s,x} - x_t^{s,x}(I)| \ge \varepsilon \} \to 0. \qquad (2.4)$$

Then we say that equation (2.1) and its solution $x_t^{s,x}$ are *regular*.

Sufficient conditions for the regularity are given in Sec. 5.2. Regular processes have many nice properties. In this subsection we prove only one of them.

To save some space we will put indices s, x at the symbols of probability and expectation meaning that they should be put inside at appropriate places. For instance (2.4) will now read

$$\sup_{s \in [0,T]} \sup_{|x| \le R} P_{s,x}\{ \sup_{t \in [0,T]} |x_t - x_t(I)| \ge \varepsilon \} \to 0.$$

Lemma 2.2. *Suppose that equation (2.1) is regular. Then for every $T, R \ge 0$, the processes $x_t^{s,x}$ and $x_t^{s,x}(I)$ are uniformly in $t \in [0, T]$ bounded in probability, uniformly with respect to $s \in [0, T]$, $|x| \le R$, and partition I, i.e.*

$$\lim_{c \to \infty} \sup_{s \in [0,T]} \sup_{|x| \le R} \sup_I P_{s,x}\{ \sup_{t \in [0,T]} |x_t(I)| \ge c \} = 0, \qquad (2.5)$$

$$\lim_{c \to \infty} \sup_{s \in [0,T]} \sup_{|x| \le R} P_{s,x}\{ \sup_{t \in [0,T]} |x_t| \ge c \} = 0. \qquad (2.6)$$

Proof. Take a partition $I = \{t(k)\}$ and use (2.2). Then for any increasing sequence of constants $c_k \geq 0$ and $x_t := x_t^{s,x}(I)$ by Doob's inequality we get

$$E \sup_{[t(k),t(k+1)]} |x_t|^2 I_{|x_{t(k)}| \leq c_k} \leq$$

$$\leq 3c_k^2 + 3E \sup_{[t(k),t(k+1)]} |\int_{t(k)}^t I_{|x_{t(k)}| \leq c_k} \sigma(s + r, x_{t(k)}) \, dw_r|^2$$

$$+ 3E(\int_{t(k)}^{t(k+1)} |b(s + r, x_{t(k)})| \, dr)^2 I_{|x_{t(k)}| \leq c_k}$$

$$\leq 3c_k^2 + 12E \int_{t(k)}^{t(k+1)} \|\sigma(s+r, x_{t(k)})\|^2 \, dr I_{|x_{t(k)}| \leq c_k} + 3(\int_{t(k)}^{t(k+1)} \sup_{|x| \leq c_k} |b(s+r,x)| \, dr)^2$$

$$\leq 3c_k^2 + 12 \int_{t(k)}^{t(k+1)} \sup_{|x| \leq c_k} \|\sigma(s + r, x)\|^2 \, dr + 3(\int_{t(k)}^{t(k+1)} \sup_{|x| \leq c_k} |b(s + r, x)| \, dr)^2$$

$$=: \xi(c_k, s + t(k), s + t(k + 1)) \leq \xi(c_k, 0, T + t(k + 1)) =: \eta(c_k, t(k + 1)).$$
$$(2.7)$$

By assumption (1.1) the function $\eta(c_k, t(k + 1))$ is finite, and by Chebyshev's inequality

$$P\{\sup_{t \leq t(k+1)} |x_t| \geq c_{k+1}\} \leq$$

$$\leq P\{\sup_{t \leq t(k)} |x_t| \geq c_k\} + P\{\sup_{[t(k),t(k+1)]} |x_t|^2 I_{|x_{t(k)}| \leq c_k} \geq c_{k+1}\}$$

$$\leq P\{\sup_{t \leq t(k)} |x_t| \geq c_k\} + c_{k+1}^{-2} \eta(c_k, t(k + 1)). \quad (2.8)$$

By induction we conclude that for every $T, R, \delta > 0$

$$P_{s,x}\{\sup_{t \leq T} |x_t(I)| \geq c\} \to 0 \qquad (2.9)$$

as $c \to \infty$ uniformly with respect to $s \leq T, |x| \leq R$, and I such that $\Delta(I) \geq \delta$. Furthermore, condition (2.4) implies that for every $\gamma > 0$ we can find $\delta(\gamma) > 0$ such that if $\Delta(I) \leq \delta(\gamma)$, then

$$P_{s,x}\{\sup_{t \leq T} |x_t - x_t(I)| \geq 1\} \leq \gamma \qquad (2.10)$$

for all $s \leq T, |x| \leq R$. This fact and the fact that (2.9) holds uniformly with respect to $s \leq T, |x| \leq R$ for *any* I immediately prove (2.6).

We also see from (2.9) that the $\sup_I$ in (2.5) can be replaced by sup over $\{I : \Delta(I) \leq \delta(\gamma)\}$ without changing the left–hand side of (2.5). By (2.10) and (2.6) this side of (2.5) is less than γ. Since γ is arbitrary, the lemma is proved.

2.2 Some properties of Euler's approximations

The first purpose of this subsection is to investigate the continuity of Euler's approximations $x_t^{s,x}(I)$ with respect to x for any fixed partition I. This is an easy task since Euler's approximations are defined by induction and we get the continuity on $[t(k), t(k+1)]$ from the explicit formula for $x_t^{s,x}(I)$ expressing it as a function of $x_{t(k)}^{s,x}(I)$.

We are also dealing with conditional expectations of $g(x_t^{s,x}(I))$ given $\mathcal{F}_{t(k)}$. The results are intuitively obvious and just amount to fixing all which depends on $w_{[0,t(k)]}$ and then averaging.

Throughout this subsection we fix a partition $I = \{t(k)\}$ of $[0,\infty)$ and do not assume that the process is regular.

Lemma 2.3. *The C-valued function $x^{s,x}(I)$ is continuous in probability with respect to (s,x).*

Proof. For the sake of simplicity of notation, for $s, r \in [0,\infty), x, y \in \mathbb{R}^d$, denote

$$x_t := x_t^{s,x}(I), \quad y_t := x_t^{r,y}(I).$$

In the same way as in the proof of Lemma 2.2 from the definition of Euler's approximation for any increasing sequence $\varepsilon_k > 0$ and $R \geq 0$ we get

$$P\{ \sup_{t \leq t(k+1)} |x_t - y_t| \geq \varepsilon_{k+1}\} \leq P\{ \sup_{t \leq t(k)} |x_t - y_t| \geq \varepsilon_k\}$$

$$+ P\{ \sup_{t \leq t(k)} |x_t - y_t| \leq \varepsilon_k, \sup_{t \leq t(k)} (|x_t| + |y_t|) \geq R\}$$

$$+ \varepsilon_{k+1}^{-2}\Big[3\varepsilon_k^2 + 12 \sup_{\substack{|u|,|v| \leq R \\ |u-v| \leq \varepsilon_k}} \int_{t(k)}^{t(k+1)} ||\sigma(s+p,u) - \sigma(r+p,v)||^2 \, dp \qquad (2.11)$$

$$+ 3\Big(\sup_{\substack{|u|,|v| \leq R \\ |u-v| \leq \varepsilon_k}} \int_{t(k)}^{t(k+1)} |b(s+p,u) - b(r+p,v)| \, dr\Big)^2\Big].$$

Now we claim that

$$F(s,r,u,v) := \int_{t(k)}^{t(k+1)} ||\sigma(s+p,u) - \sigma(r+p,v)||^2 \, dp$$

is a continuous function of (s,r,u,v). To prove this, let $(s_n, r_n, u_n, v_n) \to (s, r, u, v)$. Then, bearing in mind the inequality $|a^2 - b^2| \leq (|a| + |b|)|a - b|$, Hölder's inequality, and condition (1.1), we get that

$$\overline{\lim_{n \to \infty}} |F(s_n, r_n, u_n, v_n) - F(s, r, u, v)|^2$$

$$\leq N \overline{\lim_{n \to \infty}} \int_{t(k)}^{t(k+1)} ||\sigma(s_n + p, u_n) - \sigma(s + p, u)||^2 \, dp$$

$$+ N \overline{\lim_{n \to \infty}} \int_{t(k)}^{t(k+1)} ||\sigma(r_n + p, v_n) - \sigma(r + p, v)||^2 \, dp := NI_1 + NI_2,$$

where, for instance,

$$I_1 \leq 2 \overline{\lim_{n \to \infty}} \int_{t(k)}^{t(k+1)} ||\sigma(s_n + p, u_n) - \sigma(s_n + p, u)||^2 \, dp$$

$$+ 2 \overline{\lim_{n \to \infty}} \int_{t(k)}^{t(k+1)} ||\sigma(s_n + p, u) - \sigma(s + p, u)||^2 \, dp. \quad (2.12)$$

The first limit in (2.12) is zero because of continuity of σ in x and because it is less than

$$\overline{\lim_{n \to \infty}} \int_0^T ||\sigma(p, u_n) - \sigma(p, u)||^2 \, dp,$$

where T is an appropriate constant. The second limit in (2.12) is zero because of L_2-continuity of square integrable functions. In the same way, $I_2 = 0$, so that F is continuous indeed.

Similarly one treats the second integral in (2.11). Consequently, if if we suppose that for a k and every $\varepsilon_k > 0$

$$J_k(\varepsilon_k) := \overline{\lim_{\substack{x \to y \\ s \to r}} } P\{ \sup_{t \leq t(k)} |x_t - y_t| \geq \varepsilon_k \} = 0, \quad (2.13)$$

then passing to the limit in (2.11) first as $(s, x) \to (r, y)$ and $\varepsilon_k \downarrow 0$ and finally letting $R \to \infty$ and using Lemma 2.2, we conclude that $J_{k+1}(\varepsilon_{k+1}) = 0$ for any $\varepsilon_{k+1} > 0$. For $k = 0$ we have (2.13) since $x_0 = x$ and $y_0 = y$, and this proves the lemma.

Corollary 2.4. *If $g = g(x)$ is a bounded continuous function on $\mathbb{R}^d$, then $E_{s,x}g(x_t(I))$ is a continuous function of (s, x, t) for any partition I.*

Further we use the operator $Q_{s,t}$ defined for $0 \leq s \leq t$ on Borel functions $g(x), x \in \mathbb{R}^d$, by the formula

$$Q_{s,t}g(x) = Eg(\xi_{s,t}^x),$$

where $\xi_{s,t}^x$ is an $\mathbb{R}^d$-valued normally distributed vector with parameters

$$\left(x + \int_s^t b(r, x) \, dr, 2 \int_s^t a(r, x) \, dr \right), \quad a = \tfrac{1}{2} \sigma \sigma^*.$$

Remark 2.5. If g is a bounded continuous function, then obviously $Q_{s,t}g(x)$ is well defined for $0 \leq s \leq t$, $x \in \mathbb{R}^d$ and is continuous in (s,t,x). It follows by standard arguments that $Q_{s,t}g(x)$ is well defined and Borel in (s,t,x) if g is Borel, positive or bounded.

Lemma 2.6. *Let g be a Borel, positive or bounded function on $\mathbb{R}^d$. Then for every $s \geq 0$, $x \in \mathbb{R}^d$, $t(n) \leq t \leq t(n+1)$ we have*

$$E_{s,x}\{g(x_t(I))|\mathcal{F}_{t(n)}\} = Eg(\eta^{s,z}_{t(n),t})|_{z=x^{s,x}_{t(n)}(I)}$$

$$= Q_{s+t(n),s+t}g(x^{s,x}_{t(n)}(I)) \quad (a.s.), \quad (2.14)$$

where

$$\eta^{s,z}_{t(n),t} = z + \int_{t(n)}^{t} \sigma(s+r,z)\,dw_r + \int_{t(n)}^{t} b(s+r,z)\,dr.$$

Proof. For simplicity let

$$x_t = x^{s,x}_t(I), \quad y = x^{s,x}_{t(n)}(I),$$

and for $p \in \mathbb{R}$, and $m = 1,2,\ldots$ let $\kappa_m(p) = 2^{-m}[2^m p]$ and for $z \in \mathbb{R}^d$ let

$$\kappa_m(z) = (\kappa_m(z^1),\ldots,\kappa_m(z^d)).$$

By $\Gamma(m)$ we denote the countable set of all values of $\kappa_m(z)$, $z \in \mathbb{R}^d$.

By using the continuity of σ and b with respect to x, for $t(n) \leq t \leq t(n+1)$, we get $x_t(m) \xrightarrow{P} x_t$ as $m \to \infty$, where

$$x_t(m) := \kappa_m(y) + \int_{t(n)}^{t} \sigma(s+r,\kappa_m(y))\,dw_r + \int_{t(n)}^{t} b(s+r,\kappa_m(y))\,dr.$$

If we multiply both parts of this equality by the indicator of the set $\{\omega : \kappa_m(y) = z\}$ with nonrandom z, then we can bring this indicator inside the integrals (a.s.). After this the integrands will not change if in the arguments of σ, b we replace $\kappa_m(y)$ by z. Having made this substitution we can pull the indicator back, thus obtaining that on every set $\{\kappa_m(y) = z\}$ (a.s.) we have $x_t(m) = \eta^{s,z}_{t(n),t}$.

Note also that $\eta^{s,z}_{t(n),t}$ is independent of $\mathcal{F}_{t(n)}$ and has a normal distribution, namely that of $\xi^z_{s+t(n),s+t}$. All these arguments show that for a bounded continuous and, say positive function g the left–hand side of (2.14) is (a.s.) equal to

$$\lim_{m\to\infty} E\{g(x_t(m))|\mathcal{F}_{t(n)}\}$$

$$= \lim_{m\to\infty} E\{\sum_{z\in\Gamma_m} I_{\kappa_m(y)=z}g(x_t(m))|\mathcal{F}_{t(n)}\}$$

$$= \lim_{m\to\infty} \sum_{z\in\Gamma_m} I_{\kappa_m(y)=z}E\{g(\eta^{s,z}_{t(n),t})|\mathcal{F}_{t(n)}\}$$

$$= \lim_{m\to\infty} Q_{s+t(n),s+t}g(\kappa_m(y)) = Q_{s+t(n),s+t}g(y),$$

where in the last step we have used Remark 2.5. We have proved (2.14) for bounded continuous $g \geq 0$. The passage from these functions to all Borel, positive or bounded ones is standard, and the lemma is proved.

By taking expectations in (2.14) and using induction we get the following.

Corollary 2.7. *We have*

$$E_{s,x}g(x_t(I)) = Q_{s+t(0),s+t(1)} \cdot \cdots \cdot Q_{s+t(n),s+t}g(x). \qquad (2.15)$$

Let us derive from Lemma 2.6 the last result of this subsection.

Lemma 2.8. *Let g be a Borel, positive or bounded function on $\mathbb{R}^d$. For an integer $n \geq 0$ define*

$$\Phi_n(s, s+r, x) = E_{s,x}g(x_r(\theta_{t(n)}I)), \qquad (2.16)$$

where $\theta_{t(n)}I = \{t(k) - t(n) : k \geq n\}$. Fix n and $t \geq t(n)$. Then for every $s \geq 0$ and $x \in \mathbb{R}^d$

$$E_{s,x}\{g(x_t(I))|\mathcal{F}_{t(n)}\} = \Phi_n(s+t(n), s+t, x_{t(n)}^{s,x}(I)) \quad (a.s.). \qquad (2.17)$$

Proof. Denote by $t(m)$ the element of I nearest to t from the left. Use (2.14) with $t(m)$ instead of $t(n)$, take conditional expectations with respect to $\mathcal{F}_{t(m-1)}$, and use again (2.14) to transform the conditional expectation. After this take conditional expectations with respect to $\mathcal{F}_{t(m-2)}$ of the terms of thus obtained formula, apply again (2.14) and proceed further in an obvious manner. For $r \in [t(k) - t(n), t(k+1) - t(n))$, $k = n, n+1, \ldots$, also denote

$$\tilde{\Phi}_n(u, u+r, x)$$
$$= Q_{u,u+t(n+1)-t(n)}Q_{u+t(n+1)-t(n),u+t(n+2)-t(n)} \cdot \cdots \cdot Q_{u+t(k)-t(n),u+r}g(x).$$

We see that the left-hand side of (2.17) (a.s.) is equal to

$$Q_{s+t(n),s+t(n+1)}Q_{s+t(n+1),s+t(n+2)} \cdot \cdots \cdot Q_{s+t(m),s+t}g(x_{t(n)}^{s,x}(I)) =$$

$$= \tilde{\Phi}_n(s+t(n), s+t, x_{t(n)}^{s,x}(I)).$$

If we apply the same argument taking $r, \theta_{t(n)}I, 0$ instead of $t, I, t(n)$, then we immediately obtain that the right-hand side of (2.16) is equal to $\tilde{\Phi}_n(s, s+r, x)$, that is $\tilde{\Phi} = \Phi$ and the lemma is proved.

2.3 Markov property

In this subsection we apply the results of Subsec. 2.2 assuming that equation (2.1) is regular.

We will need a very important operator. For $0 \le s \le t$ define

$$G_{s,t}u(x) = E_{s,x}u(x_{t-s}).$$

This definition make sense, for instance, if u is Borel and bounded.

Lemma 2.9. *For every $T \in (0, \infty)$*

(i) the C-valued function $x^{s,x}$ is continuous in probability with respect to (s, x);

(ii) if u is bounded and continuous, the function $G_{s,t}u(x)$ is bounded and continuous function of (s, t, x) on the set $0 \le s \le t, x \in \mathbb{R}^d$. In particular, the process $x_t^{s,x}$ is a Feller process;

(iii) if u is Borel, bounded or positive, then $G_{s,t}u(x)$ is a Borel function of (s, t, x).

Actually, there is nothing to prove since the first two statements follow immediately from Lemma 2.3 and Definition 2.1, and the third one is a standard consequence of (ii).

Now let us prove the main results of this section.

Theorem 2.10 (Markov property). *Let u be a Borel, positive or bounded, function on $\mathbb{R}^d$ and $x \in \mathbb{R}^d, T, s \ge 0, t \in [0, T]$. Then*

$$E_{s,x}\{u(x_T)|\mathcal{F}_t\} = G_{s+t,s+T}u(x_t^{s,x}) \quad (a.s.). \tag{2.18}$$

Proof. We only need to consider bounded and continuous u. In this case (2.18) follows from Lemma 2.8 if we take a sequence of partitions $I(n)$ such that $\Delta(I(n)) \to 0$ and, in addition, $t \in I(n)$ for all n. The theorem is proved.

Corollary 2.11. *If u is bounded and continuous , then, for any $T \in [0, \infty)$, $G_{s+t,s+T}u(x_t^{s,x})$ is a bounded continuous martingale for $t \in [0, T]$.*

Corollary 2.12. *Equation (2.18) implies that, for $r, s \ge 0$, and $t \ge r$,*

$$G_{s,s+t}u = G_{s,s+r}G_{s+r,s+t}u.$$

Theorem 2.13 (Strong Markov property). *Let u be a Borel, positive or bounded, function on $\mathbb{R}^d$ and $x \in \mathbb{R}^d, T, s \ge 0$. Then equation (2.18) holds for any stopping time $t \le T$.*

Indeed, for bounded continuous u the assertion follows from Corollary 2.11 and Doob's optional sampling theorem. For general u, it is derived from this particular case by a standard measure theoretic argument.

3. Conditional version of Kolmogorov's equation

In this section we keep the assumption that σ and b are nonrandom and replace condition (1.1) with the following stronger condition: For any $R \in [0, \infty)$ there exists a finite constant $K(R)$ such that

$$||\sigma(t,x)||^2 + |b(t,x)| \le K(R) \quad \forall t \ge 0, |x| \le R. \tag{3.1}$$

In this section we also assume that, for any $s \ge 0$ and $x \in \mathbb{R}^d$ equation (2.1) has a solution, which we denote $x_t^{s,x}$ as usual.

First we want to say several words about forward Kolmogorov's equation. Take a test function ζ, that is take a $\zeta \in C_0^\infty(\mathbb{R}^d)$, fix $s \ge 0$ and $x \in \mathbb{R}^d$, and define $x_t = x_t^{s,x}$ and

$$u(t) = E_{s,x}\zeta(x_t) = \int_{\mathbb{R}^d} \zeta(y)p(s,x,s+t,y)\,dy, \tag{3.2}$$

where $p(s,x,s+t,y)$ is the generalized function of y acting on test function ζ by formula (3.2). Observe that $p(s,x,s+t,y)$ acts as a measure. Therefore, expressions like

$$\int_{\mathbb{R}^d} f(y)p(s,x,s+t,y)\,dy$$

are well defined for any bounded Borel f. By Itô's formula

$$\zeta(x_t) = \zeta(x) + \int_0^t L\zeta(s+r,x_r)\,dr + \int_0^t \zeta_{x^i}(x_r)\sigma^{ik}(s+r,x_r)\,dw_r^k, \tag{3.3}$$

where, with $a = \frac{1}{2}\sigma\sigma^*$ and $D_t u = \partial u/\partial t$,

$$Lu(t,y) := D_t u(t,y) + a^{ij}(t,y)u_{x^i x^j}(t,y) + b^i(t,y)u_{x^i}(t,y).$$

The second integral in (3.2) is a martingale, which follows from (3.1) and from compactness of the support of ζ. Hence

$$u(t) = \int_0^t \int_{\mathbb{R}^d} p(s,x,s+r,y)L\zeta(s+r,y)\,dy\,dr,$$

which after being multiplied by $\chi'(s+t)$, where $\chi(s+\cdot) \in C_0^\infty(0,\infty)$, and integrating by parts implies that

$$\int_0^\infty \int_{\mathbb{R}^d} p(s,x,s+t,y)\chi'(s+t)\zeta(y)\,dy\,dt$$

$$= -\int_0^\infty \int_{\mathbb{R}^d} p(s,x,s+t,y)\chi(s+t)L\zeta(s+t,y)\,dy\,dt,$$

$$\int_0^\infty \int_{\mathbb{R}^d} p(s,x,s+t,y)L[\chi\zeta](s+t,y)\,dy\,dt = 0.$$

It is easy to understand that one can replace here $\chi\zeta$ with any function of class $C_0^\infty((0,\infty) \times \mathbb{R}^d)$. Once this has been done, we see that by definition

$$L^* p(s, x, \cdot, \cdot)(t, y) = 0 \quad \text{on} \quad (s, \infty) \times \mathbb{R}^d$$

in the sense of generalized functions on $(s, \infty) \times \mathbb{R}^d$, where L^* is the formal adjoint to L applied with respect to the variables (t, y). Thus we got what is called *the forward Kolmogorov's equation*. This equation has usual sense if we assume that $p(s, x, t, y)$ is once differentiable in t and twice differentiable in y with the derivatives been summable (or better say been Sobolev derivatives), $a(t, y)$ is twice differentiable in y and $b(t, y)$ is once differentiable in x with the derivatives been bounded. Indeed, in this case $L^* p$ is a usual, rather than generalized, function.

The backward Kolmogorov's equation is obtained in the following way. Let $T \in [0, \infty)$ be a fixed number and g a Borel function on $\mathbb{R}^d$. Under certain conditions on g the function

$$v(s, x) := G_{s,T} g(x) \tag{3.4}$$

is well defined for $s \in [0, T]$ and $x \in \mathbb{R}^d$. If we assume that v is a smooth function, then, upon applying Itô's formula, we get that

$$v(s + t, x_t^{s,x}) = v(s, x) + \int_0^t Lv(s + r, x_r^{s,x}) \, dr + m_t \quad t \le T - s, \tag{3.5}$$

where m_t is a local martingale starting at zero. Sometimes (see Corollary 2.11) for any $s \in [0, T]$, the process $v(s + t, x_t^{s,x})$ itself is a martingale in $t \in [0, T - s]$. Then the integral in (3.5) is a continuous local martingale and since it has bounded variation, it is zero. Hence, $Lv(s + r, x_r^{s,x}) = 0$ for almost all $r \in [0, T - s]$, and this naturally leads to the guess that

$$Lv(s, x) = 0 \tag{3.6}$$

for $s < T, x \in \mathbb{R}^d$. Equation (3.6) is called (backward) Kolmogorov's equation. Below we formalize the last argument assuming certain smoothness of v.

First, it turns out to be useful to formalize the property which says that $v(s + t, x_t^{s,x})$ is a martingale.

Definition 3.1. Let Q be a bounded domain in $(0, T) \times \mathbb{R}^d$ and u be a bounded Borel function defined in Q. For $(s, x) \in Q$ let

$$\tau^{s,x} = \tau_Q^{s,x} = \inf \{t \ge 0 : (s + t, x_t^{s,x}) \notin Q\}$$

be the first exit time of $(s + t, x_t^{s,x})$ from Q.

(i) We say that u is *superharmonic (relative to Q)* at a given point $(s_0, x_0) \in Q$ if

$$u(s_0, x_0) \ge E_{s_0, x_0} u(s_0 + \tau \wedge t, x_{\tau \wedge t}) + o(t) \tag{3.7}$$

as $t \downarrow 0$ (which is true with $o(t) \equiv 0$ for $u \equiv v$ by Theorem 2.13).

(ii) We say that u is *subharmonic at a given point* $(s_0, x_0) \in Q$ if

$$u(s_0, x_0) \leq E_{s_0, x_0} u(s_0 + \tau \wedge t, x_{\tau \wedge t}) + o(t) \qquad (3.8)$$

as $t \downarrow 0$ (which is true with $o(t) \equiv 0$ for $u \equiv v$ again by Theorem 2.13).

(iii) We say that u is *superharmonic (subharmonic) in Q* if it is superharmonic (subharmonic) at any point $(s_0, x_0) \in Q$ and $o(t)$ in (3.7) (respectively in (3.8)) is uniform with respect to (s_0, x_0) lying in any closed subset of Q.

(iv) We say that u is *harmonic* if it is superharmonic and subharmonic.

The most important example of superharmonic and subharmonic functions is the function v (Theorem 2.13), and the reader may save some memory if he/she always thinks of v when we state and prove various assertions about superharmonic or subharmonic functions.

It turns out that the above notions are local properties.

Theorem 3.2. *(i) If a function u is superharmonic relative to Q at $(s_0, x_0) \in Q$, then for any subdomain $Q' \subset Q$ containing (s_0, x_0) it is superharmonic relative to Q' at (s_0, x_0).*

(ii) If a bounded Borel function u is defined in Q and $(s_0, x_0) \in Q$ and there is a subdomain $Q' \subset Q$ containing (s_0, x_0) such that u is superharmonic relative to Q' at (s_0, x_0), then it is superharmonic relative to Q at (s_0, x_0).

(iii) Similar statements hold for functions subharmonic at (s_0, x_0) or superharmonic or subharmonic in Q.

To prove the theorem we need the following lemma.

Lemma 3.3. *For $s \geq 0$ and $x \in \mathbb{R}^d$ define*

$$\gamma^{s,x}(\delta) = \inf\{t \geq 0 : |x_t^{s,x} - x| \geq \delta\}. \qquad (3.9)$$

Then for every $R > 0$, $s \geq 0$, $t \in [0, 1]$, $|x| \leq R$, $n \geq 1$, and $\delta > 0$ we have

$$P_{s,x}\{\sup_{r \leq t} |x_r - x| \geq \delta\} \leq \delta^{-n} E_{s,x} \sup_{r \leq t \wedge \gamma(\delta)} |x_r - x|^n \leq N\delta^{-n} t^{n/2}, \qquad (3.10)$$

where $N = N(n, K(R + \delta))$.

Proof. Note that in the first expression in (3.10) we can replace t by $t \wedge \gamma(\delta)$. Therefore the first inequality follows by Chebyshev's inequality. Then obviously the middle expression times δ^n is less than

$$E_{s,x}[\sup_{r \leq t} |\int_0^r I_{u \leq \gamma(\delta)} \sigma(s + u, x_u)\, dw_u| + \int_0^{t \wedge \gamma(\delta)} |b(s + r, x_u)|\, du]^n.$$

It remains only to apply the inequalities $(a + b)^n \leq 2^n(a^n + b^n)$ and $t^n \leq t^{n/2}$ and the Burkholder-Davis-Gundy inequality which shows that

$$E_{s,x} \sup_{r \le t} \Big| \int_0^r I_{u \le \gamma(\delta)} \sigma(s+u, x_u)\, du \Big|^n$$

$$\le N E_{s,x} \Big(\int_0^t I_{u \le \gamma(\delta)} ||\sigma(s+u, x_u)||^2\, du \Big)^{n/2} \le N K^{n/2} (R+\delta) t^{n/2}.$$

The lemma is proved.

Proof of Theorem 3.2. Let $(s_0, x_0) \in Q' \subset Q$ and let δ be the distance from (s_0, x_0) to the boundary of Q'. By Lemma 3.3 for $0 \le t \le 1 \wedge (\delta/2)$ it holds that

$$P_{s_0, x_0} \{\tau_Q \wedge t \ne \tau_{Q'} \wedge t\} \le P_{s_0, x_0} \{\tau_{Q'} < t\}$$

$$\le P_{s_0, x_0} \{t^2 + \sup_{r \le t} |x_t - x_0|^2 \ge \delta^2\} \le N \delta^{-4} t^2,$$

where and N is independent of (s_0, x_0) and Q'. This and boundedness of u obviously imply all our assertions. The theorem is proved.

As yet another application of Lemma 3.3 we derive the following.

Theorem 3.4. *If u is subharmonic at (s_0, x_0) and v is a bounded Borel function in Q such that*

$$u(s, x) - v(s, x) = o(|s - s_0| + |x - x_0|^2)$$

as $|s - s_0| + |x - x_0|^2 \to 0$, then v is subharmonic at (s_0, x_0).

For the proof, it suffices to denote $u(s, x) - v(s, x) = \varepsilon(s - s_0, x - x_0)(|s - s_0| + |x - x_0|^2)$ and observe that in the inequality

$$|E_{s_0, x_0} u(s_0 + \tau \wedge t, x_{\tau \wedge t}) - E_{s_0, x_0} v(s_0 + \tau \wedge t, x_{\tau \wedge t})|$$

$$\le \big(E_{s_0, x_0} \varepsilon^2 (\tau \wedge t, x_{\tau \wedge t} - x_0) \big)^{1/2} \big(E_{s_0, x_0} (\tau \wedge t + |x_{\tau \wedge t} - x_0|^2) \big)^{1/2}$$

the first factor on the right goes to zero as $t \downarrow 0$ because $(\tau \wedge t, x_{\tau \wedge t} - x_0) \to 0$ and ε is bounded, whereas the second one is less than Nt by Lemma 3.3 (applied with δ bigger than the diameter of Q).

Theorem 3.2 allows us to localize investigating differentiability properties of sub- and superharmonic functions by reducing general domains to cylinders lying inside. The following result is applicable to any cylinder and not only to $Q_R := (0, T) \times B_R$.

Theorem 3.5. *Let $Q = Q_R$ and let u be a function in Q such that, for any $s \in (0, T)$, $u(s, x)$ is twice continuously differentiable in B_R. Also assume that the first and second order derivatives of u with respect to x are bounded on Q. Then there exists a constant N such that, for any $x \in B_R$,*

(i) $u(s, x) + Ns$ is an increasing function of $s \in (0, T)$ provided that u is subharmonic in Q,

(ii) $u(s, x) - Ns$ is a decreasing function of $s \in (0, T)$ provided that u is superharmonic in Q,

(iii) $|u(t, x) - u(s, x)| \le N|t - s|$ provided that u is harmonic in Q.

Proof. Assertions (ii) and (iii) trivially follow from (i). To prove (i), apply Itô's formula to $u(s + t, x_r^{s,x})$ as a function of r. Then it is seen that, for $s \in (0, T)$ and $t \in [0, T - s)$ and $|x| < R$,

$$u(s, x) \leq E_{s,x} u(s + t, x_{t \wedge \tau}) + o(t)$$

$$= u(s + t, x) + E_{s,x} \int_0^{t \wedge \tau} L[u(s + t, \cdot)](s + r, x_r) \, dr + o(t)$$

$$\leq u(s + t, x) + Nt + o(t).$$

Now assertion (i) follows easily due to uniformity of $o(t)$ with respect to s. The theorem is proved.

We will be using this theorem while checking the assumptions of the following theorem.

Denote by $B^{1,2}(Q_R)$ the set of all bounded continuous functions $u(s, x)$ defined on Q_R which are Lipschitz continuous in s with constant independent of x, are twice differentiable with respect to x for any s with u_x and u_{xx} being bounded. Remember that Lipschitz continuous functions are almost everywhere differentiable (Rademacher's theorem), so that if $u \in B^{1,2}(Q_R)$, then $D_s u(s, x)$ exists in Q_R (a.e.). It turns out that, if $u \in B^{1,2}(Q_R)$, then, for almost any point $(s, x) \in Q_R$,

$$u(s+t, x+y) = u(s, x) + u_{x^i}(s, x) y^i + \tfrac{1}{2} u_{x^i x^j}(s, x) y^i y^j + D_s u(s, x) t + o(|t| + |y|^2)$$

as $|t| + |y|^2 \to 0$. This fact is very easy to prove if, for instance, u_x and u_{xx} are continuous as functions of (s, x). A more general result can be found in Appendix of [KR].

The following is our conditional version of Kolmogorov's equation.

Theorem 3.6. *Let $u \in B^{1,2}(Q_R)$. Then*
(i) $Lu \geq 0$ (a.e.) in Q_R if u is subharmonic in Q_R,
(ii) $Lu \leq 0$ (a.e.) in Q_R if u is superharmonic in Q_R,
(iii) $Lu = 0$ (a.e.) in Q_R if u is harmonic in Q_R.

Proof. Again we only need to prove (i). Under the assumptions in (i), by Theorem 3.4, for almost any $(s, x) \in Q_R$,

$$u(s, x) \leq u(s, x) + u_{x^i}(s, x) E_{s,x}(x_{t \wedge \tau}^i - x^i)$$
$$+ \tfrac{1}{2} u_{x^i x^j}(s, x) E_{s,x}(x_{t \wedge \tau}^i - x^i)(x_{t \wedge \tau}^j - x^j) + D_s u(s, x) t + o(t).$$

Therefore, it only remains to prove that, for almost all $(s, x) \in Q_R$,

$$H^i(s, x, t) := E_{s,x}(x_{t \wedge \tau}^i - x^i) = b^i(s, x) t + o(t), \tag{3.11}$$

$$H^{ij}(s, x, t) := E_{s,x}(x_{t \wedge \tau}^i - x^i)(x_{t \wedge \tau}^j - x^j) = 2a^{ij}(s, x) t + o(t). \tag{3.12}$$

By Itô's formula and Lemma 3.3, for any $\delta > 0$,

$$|H^i(s, x, t) - \int_0^t b^i(s + r, x)\, dr| =$$

$$|E_{s,x} \int_0^{t \wedge \tau} [b^i(s + r, x_r) - b^i(s + r, x)]\, dr| + o(t)$$

$$\leq \int_0^t h(\delta, s + r, x)\, dr + o(t), \quad (3.13)$$

where

$$h(\delta, t, x) := \sup_{|y - x| \leq \delta} |b(t, y) - b(t, x)|.$$

Since b is continuous in x, $h(\delta, t, x)$ is Borel in t and $h(\delta, t, x) \to 0$ as $\delta \downarrow 0$. Furthermore, by using the Lebesgue differentiation theorem we get that, for almost all (s, x),

$$\lim_{t \downarrow 0} \frac{1}{t} \int_s^{s+t} h(\delta, r, x)\, dr = h(\delta, s, x)$$

for all rational $\delta > 0$. Thus, for almost all (s, x) and all rational $\delta > 0$,

$$\overline{\lim_{t \downarrow 0}} \frac{1}{t} |H^i(s, x, t) - \int_0^t b^i(s + r, x)\, dr| \leq h(\delta, s, x).$$

Here the left–hand side is independent of δ and the right-hand side goes to zero as $\delta \downarrow 0$. Therefore, the last limit is zero, and to get (3.11) it only remains to invoke again the Lebesgue differentiation theorem.

Next,

$$H^{ij}(s, x, t) = E_{s,x} \int_0^{t \wedge \tau} [2a^{ij}(s+r, x_r) + (x_r^i - x^i)b^j(s+r, x_r) + (x_r^j - x^j)b^i(s+r, x_r)]\, dr$$

$$= 2E_{s,x} \int_0^{t \wedge \tau} a^{ij}(s + r, x_r) + o(t)$$

and one gets (3.12) similarly to (3.11). The theorem is proved.

Remark 3.7. It is known that the equation $Lu = 0$ (a.e.) in Q_R with boundary conditions $u = g$ on $([0, T] \times S_R) \cup (T \times B_R)$ ($S_R := \{x : |x| = R\}$) can only have one solution in $B^{1,2}(Q_R) \cap C(\bar{Q}_R)$ and even in a much wider class of functions (see, for instance, [KR]).

4. Differentiability of solutions of stochastic equations with respect to initial data

Here we want to derive some results which would guarantee that under some smoothness assumptions on σ and b the function v (see (3.4)) is twice continuously differentiable in x.

4.1 Estimating moments of solutions of Itô's equations

The only assumptions in this subsection are those stated before Remark 1.1. Fix some real–valued, $\mathcal{F}_t$–adapted, and $\mathcal{F} \otimes \mathcal{B}([0,\infty))$–measurable processes c_t, f_t, g_t such that $f_t \geq 0$, $g_t \geq 0$, and assume that

$$\varphi_t := \int_0^t c_s \, ds, \qquad \int_0^t (f_s + g_s^2) \, ds$$

are finite for any ω and finite t. Also, fix some numbers $\gamma \in (0,1)$ and $K \geq 0$. Denote by $\mathfrak{M}$ the the set of all finite stopping times.

For two cadlag processes ξ_t and η_t we write $d\xi_t \leq d\eta_t$ if and only if the process $\xi_t - \eta_t$ is decreasing.

The following result which we give without proof turns out to be extremely useful in many applications

Lemma 4.1. *Let ξ_t and η_t be nonnegative processes.*
(i) Let m_t be a local martingale starting at zero and let

$$\xi_t \leq \eta_t + m_t$$

for all t. Then for any $\tau \in \mathfrak{M}$

$$E\{\xi_\tau | \mathcal{F}_0\} \leq E\{\sup_{t \leq \tau} \eta_t | \mathcal{F}_0\}.$$

(ii) Let ξ_t and η_t be continuous and $\mathcal{F}_t$–adapted, and let

$$E\{\xi_\tau | \mathcal{F}_0\} \leq E\{\eta_\tau | \mathcal{F}_0\} \quad (a.s.) \tag{4.1}$$

for any $\tau \in \mathfrak{M}$. Then for any $\tau \in \mathfrak{M}$ we have

$$E\{\sup_{t \leq \tau} \xi_t^\gamma | \mathcal{F}_0\} \leq \tfrac{2-\gamma}{1-\gamma} E\{\sup_{t \leq \tau} \eta_t^\gamma | \mathcal{F}_0\} \quad (a.s.). \tag{4.2}$$

Lemma 4.2. *Let ξ_t be a nonnegative cadlag process and let ζ_t be a one-dimensional process having stochastic differential*

$$d\zeta_t = b_t \, dt + \sigma_t \, dw_t,$$

where b_t and σ_t are $\mathcal{F}_t$–adapted and $\mathcal{F} \otimes \mathcal{B}([0,\infty))$–measurable processes of appropriate dimensions. Assume that for all $t > 0$,

$$d\xi_t \leq b_t \, dt + \sigma_t \, dw_t.$$

and $b_t \leq c_t \xi_t + f_t$ for all ω and t. Then for all $\tau \in \mathfrak{M}$

$$E\xi_\tau e^{-\varphi_\tau} \leq E[\xi_0 + \int_0^\tau e^{-\varphi_t} f_t \, dt],$$

$$E \sup_{t \leq \tau} \xi_t^\gamma e^{-\gamma \varphi_t} \leq \tfrac{2-\gamma}{1-\gamma} E[\xi_0 + \int_0^\tau e^{-\varphi_t} f_t \, dt]^\gamma. \tag{4.3}$$

Proof. To prove (4.3), notice that by Itô's formula

$$\xi_t e^{-\varphi_t} \le \xi_0 + \int_0^t e^{-\varphi_s} f_s\, ds + \int_0^t e^{-\varphi_s} \sigma_s\, dw_s =: \eta_t. \tag{4.4}$$

Since ξ, f are nonnegative, by Lemma 4.1 (i) for any $\tau \in \mathfrak{M}$, we get

$$E\{\eta_\tau | \mathcal{F}_0\} \le E\{\xi_0 + \int_0^\tau e^{-\varphi_s} f_s\, ds | \mathcal{F}_0\} \tag{4.5}$$

and this and (4.4) prove the first inequality in (4.3). To prove the second one we apply Lemma 4.1 (ii) to (4.5) and again use (4.4). The lemma is proved.

Exercise 4.3. Prove (4.4).

Remark 4.4. Assertions like in Lemma 4.2 are *equivalent* to their "$\mathcal{F}_0$-conditional versions". By this we mean that under the condition of Lemma 4.2 it holds that (a.s.)

$$E\{\xi_\tau e^{-\varphi_\tau}|\mathcal{F}_0\} \le \xi_0 + E\{\int_0^\tau e^{-\varphi_t} f_t\, dt|\mathcal{F}_0\},$$

$$E\{\sup_{t\le\tau} \xi_t^\gamma e^{-\gamma\varphi_t}|\mathcal{F}_0\} \le \tfrac{2-\gamma}{1-\gamma} E\{[\xi_0 + \int_0^\tau e^{-\varphi_t} f_t\, dt]^\gamma|\mathcal{F}_0\}.$$

Indeed, this conditional statements obviously imply the corresponding statements of Lemma 4.2. On the other hand, one can apply Lemma 4.2 to the processes $\xi_t I_A$, $f_t I_A$, $g_t I_A$ where A is any element of $\mathcal{F}_0$. Then in (4.3) one gets the factor I_A inside all expectations, which due to the arbitrariness of A is equivalent to the above "conditional" inequalities.

Lemma 4.5. *Let ξ_t be a process as in the beginning of Lemma 4.2 and let $c_t \ge c_t I_{\xi_t \ne 0}$ (for instance, $c_t \ge 0$). Take a number $p > 0$ and assume that for all ω and t*

$$p\xi_t b_t + \tfrac{1}{2} p(p-1)|\sigma_t|^2 \le c_t \xi_t^2.$$

Also assume that $b_t \le f_t \xi_t$ and $|\sigma_t|^2 \le f_t \xi_t^2$. Then for any $\tau \in \mathfrak{M}$

$$E\xi_\tau^p e^{-\varphi_\tau} \le E\xi_0^p, \quad E\sup_t \xi_t^{\gamma p} e^{-\gamma\varphi_t} \le N(\gamma) E\xi_0^{\gamma p}.$$

Proof. Since $\xi_t = \zeta_t - A_t$ where A_t is an increasing process and $\xi_t \ge 0$, we can apply Itô's formula to $(\xi_t + \varepsilon)^p$ for any $\varepsilon > 0$. Then we get

$$d(\xi_t + \varepsilon)^p \le (\xi_t + \varepsilon)^{p-2}\left(p(\xi_t+\varepsilon)b_t + \tfrac{1}{2}p(p-1)|\sigma_t|^2\right) dt$$

$$+p(\xi_t+\varepsilon)^{p-1}\sigma_t\, dw_t \le (\xi_t+\varepsilon)^{p-2} c_t \xi_t^2\, dt$$

$$+(\xi_t+\varepsilon)^{p-2}\varepsilon p b_t\, dt + p(\xi_t+\varepsilon)^{p-1}\sigma_t\, dw_t$$

$$\le I_{\xi_t\ne0}(\xi_t+\varepsilon)^p[\eta_{\varepsilon t}^2 c_t + p f_t \eta_{\varepsilon t}\varepsilon/(\xi_t+\varepsilon)]\, dt + p(\xi_t+\varepsilon)^p \eta_{\varepsilon t} I_{\xi_t\ne0}\xi_t^{-1}\sigma_t\, dw_t,$$

where $\eta_{\varepsilon t} = \xi_t/(\xi_t + \varepsilon)$. By letting $\varepsilon \downarrow 0$ and using the dominated convergence theorem and the fact that $|I_{\xi_t \neq 0} \xi_t^{-1} \sigma_t|^2 \leq f_t$, which is integrable, we obtain

$$d\xi_t^p \leq \xi_t^p c_t \, dt + p\xi_t^p \left(I_{\xi_t \neq 0} \dot{\xi}_t^{-1} \sigma_t\right) dw_t.$$

This inequality and Lemma 4.2 (with $f \equiv 0$ there) immediately imply our assertion. The lemma is proved.

Theorem 4.6. *Let x_t be a solution of equation (1.4). Assume that*

$$||\sigma(t, x_t)|| + |b(t, x_t)| \leq K(1 + |x_t|)$$

for all $\omega \in \Omega, t \in [0, T]$. Then for any $p > 0$ and $t \in [0, T]$

$$\begin{aligned} E \sup_{s \leq t} |x_s - x_0|^p &\leq Nt^{p/2}e^{Nt}(1 + E|x_0|^p), \\ E \sup_{s \leq t} |x_s|^p &\leq Ne^{Nt}(1 + E|x_0|^p), \end{aligned} \tag{4.6}$$

where $N = N(p, K)$.

Proof. By Itô's formula for $\eta_t := x_t - x_0$ and $\xi_t := |\eta_{t \wedge T}|^2 + \varepsilon(|x_0|^2 + 1)$, where $\varepsilon > 0$ is a number, we have

$$d\xi_t = \{2(\eta_t, b(t, x_t)) + ||\sigma(t, x_t)||^2\}I_{t \leq T} \, dt + 2\eta_t^* \sigma(t, x_t)I_{t \leq T} \, dw_t.$$

Here for $\varepsilon \in (0, 1)$ and $t \leq T$

$$1 + |x_t|^2 \leq 1 + 2|\eta_t|^2 + 2|x_0|^2 \leq 4\varepsilon^{-1}\xi_t,$$

$$|\sigma^*(t, x_t)\eta_t|^2 \leq N(1 + |x_t|^2)|\eta_t|^2 \leq N(1 + |x_t|^2)\xi_t \leq N\varepsilon^{-1}\xi_t^2,$$

$$2(\eta_t, b(t, x_t)) + ||\sigma(t, x_t)||^2 \leq |\eta_t|^2 + |b(t, x_t)|^2 + ||\sigma(t, x_t)||^2 \leq N\varepsilon^{-1}\xi_t,$$

where $N = N(K)$. Hence for $c_t = N\varepsilon^{-1}$ from the first assertion of Lemma 4.5 we get

$$E \sup_t \xi_t^{p/2}e^{-N\varepsilon^{-1}t} \leq N\varepsilon^{p/2}(1 + E|x_0|^p),$$

where $N = N(p, K)$. It follows that

$$E \sup_{s \leq t} |x_s - x_0|^p \leq N\varepsilon^{p/2}e^{N\varepsilon^{-1}t}(1 + E|x_0|^p). \tag{4.7}$$

If $t \leq 1$, we take $\varepsilon = t$ and if $t \geq 1$ we take $\varepsilon = 1$. In both cases from (4.7) we obtain the first inequality in (4.6). Since the second one is an obvious corollary of the first one, the theorem is proved.

The second inequality in (4.6) can be generalized.

Theorem 4.7. *Let x_t be a solution of equation (1.4). Assume that*

$$\|\sigma(t, x_t)\| \leq K|x_t| + g_t, \quad |b(t, x_t)| \leq K|x_t| + f_t$$

for all $\omega \in \Omega, t \in [0, T]$. Then for any $p > 0$ and $t \in [0, T]$

$$E \sup_{s \leq t} |x_t|^p \leq N e^{Nt} E\big(|x_0|^p + (\int_0^t f_s \, ds)^p + (\int_0^t g_s^2 \, ds)^{p/2}\big), \qquad (4.8)$$

where $N = N(p, K)$.

Proof. For any integer $n \geq 2$ by Itô's formula

$$d|x_{t \wedge T}|^n = \{n|x_t|^{n-2}(x_t, b(t, x_t)) + \tfrac{n}{2}[(n-2)|x_t|^{n-4}|\sigma^*(t, x_t)x_t|^2$$

$$+|x_t|^{n-2}\|\sigma(t, x_t)\|^2]\}I_{t \leq T} \, dt + n|x_t|^{n-2}x_t^*\sigma(t, x_t)I_{t \leq T} \, dw_t.$$

Here

$$n|x_t|^{n-2}(x_t, b(t, x_t)) + \tfrac{n}{2}[(n-2)|x_t|^{n-4}|\sigma^*(t, x_t)x_t|^2 + |x_t|^{n-2}\|\sigma(t, x_t)\|^2]$$

$$\leq m(|x_t|^n + |x_t|^{n-1}f_t + |x_t|^{n-2}g_t^2),$$

where the constant $m \geq 0$ depends only on n and K. Hence by Lemma 4.2 for any $\tau \in \mathfrak{M}$

$$E \sup_{t \leq \tau} |x_{t \wedge T}|^{n/2} e^{-mt/2} \leq N E|x_0|^{n/2} + N E\big(\int_0^{\tau \wedge T} |x_s|^{n-1} f_s e^{-ms} \, ds\big)^{1/2}$$

$$+ N E\big(\int_0^{\tau \wedge T} |x_s|^{n-2} g_s^2 e^{-ms} \, ds\big)^{1/2}.$$

Furthermore, by Young's inequality

$$E\big(\int_0^{\tau \wedge T} |x_s|^{n-1} f_s e^{-ms} \, ds\big)^{1/2}$$

$$\leq E \sup_{t \leq \tau} |x_{t \wedge T}|^{(n-1)/2} e^{-m(n-1)t/(2n)} \big(\int_0^{\tau \wedge T} f_s \, ds\big)^{1/2}$$

$$\leq \tfrac{1}{4} E \sup_{t \leq \tau} |x_{t \wedge T}|^{n/2} e^{-mt/2} + N E\big(\int_0^{\tau \wedge T} f_s \, ds\big)^{n/2},$$

$$E\big(\int_0^{\tau \wedge T} |x_s|^{n-2} g_s^2 e^{-ms} \, ds\big)^{1/2}$$

$$\leq \tfrac{1}{4} E \sup_{t \leq \tau} |x_{t \wedge T}|^{n/2} e^{-mt/2} + N E\big(\int_0^{\tau \wedge T} g_s^2 \, ds\big)^{n/4}.$$

It follows that

$$E \sup_{t \leq \tau} |x_{t \wedge T}|^{n/2} e^{-mt/2}$$

$$\leq NE|x_0|^{n/2} + NE \Big(\int_0^{\tau \wedge T} f_s \, ds \Big)^{n/2} + NE \Big(\int_0^{\tau \wedge T} g_s^2 \, ds \Big)^{n/4} \quad (4.9)$$

with $N = N(n, K)$ if the left–hand side is finite. Since the process x_t is continuous and defined for all t, any $\tau \in \mathfrak{M}$ can be approximated by stopping times for which the left–hand side is finite. Therefore, (4.9) holds for all $\tau \in \mathfrak{M}$. This by Remark 4.4 and Lemma 4.1 implies that for any $\gamma \in (0, 1)$ and $\tau \in \mathfrak{M}$

$$E \sup_{t \leq \tau} |x_{t \wedge T}|^{n\gamma/2} e^{-mt\gamma/2}$$

$$\leq NE|x_0|^{n\gamma/2} + NE \Big(\int_0^{\tau \wedge T} f_s \, ds \Big)^{n\gamma/2} + NE \Big(\int_0^{\tau \wedge T} g_s^2 \, ds \Big)^{n\gamma/4}.$$

By choosing $n\gamma = 2p$ and $\tau = T$, we easily get (4.8). The theorem is proved.

Exercise 4.8. By considering $x_t - x_0$ instead of x_t derive Theorem 4.6 from Theorem 4.7.

4.2 Smoothness of solutions depending on a parameter

Here we apply the results of Sec. 1 to equations depending on a parameter and show that the solutions are smooth as functions of the parameter provided the coefficients are smooth.

Let Q be a domain in a Euclidean space $\mathbb{R}^k$ and let functions $\sigma(t, x, q)$ and $b(t, x, q)$ be defined on $\Omega \times [0, \infty) \times \mathbb{R}^d \times Q$. We assume that σ takes values in the set of $d \times d_1$ matrices and b is $\mathbb{R}^d$–valued. We also assume that they are $\mathcal{F}_t$–adapted and $\mathcal{F} \otimes \mathcal{B}([0, \infty))$–measurable for any x, q. As far as their dependence on x, q is concerned we fix an integer $n \geq 1$ and assume that σ and b are n times continuously differentiable with respect to (x, q) for any (ω, t) and for any finite $T, R > 0$

$$\sum \int_0^T \sup_{|x| \leq R, q \in Q} \{\|D^i \sigma(t, x, q)\|^2 + |D^i b(t, x, q)|\} \, dt < \infty, \quad (4.10)$$

where D^i stands for arbitrary derivative of order i with respect to (x, q) and the summation is taken over all derivatives of all orders $i \leq n$.

We will be dealing with solutions of the following equation

$$x_t = x_0(q) + \int_0^t \sigma(s, x_s, q) \, dw_s + \int_0^t b(s, x_s, q) \, ds \quad t \geq 0. \quad (4.11)$$

If there is no stochastic term in the equation, then from the theory of ODE one knows that the solutions are differentiable with respect to q if this is true for the initial data. Also the partial derivatives satisfy a linear system which is obtained by formal differentiation of (4.11). Here we show that the same is true in the general situation as well. In particular, the first and the second directional derivatives in the direction of a vector $\kappa \in \mathbb{R}^k$ of the solution of (4.11) satisfy

$$d\xi_t = \partial(\xi_t, \kappa)\sigma(t, x_t, q)\, dw_t + \partial(\xi_t, \kappa)b(t, x_t, q)\, dt, \qquad (4.12)$$

$$d\eta_t = [\partial(\eta_t, 0)\sigma(t, x_t, q) + \partial^2(\xi_t, \kappa)\sigma(t, x_t, q)]\, dw_t$$
$$+ [\partial(\eta_t, 0)b(t, x_t, q) + \partial^2(\xi_t, \kappa)\sigma(t, x_t, q)]\, dt, \quad (4.13)$$

where for any function $u(x, q)$, $\xi \in \mathbb{R}^d$, and $\kappa \in \mathbb{R}^k$

$$\partial(\xi, \kappa)u(x, q) = u_{x^i}(x, q)\xi^i + u_{q^i}(x, q)\kappa^i,$$

$$\partial^2(\xi, \kappa)u(x, q) = u_{x^i x^j}(x, q)\xi^i \xi^j + 2u_{x^i q^j}(x, q)\xi^i \kappa^j + u_{q^i q^j}(x, q)\kappa^i \kappa^j.$$

Definition 4.9. Given a random process $x_t(r)$ defined for $t \geq 0$ and $r \in (r_1, r_2)$, where $r_1 < r_2$, and a point $r_0 \in (r_1, r_2)$, we say that $x_t(r)$ is t-*uniformly continuous at point $r = r_0$ in probability* if

$$P\text{-}\lim_{r \to r_0} \sup_{t \leq T} |x_t(r) - x_t(r_0)| = 0 \quad \forall T \in [0, \infty).$$

Also, if a random process ξ_t is defined for $t \in [0, \infty)$, we call ξ_t a t-*uniform derivative of $x_t(r)$ in probability* with respect to r at $r = r_0$ if

$$P\text{-}\lim_{\varepsilon \to 0} \sup_{t \leq T} \left| \varepsilon^{-1}[x_t(r_0 + \varepsilon) - x_t(r_0)] - \xi_t \right| = 0 \quad \forall T \in [0, \infty).$$

We write

$$\xi_t = P\text{-}\partial x_t(r)/\partial r|_{r=r_0}$$

and notice that two different processes ξ_t and ξ_t' can satisfy this equality only if they are indistinguishable.

In an obvious way one defines t-uniform continuity (on (r_1, r_2)), higher order t-uniform derivatives in probability and also partial derivatives and directional derivatives if x_t depends on a multidimensional parameter. In the case of processes like solutions $x_t = x_t(q)$ of (4.11), if $\kappa \in \mathbb{R}^k$ and $q \in Q$, we denote

$$x_{t(\kappa)}(q) = P\text{-}\partial x_t(q + r\kappa)/\partial r|_{r=0}, \quad x_{t(\kappa)(\kappa)}(q) = P\text{-}\partial x_{t(\kappa)}(q + r\kappa)/\partial r|_{r=0}.$$

Notice that it makes sense to speak about t-uniform continuity in probability of t-uniform derivatives in probability.

The above definitions make sense and will be also used in the case of random vectors rather than random processes depending on q.

Theorem 4.10. *Let $x_t(q)$ be a process which satisfies (4.11) for any $q \in Q$. Assume that $x_0(q)$ has all derivatives in probability of all orders $\leq n$ in Q. Then the process $x_t(q)$ has all t–uniform derivatives in probability of all orders $\leq n$ in Q.*

Also, for any $q \in Q$ and $\kappa \in \mathbb{R}^k$, the process $x_{t(\kappa)}(q)$ satisfies (4.12) (where $x_t = x_t(q)$) and $x_{t(\kappa)(\kappa)}(q)$ satisfies (4.13) if $n \geq 2$.

Finally, if the derivatives of $x_0(q)$ are continuous in probability, the derivatives of $x_t(q)$ are t–uniformly continuous in probability.

Proof. First of all notice that the functions $\partial(\xi, \kappa)\sigma$ and $\partial(\xi, \kappa)b$ are affine functions of ξ. Furthermore, for any T, ω, i, j

$$\int_0^T \{\|\sigma_{x^i}(t, x_t(q), q)\|^2 + \|\sigma_{q^j}(t, x_t(q), q)\|^2$$
$$+ |b_{x^i}(t, x_t(q), q)| + |b_{q^j}(t, x_t(q), q)|\} \, dt < \infty$$

which follows from (4.10) and the fact that any trajectory of $x_t(q)$ on $[0, T]$ is bounded. By Theorem 1.2 for any fixed $q \in Q$ and $\kappa \in \mathbb{R}^k$, there exists a unique solution ξ_t of (4.12) with initial data $\xi_0 = x_{0(\kappa)}(q)$.

Next, for r small enough so that $q + r\kappa \in Q$ let

$$\xi_t^{(r)} = \xi_t^{(r)}(q) = r^{-1}[x_t(q + r\kappa) - x_t(q)],$$
$$\sigma^{(r)}(t, \zeta) = r^{-1}[\sigma(t, x_t(q + r\kappa), q + r\kappa) - \sigma(t, x_t(q), q)]$$

and introduce $b^{(r)}(t, \zeta)$ similarly. Also let

$$\gamma^{(r)}(R) = \inf \{t \geq 0 : |\xi_t^{(r)}| \geq R\} \wedge T.$$

It is not hard to see that $\xi_t^{(r)}$ satisfies the following analogue of (1.16)

$$d\xi_t^{(r)} = \sigma^{(r)}(t, \xi_t^{(r)}) \, dw_t + b^{(r)}(t, \xi_t^{(r)}) \, dt.$$

Moreover, for $t < \gamma^{(r)}(R)$ obviously

$$|x_t(q + r\kappa) - x_t(q)| \leq rR$$

which in view of (4.10) and the continuity of the first derivatives of σ and b by Hadamard's formula and by the dominated convergence theorem implies that, for $(x_t(q, s, r), q(s, r)) := (sx_t(q+r\kappa) + (1-s)x_t(q), s(q+r\kappa) + (1-s)q)$, we have

$$\int_0^{\gamma^{(r)}(R)} \|\sigma^{(r)}(t, \xi_t^{(r)}) - \partial(\xi_t^{(r)}, \kappa)\sigma(t, x_t(q), q)\|^2 \, dt$$

$$= \int_0^{\gamma^{(r)}(R)} \| \int_0^1 \partial(\xi_t^{(r)}, \kappa)\sigma(t, x_t(q, s, r), q(s, r)) \, ds$$

$$- \partial(\xi_t^{(r)}, \kappa)\sigma(t, x_t(q), q)\|^2 \, dt$$

$$\leq \int_0^{\gamma^{(r)}(R)} \sup_{s\in(0,1)} \sup_{|\zeta|\leq R} ||\partial(\zeta,\kappa)\sigma(t,x_t(q,s,r),q(s,r))$$

$$- \partial(\zeta,\kappa)\sigma(t,x_t(q),q)]||^2 \, dt \to 0$$

as $r \to 0$ for every R, T, and ω. The same is true for b. By Theorem 1.8 this implies $\xi_t = x_{t(\kappa)}(q)$ and proves the first assertion of our theorem for $n = 1$.

Notice that equation (4.12) combined with (4.11) composes a system which can be considered as a stochastic equation relative to the couple x_t, ξ_t. By the previous result the solution of this system–equation is t–uniformly differentiable if $n \geq 2$ and as above one can write an equation for its derivative. This equation turns out to be system (4.12)–(4.13). One can obviously keep going to higher values of n, so that the first two assertions of the theorem are proved.

To prove the last assertion, we first prove that $x_t(q)$ is t-uniformly continuous in probability. Fix $q_0 \in Q$ and let a cut–off function $\zeta \in C_0^\infty(Q)$ be such that $\zeta = 1$ in a neighborhood of q_0. Observe that for q close to q_0 the couple $(x_t(q), q)$ satisfies the system

$$dx_t = \sigma(t,x_t,q_t)\zeta(q_t) \, dw_t + b(t,x_t,q_t)\zeta(q_t) \, dt, \quad dq_t = 0.$$

The coefficients of this system satisfy (1.2) with $K_t(R)$ equal to a constant, depending only on d and k, times

$$\sum_{i,j} \sup_{|x|\leq R, |q|\leq R} \left(||\sigma_{x^i}(t,x,q)\zeta(q)||^2 + ||(\sigma(t,x,q)\zeta(q))_{q^j}||^2 \right.$$

$$\left. + |b_{x^i}(t,x,q)\zeta(q)| + |(b(t,x,q)\zeta(q))_{q^j}| \right),$$

which is integrable in t owing to condition (4.10). By Theorem 1.7 we conclude that $x_t(q)$ is t-uniformly continuous in probability indeed.

Now let $q_n \to q_0 \in Q$. Observe that $\xi_t^n := x_{t(\kappa)}(q_n)$ and $\xi_t := x_{t(\kappa)}(q_0)$ satisfy

$$d\xi_t^n = \sigma^n(t,\xi_t^n) \, dw_t + b^n(t,\xi_t^n) \, dt, \quad d\xi_t = \sigma(t,\xi_t) \, dw_t + b(t,\xi_t) \, dt$$

where

$$\sigma^n(t,\xi) = \partial(\xi,\kappa)\sigma(t,x_t(q_n),q_n), \quad b^n(t,\xi) = \partial(\xi,\kappa)b(t,x_t(q_n),q_n),$$

$$\sigma(t,\xi) = \partial(\xi,\kappa)\sigma(t,x_t(q_0),q_0), \quad b(t,\xi) = \partial(\xi,\kappa)b(t,x_t(q_0),q_0).$$

Define

$$\gamma^n(R) = \inf\{t \geq 0 : |\xi_t^n| \geq R\} \wedge T,$$

and notice that by virtue of t–uniform continuity of $x_t(q)$ (and condition (4.10)) we have

$$\int_0^{\gamma^n(R)} \left[||\sigma^n - \sigma||^2 + |b^n - b|\right](t, \xi_t^n)\, dt$$

$$\leq NR \sum_{i,j} \int_0^{\gamma^n(R)} \left[||\sigma_{x^i}(t, x_t(q_n), q_n) - \sigma_{x^i}(t, x_t(q_0), q_0)||^2\right.$$

$$+ ||\sigma_{q^j}(t, x_t(q_n), q_n) - \sigma_{q^j}(t, x_t(q_0), q_0)||^2$$

$$+ |b_{x^i}(t, x_t(q_n), q_n) - b_{x^i}(t, x_t(q_0), q_0)|$$

$$\left. + |b_{q^j}(t, x_t(q_n), q_n) - b_{q^j}(t, x_t(q_0), q_0)|\right] dt \to 0,$$

where N depends only on d and k. Also $\xi_0^n \to \xi_0$ by assumption.

Hence, by Theorem 1.8 we get that $\xi_t^n \to \xi_t$ t-uniformly in probability, which means that the first t-uniform derivatives of $x_t(q)$ are t-uniformly continuous in probability. By applying this conclusion to the couple $(x_t(q), x_{t(\kappa)}(q))$, which satisfies (4.11)–(4.12), we see that its first t-uniform derivatives are t-uniformly continuous in probability if $n \geq 2$. This means that the second t-uniform derivatives of $x_t(q)$ are t-uniformly continuous in probability if $n \geq 2$. We apply again this result to the couple $(x_t(q), x_{t(\kappa)}(q))$ and in an obvious way by induction we get that the nth t-uniform derivatives of $x_t(q)$ are t-uniformly continuous in probability. The theorem is proved.

Remark 4.11. Without assuming that the nth derivatives of $x_0(q)$ are continuous in probability we still have that the $n - 1$st t–uniform derivatives of $x_t(q)$ are t-uniformly continuous in probability.

Indeed, above we have seen that $x_t(q)$ is t-uniformly continuous in probability (for any $n \geq 1$). By the same reason the couple $(x_t(q), x_{t(\kappa)}(q))$ is t-uniformly continuous in probability if $n \geq 2$. Hence, the first t–uniform derivatives of $x_t(q)$ are t-uniformly continuous in probability if $n \geq 2$. By applying this conclusion to the couple $(x_t(q), x_{t(\kappa)}(q))$, we see that its first t–uniform derivatives are t-uniformly continuous in probability if $n \geq 3$. This means that the second t–uniform derivatives of $x_t(q)$ are t-uniformly continuous in probability if $n \geq 3$. By induction we get our conclusion.

Exercise 4.12. Under the conditions of Theorem 4.10 take a real–valued function $f(t, x, q)$ satisfying the same conditions as $\sigma(t, x, q)$ and $b(t, x, q)$. Prove that the process

$$\int_0^t f(s, x_s(q), q)\, ds$$

is n times t-uniformly differentiable in probability with respect to q. (Hint: denote this process by $x_t^{d+1}(q)$ and consider $(x_t^1, ..., x_t^{d+1})$.)

Exercise 4.13. In the above exercise also take a real–valued function $c_t(x, q)$ satisfying the same conditions as $\sigma(t, x, q)$ and $b(t, x, q)$. Prove that the process

$$\int_0^t f(s, x_s(q), q) \exp\left(\int_0^s c(r, x_r(q), q)\, dr\right) ds$$

is n times t-uniformly differentiable in probability with respect to q.

Remark 4.14. In [KR92] a version of Theorem 4.10 is given in which equation has solutions which stay in a domain for all times and the coefficients may blow up when the process approaches the boundary.

4.3 Estimating moments of derivatives of solutions

Let $x_t(q)$ be a solution of equation (4.11). It is convenient in this subsection to assume that the domain $Q \subset \mathbb{R}^k$ (where q varies) is convex. Fix some constants $T, K, p, m > 0$. In Subsec. 4.2 we found some conditions under which $x_t(q)$ is differentiable in probability. In the future we need $x_t(q)$ to be differentiable in the mean because we are going to differentiate expectations. Here we give some sufficient conditions for that to happen. These conditions also provide some estimates of derivatives.

Theorem 4.15. *Assume that*

$$||\sigma(t, x_1, q_1) - \sigma(t, x_2, q_2)|| + |b(t, x_1, q_1) - b(t, x_2, q_2)| \leq K[|x_1 - x_2| + |q_1 - q_2|]$$

for all $t \in [0, T]$, $x_i \in \mathbb{R}^d$, $q_i \in Q$, and ω. Then for any $t \in [0, T]$, $q \in Q$, $r > 0$, and unit $\kappa \in \mathbb{R}^k$ such that $q + r\kappa \in Q$ we have

$$E \sup_{s \leq t} r^{-p} |x_s(q + r\kappa) - x_s(q)|^p \leq N e^{Nt} (1 + E r^{-p} |x_0(q + r\kappa) - x_0(q)|^p),$$

where $N = N(p, K)$.

Proof. Denote $\xi_t^{(r)} = r^{-1}[x_t(q + r\kappa) - x_t(q)]$. Obviously $\xi_t^{(r)}$ satisfies

$$d\xi_t^{(r)} = \sigma^{(r)}(t, \xi_t^{(r)}) \, dw_t + b^{(r)}(t, \xi_t^{(r)}) \, dt,$$

where

$$\sigma^{(r)}(t, \xi_t^{(r)}) := r^{-1}[\sigma(t, x_t(q + r\kappa), q + r\kappa) - \sigma(t, x_t(q), q)]$$

and $b^{(r)}(\xi_t^{(r)})$ is defined similarly. Since

$$||\sigma^{(r)}(t, \xi_t^{(r)})|| + |b^{(r)}(t, \xi_t^{(r)})| \leq K(1 + |\xi_t^{(r)}|),$$

the assertion of our theorem follows from Theorem 4.6. The theorem is proved.

Corollary 4.16. *If $E|x_0(q + r\kappa) - x_0(q)|^p \to 0$ as $r \to 0$, then*

$$E \sup_{s \leq T} |x_s(q + r\kappa) - x_s(q)|^p \to 0.$$

By combining Theorem 4.15 with Theorem 4.10 and well–known results bearing on convergence in probability under boundedness of moments, we arrive at the following corollary.

Corollary 4.17. *If $\sigma(t, x, q)$ and $b(t, x, q)$ are continuously differentiable with respect to (x, q) and if $x_0(q)$ is differentiable in the mean of order p in Q, then for any $0 < p_1 < p$*

$$\lim_{r \to 0} E \sup_{s \le T} |x_{s(\kappa)}(q) - r^{-1}[x_s(q + r\kappa) - x_s(q)]|^{p_1} = 0,$$

$$E \sup_{s \le T} |x_{t(\kappa)}(q)|^p \le N e^{NT}(1 + E|x_{0(\kappa)}(q)|^p).$$

Theorem 4.18. *Let the assumption of Theorem 4.15 be satisfied. Let $x_0(q)$ be differentiable in the mean of order $2p\delta$ in Q, where δ is a constant, $\delta \in (1, \infty)$. Moreover, let $\sigma(t, x, q)$ and $b(t, x, q)$ be twice continuously differentiable with respect to (x, q) and the absolute values of any their second order partial derivative with respect to (x, q) be less than $K(1 + |x|^m)$. Then for any $t \in [0, T]$, $q \in Q$, $r > 0$, and unit $\kappa \in \mathbb{R}^k$ such that $q + r\kappa \in Q$ we have*

$$E \sup_{s \le t} r^{-p} |x_{s(\kappa)}(q + r\kappa) - x_{s(\kappa)}(q)|^p$$

$$\le N e^{Nt} \left(E r^{-p} |x_{0(\kappa)}(q + r\kappa) - x_{0(\kappa)}(q)|^p + J \right), \quad (4.14)$$

where $N = N(\delta, p, K, d, d_1, k, m)$, $J = J_1 J_2$, and for $\delta' := \delta/(\delta - 1)$

$$J_1^\delta := 1 + E r^{-2p\delta} |x_0(q + r\kappa) - x_0(q)|^{2p\delta} + E|x_{0(\kappa)}(q)|^{2p\delta},$$

$$J_2^{\delta'} := 1 + E|x_0(q)|^{pm\delta'} + E|x_0(q + r\kappa)|^{pm\delta'}.$$

Proof. By Theorem 4.10 the process $\xi_t(q) := x_{t(\kappa)}(q)$ satisfies (4.12). Hence, for $\eta_t^{(r)} := r^{-1}[\xi_t(q + r\kappa) - \xi_t(q)]$ we have

$$d\eta_t^{(r)} = r^{-1}[\partial(\xi_t(q+r\kappa), \kappa)\sigma(t, x_t(q+r\kappa), q+r\kappa) - \partial(\xi_t(q), \kappa)\sigma(t, x_t(q), q)] \, dw_t$$

$$+ r^{-1}[\partial(\xi_t(q + r\kappa), \kappa)b(t, x_t(q + r\kappa), q + r\kappa) - \partial(\xi_t(q), \kappa)b(t, x_t(q), q)] \, dt,$$

$$d\eta_t^{(r)} = [\sigma^{(r)}(t, \eta_t^{(r)}) + g_t^{(r)}] \, dw_t + [b^{(r)}(t, \eta_t^{(r)}) + f_t^{(r)}] \, dt, \quad (4.15)$$

where

$$\sigma^{(r)}(t, \eta_t^{(r)}) = \partial(\eta_t^{(r)}, 0)\sigma(t, x_t(q + r\kappa), q + r\kappa),$$

$$b^{(r)}(t, \eta_t^{(r)}) = \partial(\eta_t^{(r)}, 0)b(t, x_t(q + r\kappa), q + r\kappa),$$

$$g_t^{(r)} = r^{-1}[\partial(\xi_t(q), \kappa)\sigma(t, x_t(q + r\kappa), q + r\kappa) - \partial(\xi_t(q), \kappa)\sigma(t, x_t(q), q)],$$

$$f_t^{(r)} = r^{-1}[\partial(\xi_t(q), \kappa)b(t, x_t(q + r\kappa), q + r\kappa) - \partial(\xi_t(q), \kappa)b(t, x_t(q), q)].$$

Furthermore,

$$\|\sigma^{(r)}(t, \eta_t^{(r)})\| + |b^{(r)}(t, \eta_t^{(r)})| \le N|\eta_t^{(r)}|,$$

$$|f_t^{(r)}| + |g_t^{(r)}| \le N(1 + |\xi_t(q)|)(1 + |\xi_t^{(r)}|)(1 + |x_t(q)|^m + |x_t(q + r\kappa)|^m).$$

Hence from (4.15) by Theorem 4.7 we get

$$E \sup_{s \leq t} |\eta_s^{(r)}|^p \leq N e^{Nt} (E |\eta_0^{(r)}|^p$$

$$+ E \sup_{s \leq t} \{(1 + |\xi_t(q)|^p)(1 + |\xi_t^{(r)}|^p)(1 + |x_t(q)|^{pm} + |x_t(q + r\kappa)|^{pm})\}).$$

Finally, we use the inequality

$$E a^p b^p c^{mp} \leq E(a^{2p} + b^{2p}) c^{mp} \leq N \big(E(a^{2p\delta} + b^{2p\delta}) \big)^{1/\delta} \big(E c^{mp\delta'} \big)^{1/\delta'},$$

Theorems 4.6 and 4.15 and Corollary 4.17 and we arrive at (4.14). The theorem is proved.

Similarly to Corollary 4.17 we get the following statement.

Corollary 4.19. *Under the assumptions of Theorem 4.18, if $x_0(q)$ is continuous in the mean of order $pm\delta'$ and differentiable in the mean of order $2p\delta$ and if $x_{0(\kappa)}(q)$ is differentiable in the mean of order p in Q, then for any $t \in [0, T]$ and $0 < p_1 < p$*

$$\lim_{r \to 0} E \sup_{s \leq T} |x_{s(\kappa)(\kappa)}(q) - r^{-1}[x_{s(\kappa)}(q + r\kappa) - x_s(q)]|^{p_1} = 0,$$

$$E \sup_{s \leq T} |x_{t(\kappa)(\kappa)}(q)|^p \leq N e^{NT} \Big(E |x_{0(\kappa)(\kappa)}(q)|^p$$

$$+ \big(1 + E|x_{0(\kappa)}(q)|^{2p\delta}\big)^{1/\delta} \big(1 + E|x_0(q)|^{pm\delta'}\big)^{1/\delta'} \Big).$$

4.4 The notions of $\mathcal{L}$–continuity and $\mathcal{L}$–differentiability

In this subsection T is a fixed number from $(0, \infty)$.

Definition 4.20. Let a real–valued random process $x_t = x_t(\omega)$ be defined for $t \in [0, T]$. We write $x_t \in \mathcal{L}$ if the process $x_t(\omega)$ is measurable in (ω, t) and for all $p > 0$

$$E \int_0^T |x_t|^p \, dt < \infty.$$

We write $x_t \in \mathcal{LB}$ if x_t is separable and for all $p > 0$

$$E \sup_{t \leq T} |x_t|^p < \infty.$$

The convergence in spaces $\mathcal{L}$ and $\mathcal{LB}$ is defined in the following way.

Definition 4.21. Let $x_t^n \in \mathcal{L}(\in \mathcal{LB})$, $n = 0, 1, 2....$ We say that $\mathcal{L}$–limit ($\mathcal{LB}$–limit) of x_t^n equals x_t^0 and we write

$$\mathcal{L}\text{-} \lim_{n \to \infty} x_t^n = x_t^0 \quad (\mathcal{LB}\text{-} \lim_{n \to \infty} x_t^n = x_t^0)$$

if for all $p > 0$

$$\lim_{n \to \infty} E \int_0^T |x_t^n - x_t^0|^p \, dt = 0 \quad (\text{respectively,} \lim_{n \to \infty} E \sup_{t \leq T} |x_t^n - x_t^0|^p = 0).$$

By Hölder's inequality in both definitions we can replace $p > 0$ with $p \geq 1$.

Having introduced the notions of $\mathcal{L}$–limit and $\mathcal{LB}$–limit, it is clear what we mean by $\mathcal{L}$–continuity and $\mathcal{LB}$–continuity at a point q_0 of a function $x_t(q)$ defined in a domain $Q \subset \mathbb{R}^k$.

Definition 4.22. Let $q_0 \in Q$, $\kappa \in \mathbb{R}^k$, $y_t \in \mathcal{L}$ (respectively $\mathcal{LB}$). Suppose that for each q from some neighborhood of q_0 a process $x_t(q) \in \mathcal{L}$ ($\mathcal{LB}$) is given. We say that y_t is a $\mathcal{L}$–derivative ($\mathcal{LB}$–derivative) of $x_t(q)$ in the direction of κ at q_0 and we write

$$y_t = \mathcal{L}\text{-}x_{t(\kappa)}(q_0) \quad \left(y_t = \mathcal{LB}\text{-}x_{t(\kappa)}(q_0) \right)$$

if

$$y_t = \mathcal{L}\text{-}\lim_{r \to 0} r^{-1}[x_t(q_0 + r\kappa) - x_t(q_0)] \quad \left(y_t = \mathcal{LB}\text{-}\lim_{r \to 0} r^{-1}[x_t(q_0 + r\kappa) - x_t(q_0)] \right).$$

We say that the process $x_t(q)$ is once $\mathcal{L}$–differentiable (respectively, $\mathcal{LB}$–differentiable) at q_0 if $x_t(q)$ has $\mathcal{L}$–derivatives ($\mathcal{LB}$–derivatives) in all directions at q_0. The process $x_t(q)$ is said to be i times ($i \geq 2$) $\mathcal{L}$–differentiable (respectively, $\mathcal{LB}$–differentiable) at q_0 if $x_t(q)$ is once $\mathcal{L}$–differentiable (respectively, $\mathcal{LB}$–differentiable) at any point of some neighborhood of q_0 and each first $\mathcal{L}$–derivative ($\mathcal{LB}$–derivative) is $i - 1$ times $\mathcal{L}$–differentiable ($\mathcal{LB}$–differentiable) at q_0.

Notice that, of course, $\mathcal{LB}$–derivatives also are t-uniform derivatives in probability.

Definitions 4.20, 4.21, and 4.22 have been given for real–valued processes. They are extended to vector- or matrix–valued processes in an obvious way.

Further, as is commonly done in calculus we write $y_t(q) = \mathcal{L}\text{-}x_{t(\kappa)}(q)$ if $y_t(q_0) = \mathcal{L}\text{-}x_{t(\kappa)}(q_0)$ for any q_0, $\mathcal{L}\text{-}x_{t(\kappa_1)(\kappa_2)}(q) = \mathcal{L}\text{-}y_{t(\kappa_2)}(q)$, where $y_t(q) = \mathcal{L}\text{-}x_{t(\kappa_1)}(q)$, etc. We say that $x_t(q)$ is i times $\mathcal{L}$–continuously $\mathcal{L}$–differentiable if all $\mathcal{L}$–derivatives of $x_t(q)$ up to and including order i are $\mathcal{L}$–continuous. We will not dwell on explaining such obvious figures of speech in the future.

In order to get familiar with the given definitions, we note a few simple properties these definitions imply. It is obvious that for random variables, the notion of $\mathcal{L}$–continuity is equivalent to that of $\mathcal{LB}$–continuity. Furthermore, $|Ex(q) - Ex(q_0)| \leq E|x(q) - x(q_0)|$. Hence the expectation of a $\mathcal{L}$–continuous random variable is continuous. Since

$$\left| r^{-1}(Ex(q_0 + r\kappa) - Ex(q_0)) - Ey \right| \leq E\left| r^{-1}(x(q_0 + r\kappa) - x(q_0)) - y \right|,$$

the derivative of $Ex(q)$ in the direction of κ at q_0 is equal to the expectation of the $\mathcal{L}$–derivative of $x(p)$ if the latter exists. Therefore, the sign of the first derivative is interchangeable with the sign of the expectation. Combining the above properties, we deduce that $(Ex(q))_{(\kappa)}$ exists and is continuous at q_0 if $x(q)$ is $\mathcal{L}$–continuously $\mathcal{L}$–differentiable at q_0 in the direction of κ. A similar situation occurs for higher order derivatives.

Since for a random variable $\tau \leq T$

$$E|x_\tau(q) - x_\tau(q_0)| \le E \sup_{t \le T} |x_t(q) - x_t(q_0)|,$$

the variable $x_\tau(q)$ is $\mathcal{L}$-continuous if $x_t(q)$ is $\mathcal{LB}$-continuous and $x_\tau(q)$ is a measurable function of ω. A similar inequality shows that for the same τ

$$\mathcal{L}\text{-}x_{\tau(\kappa)}(q) = (\mathcal{LB}\text{-}x_{t(\kappa)}(q))|_{t=\tau} \qquad (4.16)$$

if $x_t(q)$ is $\mathcal{LB}$-differentiable in the direction of κ and $x_\tau(q)$ and the right-hand side of (4.16) are random variables. This argument allows us to derive properties of $\mathcal{L}$-continuity and $\mathcal{L}$-differentiability of the random variable $x_\tau(q)$ from those of the process $x_t(q)$. In addition, (4.16) shows that substituting τ for t is interchangeable with derivating.

Next, suppose that $x(t,q) := x_t(q)$ is continuous with respect to t and $\mathcal{LB}$-continuous at q_0. Also, let $\tau(q)$ be random functions with values in $[0,T]$ continuous in probability at q_0. Then we assert that $x(\tau(q), q_0)$ and $x(\tau(q), q)$ are $\mathcal{L}$-continuous at q_0.

In fact, $|x(\tau(q), q_0) - x(\tau(q_0), q_0)|^p \to 0$ in probability as $q \to q_0$ and in addition this quantity is bounded by $2^p \sup_{t \le T} |x(t, q_0)|^p$, which is summable and independent of q. Therefore, $E|x(\tau(q), q_0) - x(\tau(q_0), q_0)|^p \to 0$, i.e., $x(\tau(q), q_0)$ is $\mathcal{L}$-continuous. The $\mathcal{L}$-continuity of $x(\tau(q), q)$ follows from that of $x(\tau(q), q_0)$ and from the inequalities

$$E|x(\tau(q), q) - x(\tau(q_0), q_0)|^p$$
$$\le 2^p E|x(\tau(q), q) - x(\tau(q), q_0)|^p + 2^p E|x(\tau(q), q_0) - x(\tau(q_0), q_0)|^p$$
$$\le 2^p E \sup_{t \le T} |x(t, q) - x(t, q_0)|^p + 2^p E|x(\tau(q), q_0) - x(\tau(q_0), q_0)|^p.$$

According to Hölder's inequality for $p \ge 1$

$$E \sup_{t \le T} \left| \int_0^t x_s(q)\, ds - \int_0^t x_s(q_0)\, ds \right|^p \le E \left(\int_0^T |x_s(q) - x_s(q_0)|\, ds \right)^p$$

$$\le T^{p-1} E \int_0^T |x_s(q) - x_s(q_0)|^p\, ds.$$

Therefore, $\int_0^t x_s(q)\, ds$ is an $\mathcal{LB}$-continuous process if the process $x_t(q)$ is $\mathcal{L}$-continuous. It is proved in a similar way that this integral has an $\mathcal{LB}$-derivative in the direction of κ and this derivative coincides with the integral of the $\mathcal{L}$-derivative of $x_t(q)$ in the direction of κ if the latter derivative exists.

Combining the assertions discussed above one can obtain numerous useful facts. We do not list them, because all they are trivial. Less trivial is the following fact from which we later derive continuity of compositions.

Lemma 4.23. *Let x_t^n be d-dimensional processes measurable with respect to (ω, t) given for $n = 0, 1, 2\ldots$. Assume that*

$$\mathcal{L}\text{-}\lim_{n \to \infty} x_t^n = x_t^0. \qquad (4.17)$$

Let $f(t, x)$ be a random function defined for $x \in \mathbb{R}^d$ and $t \in [0, T]$. Assume that f is measurable with respect to (ω, t) for any x and is continuous in x for any ω, t. Also let

$$|f(t, x)| \leq K(1 + |x|^m) \tag{4.18}$$

for all ω, t, x where K, m are some constants. Then

$$\mathcal{L}\text{-}\lim_{n \to \infty} f(t, x_t^n) = f(t, x_t^0).$$

Proof. It follows from (4.17) that $x_t^n \to x_t^0$ in measure $dP \times dt$. The convergence in measure is related in a very well–known way to the convergence of subsequences almost everywhere, whence and from the continuity of $f(t, x)$ in x we get that $f(t, x_t^n) \to f(t, x_t^0)$ in measure $dP \times dt$. It only remains to show that for any $p \geq 1$

$$\overline{\lim_{n \to \infty}} \, E \int_0^T |f(t, x_t^n)|^p \, dt < \infty.$$

Owing to (4.18) it suffices to show

$$\sup_n E \int_0^T |x_t^n|^p \, dt < \infty$$

for any $p \geq 1$. But the last inequality follows obviously from the fact that $x_t^n \to x_t^0$ in $L_p(\Omega \times [0, T])$ (for all p). The lemma is proved.

Corollary 4.24. *If we are given one–dimensional processes x_t^n and y_t^n defined for $n = 0, 1, 2...$ and such that $\mathcal{L}\text{-}\lim_{n \to \infty} x_t^n = x_t^0$ and $\mathcal{L}\text{-}\lim_{n \to \infty} y_t^n = y_t^0$, then*

$$\mathcal{L}\text{-}\lim_{n \to \infty} x_t^n y_t^n = x_t^0 y_t^0.$$

Indeed, the two–dimensional process (x_t^n, y_t^n) has $\mathcal{L}$–limit equal to (x_t^0, y_t^0). In addition, the function $f(x, y) := xy$ satisfies the growth condition $|f(x, y)| \leq |(x, y)|^2 = x^2 + y^2$. Hence $\mathcal{L}\text{-}\lim_{n \to \infty} f(x_t^n, y_t^n) = f(x_t^0, y_t^0)$.

4.5 Differentiability of certain expectations depending on a parameter

Here we continue the investigation of Subsec. 4.4 with more emphasis on differentiability of composite functions and their expectations.

Take some constants $T, K, m > 0$.

Lemma 4.25. *Let the assumptions of Lemma 4.23 be satisfied and let $x_t^n(u)$ be random d-vectors defined for $n = 1, 2, 3..$, $u \in [0, 1]$ $t \leq T$. Assume that $x_t^n(u)$ are continuous in u, measurable in (ω, t) and satisfy*

$$|x_t^n(u) - x_t^0| \leq |x_t^n - x_t^0|.$$

Then

$$\mathcal{L}\text{-}\lim_{n \to \infty} \int_0^1 f(t, x_t^n(u)) \, du = f(t, x_t^0).$$

Proof. By Hölder's inequality for $p \geq 1$

$$E \int_0^T | \int_0^1 f(t, x_t^n(u)) \, du - f(t, x_t^0)|^p \, dt = E \int_0^T | \int_0^1 [f(t, x_t^n(u)) - f(t, x_t^0)] \, du|^p \, dt$$

$$\leq E \int_0^T \int_0^1 |f(t, x_t^n(u)) - f(t, x_t^0)|^p \, du \, dt.$$

The last expression goes to zero as $n \to \infty$. Indeed, as in the proof of Lemma 4.23 we have $f(t, x_t^n(u)) \to f(t, x_t^0)$ in measure $dP \times dt \times du$ and

$$\sup_n E \int_0^T \int_0^1 |f(t, x_t^n(u)) - f(t, x_t^0)|^{2p} \, du \, dt$$

$$\leq N(1 + \sup_{n,u} E \int_0^T |x_t^n(u)|^{2mp} \, dt + E \int_0^T |x_t^0|^{2mp} \, dt)$$

$$\leq N(1 + \sup_n E \int_0^T |x_t^n|^{2mp} \, dt + E \int_0^T |x_t^0|^{2mp} \, dt) < \infty.$$

The lemma is proved.

In the following few results we consider composite functions like $f(t, x_t(q))$ of functions $f(t, x)$ and $x_t(q)$, so that $f(t, x)$ does not depend explicitly on the parameter q, which varies in a domain $Q \subset \mathbb{R}^k$. This case is not less general than when $f = f(t, q, x)$ because one can always consider the couple $(q, x_t(q))$ as one process.

Theorem 4.26. *Let an $\mathbb{R}^d$-valued random process $x_t(q)$, $t \leq T$ be defined for $q \in Q$ and let a real valued process $f(t, x)$, $t \leq T$ be defined for any $x \in \mathbb{R}^d$. Assume that $f(t, x)$ is measurable with respect to (ω, t) for any x. Take a point $q_0 \in Q$.*

(i) Let $f(t, x)$ be continuous with respect to x for any ω, t and let (4.18) hold and let $x_t(q)$ be $\mathcal{L}$-continuous at q_0. Then the process $f(t, x_t(q))$ is $\mathcal{L}$-continuous at q_0.

(ii) Let $f(t, x)$ be i times continuously differentiable with respect to x for any ω, t and let f and all its derivatives up to and including order i satisfy (4.18). Finally, let $x_t(q)$ be i times ($\mathcal{L}$-continuously) $\mathcal{L}$-differentiable at q_0. Then the process $f(t, x_t(q))$ is i times (respectively, $\mathcal{L}$-continuously) $\mathcal{L}$-differentiable at q_0 as well. In addition,

$$\mathcal{L}\text{-}(f(t, x_t(q)))_{(\kappa)} = f_{(y_t(q))}(t, x_t(q)),$$
$$\mathcal{L}\text{-}(f(t, x_t(q)))_{(\kappa)(\kappa)} = f_{(z_t(q))}(t, x_t(q)) + f_{(y_t(q))(y_t(q))}(t, x_t(q)), \qquad (4.19)$$

where

$$y_t(q) = \mathcal{L}\text{-}x_{t(\kappa)}(q), \qquad z_t(q) = \mathcal{L}\text{-}x_{t(\kappa)(\kappa)}(q),$$

whenever the existence of the left-hand sides in (4.19) is asserted.

Proof. To prove (i) it suffices to take any sequence $q_n \to q_0$, let $x_t^n = x_t(q_n)$, and to use Lemma 4.23.

To prove (ii) first assume that $i = 1$. Observe that $f(t, x, y) := f_{t(y)}(x)$ is a continuous function of $(x, y) \in \mathbb{R}^{2d}$ and

$$|f(t, x, y)| \le N(1 + |x|^m)|y| \le N(1 + |(x, y)|^{m+1}).$$

Next, take $\kappa \in \mathbb{R}^k$ and a sequence of numbers $r_n \to 0$ and define

$$x_t^n(u) = u x_t(q_0 + r_n \kappa) + (1 - u) x_t(q_0), \quad y_t^n = r_n^{-1}(x_t(q_0 + r_n \kappa) - x_t(q_0)).$$

By the Newton–Leibniz rule we have

$$r_n^{-1}\left[f(t, x_t(q_0 + r_n \kappa)) - f(t, x_t(q_0))\right] = r_n^{-1} \int_0^1 \frac{\partial}{\partial u} f(t, x_t^n(u))\, du$$

$$= \int_0^1 f(t, x_t^n(u), y_t^n)\, du.$$

Here

$$|x_t^n(u) - x_t(q_0)| \le |x_t(q_0 + r_n \kappa) - x_t(q_0)|,$$

so that by Lemma 4.25 applied to $x_t^n(u)$ and $y_t^n(u) := y_t^n$ we have

$$\mathcal{L}\text{-} \lim_{n \to \infty} \int_0^1 f(t, x_t^n(u), y_t^n)\, du = f(t, x_t(q_0), y_t(q_0)).$$

Therefore,

$$\mathcal{L}\text{-} \lim_{r \to 0} r^{-1}[f(t, x_t(q_0 + r\kappa)) - f(t, q_0)] = f(t, x_t(q_0), y_t(q_0)).$$

Finally, by (i), $f(t, x_t(q_0), y_t(q_0))$ is $\mathcal{L}$–continuous with respect to q_0 if $x_t(q_0)$ is $\mathcal{L}$–continuously $\mathcal{L}$–differentiable with respect to q_0. This proves assertion (ii) for $i = 1$ and also proves the first formula in (4.19) which we choose to write in the following way

$$\mathcal{L}\text{-}(f(t, x_t(q)))_{(\kappa)} = f(t, x_t(q), y_t(q)).$$

For general $i \ge 1$ we use the induction on i. Assume that (ii) is proved for $i \le j$ and for any processes $f(t, x)$ and $x_t(q)$ satisfying the assumptions in (ii). Let our couple $f(t, x), x_t(q)$ satisfy the conditions in (ii) for $i = j+1$. We take the derivative $\mathcal{L}\text{-}(f(t, x_t(q)))_{(\kappa)}$ and prove that it is j times $\mathcal{L}$–differentiable at q_0.

Let us write this derivative as $f(t, x_t(q), y_t(q))$. Notice that the process $(x_t(q), y_t(q))$ is j times $\mathcal{L}$–differentiable at q_0 by assumption and the function $f(t, x, y)$ is j times continuously differentiable with respect to the couple (x, y). Also the absolute values of the partial derivatives of $f(t, x, y)$ of orders up to and including j are less than $N(1 + |(x, y)|^{m+1})$. Therefore, by the induction assumption, $f(t, x_t(q), y_t(q))$ is j times $\mathcal{L}$–differentiable at q_0. Since

κ is arbitrary, $f(t, x_t(q))$ is, by definition (!), $j + 1$ times $\mathcal{L}$–differentiable at q_0.

Similarly one proves $\mathcal{L}$–continuity of $\mathcal{L}$–derivatives of $f(t, x_t(q))$ at q_0 if $\mathcal{L}$–derivatives of $x_t(q)$ are $\mathcal{L}$–continuous at q_0. Finally, in conjunction with the first formula in (4.19)

$$\mathcal{L}\text{-}\big(f(t, x_t(q))\big)_{(\kappa)(\kappa)} = \mathcal{L}\text{-}\big(f(t, x_t(q), y_t(q))\big)_{(\kappa)} = f_{(y_t(q), z_t(q))}(t, x_t(q), y_t(q)),$$

which after simple transformations yields the second formula in (4.19). The theorem is proved.

Remark 4.27. Theorem 4.26 allows one to prove the $\mathcal{L}$–continuity and $\mathcal{L}$–differentiability of various expressions containing random processes.

For example, arguing in the same way as in Corollary 4.24, one proves that if $x_t(q)$ and $y_t(q)$ are real–valued i times $\mathcal{L}$–differentiable processes, the product $x_t(q)y_t(q)$ is i times $\mathcal{L}$–differentiable as well.

Another important example is given by the process $\exp(-x_t(q))$ which is i times $\mathcal{L}$–differentiable if $x_t(q)$ is i times $\mathcal{L}$–differentiable and $x_t(q)$ is nonnegative or just bounded from below. Indeed, notwithstanding that the function $\exp(-x)$ grows more rapidly than any polynomial as $x \to -\infty$ while considering nonnegative $x_t(q)$ we may write $\exp(-x_t(q)) = f(x_t(q))$, where $f(x)$ is any smooth function coinciding with $\exp(-x)$ for $x \geq 0$ and vanishing for $x \leq -1$.

Combining these examples with the known properties of integrals of $\mathcal{L}$–continuous and $\mathcal{L}$–differentiable functions we arrive at the following results.

Theorem 4.28. *Let the processes* $x_t(q), f_1(t, x), f_t(t, x)$ *satisfy the assumptions of Theorem 4.26 (i) (respectively, Theorem 4.26 (ii)) and let* $f_1(t, x) \geq 0$. *Then the process*

$$f_2(t, x_t(q)) \exp\Big\{ - \int_0^t f_1(t, x_s(q))\, ds \Big\} \tag{4.20}$$

is $\mathcal{L}$–continuous (respectively, i times (respectively, $\mathcal{L}$–continuously) $\mathcal{L}$–differentiable) at q_0.

Theorem 4.29. *Let the processes* $x_t(q), f(t, x)$ *satisfy the assumptions of Theorem 4.26 (i) (respectively, Theorem 4.26 (ii)) and let* $f(t, x) \geq 0$. *Let the random variable* $\tau(\omega) \in [0, T]$ *and let the time independent processes* $y(q), g(x)$ *satisfy the assumptions of Theorem 4.26 (i) (respectively, Theorem 4.26 (ii)). Then the random variable*

$$g(y(q)) \exp\Big\{ - \int_0^\tau f(s, x_s(q))\, ds \Big\} \tag{4.21}$$

is $\mathcal{L}$–continuous (respectively, i times (respectively, $\mathcal{L}$–continuously) $\mathcal{L}$–differentiable) at q_0.

Remark 4.30. Formulas 4.19 and our discussion concerning $\mathcal{L}$–derivatives of integrals and derivatives of expectations show that while computing $\mathcal{L}$–derivatives of (4.20) and (4.21) or of their expectations we can apply usual formulas from calculus.

Let us apply these results to the situation in which $x_t(q)$ is a solution of Itô's equation (4.11).

Theorem 4.31. *Let Q be convex and $x_0(q)$ be once (respectively twice) $\mathcal{L}$–continuously $\mathcal{L}$–differentiable in Q. Fix a point $q_0 \in Q$ and assume that $\sigma(t, 0, q_0)$ and $b(t, 0, q_0)$ are bounded. Also assume that the first order derivatives of $\sigma(t, x, q)$ and $b(t, x, q)$ with respect to (x, q) are continuous in (x, q) for any ω, t and are bounded functions of ω, t, x, q.*

Let $c(t, x, q)$, $f(t, x, q)$, and $g(x, q)$ be real–valued functions defined on $\Omega \times [0, T] \times \mathbb{R}^d \times Q$. Assume that they are once (respectively, they and $\sigma(t, x, q)$ and $b(t, x, q)$ are twice) continuously differentiable with respect to (x, q) for any ω, t and their absolute values together with the absolute values of their partial derivatives (respectively, up to and including the second order derivatives) with respect to (x, q) are less than $K(1 + |x|^m)$ for any ω, t, x, q. Finally, let $c(t, x, q) \geq 0$ and let τ be a stopping time, $\tau \leq T$.

Then $x_t(q)$ is once (twice) $\mathcal{LB}$–continuously $\mathcal{LB}$–differentiable in Q and the function

$$F(q) := E \int_0^\tau f(t, x_t(q), q) e^{-\int_0^t c(s, x_s(q), q)\, ds}\, dt + E g(x_\tau(q), q) e^{-\int_0^\tau c(s, x_s(q), q)\, ds}$$

is once (twice) continuously differentiable in Q.

Proof. By Theorem 4.6 we have $x_t(q) \in \mathcal{LB}$ for any $q \in Q$. By Theorem 4.10 the process $x_t(q)$ has one (respectively, two) t–uniform derivatives in probability which are t–uniformly continuous in probability.

By Corollaries 4.17 and 4.19 the t–uniform derivatives in probability of $x_t(q)$ also are its $\mathcal{LB}$–derivatives. From the estimates of these derivatives given in Corollaries 4.17 and 4.19 and from the above mentioned t–uniform continuity of the derivatives in probability, it follows that the derivatives are $\mathcal{LB}$–continuous.

This proves that $x_t(q)$ is once (respectively, twice) $\mathcal{LB}$–continuously $\mathcal{LB}$–differentiable. Our assertion about the function $F(q)$ follows now from Theorems 4.28 and 4.29 and Remark 4.30 applied to the two–component process $(x_t(q), q)$. The theorem is proved.

An important particular case of this theorem occurs when $Q = \mathbb{R}^d$, $x_0(q) = q$, $\sigma(t, x, q) = \sigma(t, x)$ and $b(t, x, q) = b(t, x)$ so that we are dealing with solutions $x_t(x)$ of the equation

$$x_t = x + \int_0^t \sigma(s, x_s)\, dw_s + \int_0^t b(s, x_s)\, ds.$$

Theorem 4.32. *Assume that $\sigma(t,x)$ and $b(t,x)$ are once (respectively, twice) continuously differentiable with respect to x for any ω, t and*

$$\|\sigma(t,0)\| + |b(t,0)| + \|\sigma_{x^i}(t,x)\| + |b_{x^i}(t,x)| \leq K$$

for any ω, t, x, i. Let $c(t,x), f(t,x)$, and $g(x)$ be real–valued functions on $\Omega \times [0,T] \times \mathbb{R}^d$. Assume that they are once (respectively, they and $\sigma(t,x)$ and $b(t,x)$ are twice) continuously differentiable with respect to x for any ω, t and their absolute values together with the absolute values of their derivatives (respectively, including the second order derivatives) with respect to x are less than $K(1+|x|^m)$ for any ω, t. Finally let $c(t,x) \geq 0$ and let τ be a stopping time, $\tau \leq T$.

Then $x_t(x)$ is once (twice) $\mathcal{LB}$–continuously $\mathcal{LB}$–differentiable with respect to x and the function

$$F(x) := E \int_0^\tau f(t,x_t(x))e^{-\int_0^t c(s,x_s(x))\,ds}\,dt + Eg(x_\tau(x))e^{-\int_0^\tau c(s,x_s(x))\,ds}$$

is once (twice) continuously differentiable in $\mathbb{R}^d$. In addition, for a constant $N = N(d,m,K)$ and all x we have

$$|F(x)| \leq Ne^{NT}(1+|x|^m), \quad |F_x(x)| \leq Ne^{NT}(1+|x|^{2m}),$$
$$(\text{respectively,} \quad |F_{xx}(x)| \leq Ne^{NT}(1+|x|^{3m})). \qquad (4.22)$$

Proof. We only need to prove the last assertion related to (4.22). The first inequality in (4.22) follows from Theorem 4.6. To prove the second one notice that for any $\xi \in \mathbb{R}^d$ and $\xi_t := \mathcal{L}\text{-}(x_t(x))_{(\xi)} = x_{t(\xi)}(x)$ we have

$$F_{(\xi)}(x) = E\mathcal{L}\text{-}\Big(\int_0^\tau f(t,x_t(x))e^{-\int_0^t c(s,x_s(x))\,ds}\,dt + g(x_\tau(x))e^{-\int_0^\tau c(s,x_s(x))\,ds} \Big)_{(\xi)}$$

$$= E\Big(\int_0^\tau [f_{(\xi_t)}(t,x_t) - f(t,x_t)\int_0^t c_{(\xi_s)}(s,x_s)\,ds]e^{-\int_0^t c(s,x_s)\,ds}\,dt$$

$$+ [g_{(\xi_\tau)}(x_\tau) - g(x_\tau)\int_0^\tau c_{(\xi_s)}(s,x_s)\,ds]e^{-\int_0^\tau c(s,x_s)\,ds} \Big), \quad (4.23)$$

where for simplicity the argument x is dropped. Hence

$$|F_{(\xi)}(x)| \leq N(1+T^2)E\sup_{t\leq T}|\xi_t|\sup_{t\leq T}(1+|x_t|^{2m})$$

$$\leq N(1+T^2)(E\sup_{t\leq T}|\xi_t|^2)^{1/2}(E\sup_{t\leq T}(1+|x_t|^{4m}))^{1/2}$$

To finish the proof of the second inequality in (4.22) it only remains to use Theorem 4.6 and Corollary 4.17, where $x_{0(\kappa)}(q)$ is to be replaced with $\xi_0 = \xi$.

While considering second order derivatives we use the fact that, for $\eta_t = x_{t(\xi)(\xi)}(x)$,

$$F_{(\xi)(\xi)}(x) = E\bigg(\int_0^T \Big[f_{(\eta_t)}(t, x_t) + f_{(\xi_t)(\xi_t)}(t, x_t) - f_{(\xi_t)}(t, x_t) \int_0^t c_{(\xi_s)}(s, x_s)\, ds$$

$$-f(t, x_t) \int_0^t \{ c_{(\eta_s)}(s, x_s) + c_{(\xi_s)(\xi_s)}(s, x_s) \}\, ds$$

$$-\{ f_{(\xi_t)}(t, x_t) - f(t, x_t) \int_0^t c_{(\xi_s)}(s, x_s)\, ds \} \int_0^t c_{(\xi_s)}(s, x_s)\, ds \Big] e^{-\int_0^t c(s, x_s)\, ds}\, dt$$

$$+\Big[g_{(\eta_\tau)}(x_\tau) + g_{(\xi_\tau)(\xi_\tau)}(x_\tau) - g_{(\xi_\tau)}(x_\tau) \int_0^T c_{(\xi_s)}(s, x_s)\, ds$$

$$-g(x_\tau) \int_0^T \{ c_{(\eta_s)}(s, x_s) + c_{(\xi_s)(\xi_s)}(s, x_s) \}\, ds$$

$$- \{ g_{(\xi_\tau)}(x_\tau) - g(x_\tau) \int_0^T c_{(\xi_s)}(s, x_s)\, ds \} \int_0^T c_{(\xi_s)}(s, x_s)\, ds \Big] e^{-\int_0^T c(s, x_s)\, ds} \bigg).$$

$$(4.24)$$

We estimate $|F_{(\xi)(\xi)}(x)|$ from above through

$$N(1 + T^3)\big(E\sup_{t \leq T}(|\eta_t| + |\xi_t|^2) \sup_{t \leq T}(1 + |x_t|^{2m}) + E\sup_{t \leq T}|\xi_t|^2 \sup_{t \leq T}(1 + |x_t|^{3m})\big)$$

and then as above we use Hölder's inequality, Theorem 4.6 and Corollary 4.17 along with Corollary 4.19, which in our situation says that (notice that $\eta_0 = 0$)

$$E\sup_{t \leq T}|\eta_t|^2 \leq Ne^{NT}(1 + |x|^{2m}).$$

This easily gives the desired result. The theorem is proved.

5. Kolmogorov's equation in the whole space

Here we want to apply the results of Subsec. 4.5 in order to give conditions in terms of the given data which could allow us to "uncondition" the results of Sec. 3.. As in Sec. 3., we consider equation (2.1) with nonrandom coefficients assuming that the conditions stated before Remark 1.1 are satisfied.

Our first goal is to give sufficient conditions for the regularity of (2.1).

5.1 Stratified equations

In this subsection we prove a theorem which will be used in the following one for deriving more or less explicit conditions on σ, b which are sufficient for (2.1) to be regular. It turns out that for further applications it is more useful to consider not equation (2.1) alone but append it by another equation. Definition 2.1 concerns (2.1) which, in fact, is a system of one–dimensional equations for each coordinate x_t^i. Let us take a system for a $(d + d_0)$–dimensional process $z_t = (x_t, y_t)$ whose first d coordinates we represent as x_t and remaining d_0 coordinates as y_t. We seek a process z_t such that x_t satisfies (2.1) and y_t satisfies the equation

$$y_t = y + \int_0^t \bar{\sigma}(s + r, z_r) \, dw_r + \int_0^t \bar{b}(s + r, z_r) \, dr, \ t \geq 0 \qquad (5.1)$$

For convenience we enumerate the following assumptions about $\bar{\sigma} = \bar{\sigma}(t, z)$ and $\bar{b} = \bar{b}(t, z)$.

Assumption 5.1. The functions $\bar{\sigma}, \bar{b}$ are Borel functions defined on $[0, \infty) \times \mathbb{R}^{d+d_0}$ with values in the set of all $d_0 \times d_1$-matrices and in $\mathbb{R}^{d_0}$ respectively. They are continuous with respect to z for every $t \geq 0$ and for every $T, R \geq 0$

$$\int_0^T \sup_{|z| \leq R} [\|\bar{\sigma}(t, z)\|^2 + |\bar{b}(t, z)|] \, dt < \infty. \qquad (5.2)$$

In other words, this assumption means that the coefficients of *system* (2.1), (5.1) considered as a single equation with respect to $z_t = (x_t, y_t)$ satisfy the conditions assumed for the coefficients of (2.1). Therefore, Definition 2.1 makes sense for this system.

Assumption 5.2. The couple $(\bar{\sigma}, \bar{b})$ is monotone and coercive with respect to y. More precisely, for every $R \in [0, \infty)$, on $[0, \infty)$ there exists a positive Borel locally integrable function $K_t(R)$ such that for all $t \in [0, \infty), x \in \mathbb{R}^d, y, y^{(1)}, y^{(2)} \in \mathbb{R}^{d_0}, R_0 \in [0, \infty)$, satisfying $|x| \leq R_0, |y^{(1)}|, |y^{(2)}| \leq R$, it holds that

$$2(y^{(1)} - y^{(2)}, \bar{b}(t, x, y^{(1)}) - \bar{b}(t, x, y^{(2)}))$$
$$+ \|\bar{\sigma}(t, x, y^{(1)}) - \bar{\sigma}(t, x, y^{(2)})\|^2 \leq K_t(R \vee R_0)|y^{(1)} - y^{(2)}|^2, \quad (5.3)$$

$$2(y, \bar{b}(t, x, y)) + \|\bar{\sigma}(t, x, y)\|^2 \leq K_t(R_0)(1 + |y|^2). \qquad (5.4)$$

Condition (5.3) is called the monotonicity condition whereas (5.4) is called the coercivity condition.

Theorem 5.3. *If equation (2.1) is regular, then, under Assumptions 5.1 and 5.2, system (2.1), (5.1) is regular too.*

Proof. First we prove that system (2.1), (5.1) has a unique solution $z_t = z_t^{s,z} = (x_t^{s,x}, y_t^{s,z})$ on $[0,T]$ for any $s \geq 0, z = (x,y) \in \mathbb{R}^d \times \mathbb{R}^{d_0}$, and $T \in [0,\infty)$. For equation (2.1) this is given by definition, and the first d components of z_t represent $x_t^{s,x}$ indeed. For equation (5.1) we get existence and uniqueness from Theorem 1.2 (with $K_t(R \vee \sup_{r \leq t} |x_r^{s,x}|)$ as $K_t(R)$).

To prove the regularity, we use the arguments from Lemma 1.5 similarly to the proof of Theorem 1.8. Take a sequence of partitions $I(n) = \{t(k,n) : k = 0, 1, 2, \ldots\}$ such that $\Delta(I(n)) \to 0$ and two converging sequences $s(n) \to s \in [0,\infty)$ and $z(n) \to z \in \mathbb{R}^{d+d_0}$. Obviously, to prove the regularity, it suffices to prove that given any such $I(n), s(n), z(n)$, we have

$$P\{\sup_{t \leq T} |z_t^{s(n),z(n)}(I(n)) - z_t^{s(n),z(n)}| \geq \varepsilon\} = 0 \quad \forall \varepsilon > 0, T \in [0,\infty).$$

Since these relations hold for x_t in place of z_t by Definition 2.1, we only need to prove that

$$\sup_{t \leq T} |y_t^{s(n),z(n)}(I(n)) - y_t^{s(n),z(n)}| \xrightarrow{P} 0. \tag{5.5}$$

Define

$$\bar{x}_t^n = x_t^{s(n),x(n)}(I(n)), \quad x_t^n = x_t^{s(n),x(n)}, \quad \bar{y}_t^n = y_t^{s(n),z(n)}(I(n)),$$

$$\bar{p}_t^n = \bar{y}_{\kappa(n,t)}^n - \bar{y}_t^n, \quad y_t^n = y_t^{s(n),z(n)} - \bar{q}_t^n,$$

$$\bar{q}_t^n = \int_0^t [\tilde{\sigma}(s(n) + r, x_r^n, y_r^{s(n),z(n)}) - \tilde{\sigma}(s(n) + r, \bar{x}_{\kappa(n,r)}^n, y_r^{s(n),z(n)})] \, dw_r$$

$$+ \int_0^t [\tilde{b}(s(n) + r, x_r^n, y_r^{s(n),z(n)}) - \tilde{b}(s(n) + r, \bar{x}_{\kappa(n,r)}^n, y_r^{s(n),z(n)})] \, dr,$$

$$\gamma_0^n(R) = \inf\{t \geq 0 : |x_t^n| + |\bar{x}_t^n| \geq R\}, \quad \gamma_1^n(R) = \inf\{t \geq 0 : |y_t^{s(n),z(n)}| \geq R\}$$

$$\gamma_2^n(R) = \inf\{t \geq 0 : |\bar{y}_t^n| + |\bar{q}_t^n| \geq R + 2\}, \quad \gamma^n(R) = \gamma_0^n(R) \wedge \gamma_1^n(R) \wedge \gamma_2^n(R),$$

$$\psi_t^n(R) = \exp(-2 \int_0^t K_{s(n)+r}(R) \, dr),$$

and notice that

$$d\bar{y}_t^n = \tilde{\sigma}(s(n) + t, \bar{x}_{\kappa(n,t)}^n, \bar{y}_t^n + \bar{p}_t^n) \, dw_t + \tilde{b}(s(n) + t, \bar{x}_{\kappa(n,t)}^n, \bar{y}_t^n + \bar{p}_t^n) \, dt,$$

$$dy_t^n = \tilde{\sigma}(s(n) + t, \bar{x}_{\kappa(n,t)}^n, y_t^n + \bar{q}_t^n) \, dw_t + \tilde{b}(s(n) + t, \bar{x}_{\kappa(n,t)}^n, y_t^n + \bar{q}_t^n) \, dt.$$

As in the proof of Theorem 1.2, we get that

$$\lim_{n \to \infty} E \int_0^{T \wedge \gamma^n(R)} |\bar{p}_t^n| \, dt = 0 \quad \forall R, T \in [0,\infty).$$

To prove similar relation for $|\bar{q}_t^n|$, observe that, from the fact that $x^n - \bar{x}^n \to 0$ in C in probability (which holds by assumption) and $x^n - x^{s,x} \to 0$ in C in probability (Lemma 2.3), we get that

$$\sup_{r \leq T} |x_r^n - \bar{x}_{\kappa(n,r)}^n| \leq \sup_{r \leq T} |x_r^n - x_{\kappa(n,r)}^n| + \sup_{r \leq T} |x_{\kappa(n,r)}^n - \bar{x}_{\kappa(n,r)}^n|$$

$$\leq 2 \sup_{r \leq T} |x_r^n - x_r^{s,x}| + \sup_{r \leq T} |x_r^{s,x} - x_{\kappa(n,r)}^{s,x}| + \sup_{r \leq T} |x_r^n - \bar{x}_r^n| \to 0$$

in probability, which along with the continuity of $\tilde{\sigma}(t,z)$ in z implies that, for $s(n) + t \leq T$,

$$\int_0^{t \wedge \gamma^n(R)} ||\tilde{\sigma}(s(n) + r, x_r^n, y_r^{s(n),z(n)}) - \tilde{\sigma}(s(n) + r, \bar{x}_{\kappa(n,r)}^n, y_r^{s(n),z(n)})||^2 \, dr$$

$$\leq \int_0^T \sup_{r \leq T, |y| \leq R} ||\tilde{\sigma}(s, x_r^n, y) - \tilde{\sigma}(s, \bar{x}_{\kappa(n,r)}^n, y)||^2 \, ds \to 0$$

in probability. It follows that the stochastic integral in the definition of $\bar{q}_t^n$ tends to zero in probability. In the same way one treats the usual integral in the definition of $\bar{q}_t^n$ and one concludes that

$$\bar{q}_{\cdot \wedge \gamma^n(R)}^n \to 0 \tag{5.6}$$

in C in probability for all R. Since $|\bar{q}_t^n| \leq R$ for $0 \leq t \leq \gamma^n(R)$, we get

$$\lim_{n \to \infty} E \int_0^{T \wedge \gamma^n(R)} |\bar{q}_t^n| \, dt = 0 \quad \forall R, T \in [0, \infty).$$

Hence, as in the proof of Lemma 1.5

$$\sup_{t \leq T \wedge \gamma^n(R)} |\bar{y}_t^n - y_t^n| \xrightarrow{P} 0 \quad \forall R, T \in [0, \infty).$$

Owing to (5.6), we have proved (5.5) with $T \wedge \gamma^n(R)$ in place of T:

$$\sup_{t \leq T \wedge \gamma^n(R)} |\bar{y}_t^n - y_t^{s(n),z(n)}| \xrightarrow{P} 0 \quad \forall R, T \in [0, \infty). \tag{5.7}$$

Furthermore, for $\tau^n(R) := \inf\{t \geq 0 : |\bar{y}_t^n| \geq R + 1\}$, we obviously have

$$P\{\gamma_2^n(R) \leq \tau^n(R) \wedge T\} \leq P\{\sup_{t \leq T \wedge \gamma^n(R)} |\bar{q}_t^n| \geq 1\},$$

where the last probability tends to zero as $n \to \infty$ again due to (5.6). Hence, (5.7) implies that

$$\sup_{t \leq T \wedge \gamma_0^n(R) \wedge \gamma_1^n(R) \wedge \tau^n(R)} |\bar{y}_t^n - y_t^{s(n),z(n)}| \xrightarrow{P} 0 \quad \forall R, T \in [0, \infty). \tag{5.8}$$

By the way, if $\tau^n(R) = T \wedge \gamma_0^n(R) \wedge \gamma_1^n(R) \wedge \tau^n(R)$, then $|\bar{y}_t^n - y_t^{s(n),z(n)}| \geq 1$ at $t = \tau^n(R)$. This and (5.8) yield

$$\sup_{t \leq T \wedge \gamma_0^n(R) \wedge \gamma_1^n(R)} |\bar{y}_t^n - y_t^{s(n),z(n)}| \xrightarrow{P} 0 \quad \forall R, T \in [0, \infty),$$

and it only remains to prove that

$$\lim_{R \to \infty} \overline{\lim_{n \to \infty}} \, P\{\gamma_0^n(R) \wedge \gamma_1^n(R) \le T\} = 0. \tag{5.9}$$

Coming back to (5.1), applying Itô's formula to $|y_t^{s(n),z(n)}|^2 \psi_t^n(R_0)$ with a fixed R_0, and using (5.4), we get that for any stopping time $\tau \le \gamma_0^n(R_0) \wedge T$

$$E|y_\tau^{s(n),z(n)}|^2 \psi_\tau^n(R_0) \le |y^n|^2,$$

$$E|y_\tau^{s(n),z(n)}|^2 \le |y^n|^2 \exp(2 \int_0^T K_{s(n)+t}(R_0) \, dt).$$

Taking here $\tau = \gamma_0^n(R) \wedge \gamma_1^n(R_0) \wedge T$ and using Chebyshev's inequality, we get

$$P\{\gamma_1^n(R) \le \gamma_0^n(R_0) \wedge T\} \le R^{-2}|y^n|^2 \exp(2 \int_0^T K_{s(n)+t}(R_0) \, dt).$$

Therefore, the left-hand side of (5.9) is less than

$$\lim_{R \to \infty} \overline{\lim_{n \to \infty}} \, P\{\gamma_0^n(R_0) \wedge \gamma_0^n(R) \le T\} = \overline{\lim_{n \to \infty}} \, P\{\gamma_0^n(R_0) \le T\}$$

for any R_0. Now to get (5.9) remember that sequences converging in probability are bounded in probability. Therefore, the above stated properties of x_t^n and $\bar{x}_t^n$ imply that

$$\lim_{R_0 \to \infty} \overline{\lim_{n \to \infty}} \, P\{\gamma_0^n(R_0) \le T\} \le \lim_{R_0 \to \infty} \sup_n P\{\sup_{s \le T}[|x_t^n| + |\bar{x}_t^n|] \ge R_0\} = 0.$$

The theorem is proved.

5.2 Sufficient conditions for regularity

First we give the following corollary of Theorem 5.3.

Corollary 5.4. *Under the hypotheses stated before Definition 2.1 assume that for every $R \ge 0$ there exists a Borel function $K_t(R)$ defined and locally integrable on $[0, \infty)$ and such that*

$$2(x - y, b(t, x) - b(t, y)) + \|\sigma(t, x) - \sigma(t, y)\|^2 \le K_t(R)|x - y|^2, \tag{5.10}$$

$$2(x, b(t, x)) + \|\sigma(t, x)\|^2 \le K_t(1)(1 + |x|^2) \tag{5.11}$$

if $t \ge 0$, $x, y \in \mathbb{R}^d$, $|x|, |y| \le R$, $R \in [0, \infty)$. Then equation (2.1) is regular.

Indeed, it suffices to replace the system (2.1), (5.1) by a system which has the first equation with coefficients identically equal to zero and the second one coinciding with (2.1) (and to modify $K_t(R)$ so that they would become increasing in R and satisfying the inequality $K_t(R) \ge K_t(1)$ for all $R \ge 0$).

Corollary 5.5. *Under the hypotheses stated before Definition 2.1 assume that there exists a constant $K \geq 0$ such that*

$$|b(t, x) - b(t, y)| + \|\sigma(t, x) - \sigma(t, y)\| \leq K|x - y|$$

if $t \geq 0$, $x, y \in \mathbb{R}^d$. Then equation (2.1) is regular.

Indeed, in this case the left–hand side of (5.10) is less than

$$2|x - y| \, |b(t, x) - b(t, y)| + K^2|x - y|^2 \leq (2 + K)K|x - y|^2,$$

whereas the left–hand side of (5.11) is less than

$$2|x| \, |b(t, x) - b(t, 0)| + 2|x| \, |b(t, 0)| + 2\|\sigma(t, x) - \sigma(t, 0)\|^2 + 2\|\sigma(t, 0)\|^2$$

$$\leq 2K|x|^2 + (1 + |x|^2)|b(t, 0)| + 2K^2|x|^2 + 2\|\sigma(t, 0)\|^2$$

$$\leq (1 + |x|^2)(2K + 2K^2 + |b(t, 0)| + 2\|\sigma(t, 0)\|^2).$$

Remark 5.6. Theorem 5.3 says that the property of equations to be regular is inherited while (2.1) is supplemented by equation (5.1) whose coefficients satisfy some conditions. It is obvious that we can add to system (2.1), (5.1) yet another equation, and if its coefficients satisfy conditions like (5.3) and (5.4), the new system, thus enlarged, will still be regular.

In the future we will see quite a few applications of this remark.

Remark 5.7. If σ, b satisfy the assumptions of Corollary 5.4 and $\bar{\sigma}, \bar{b}$ satisfy Assumptions 5.1 and 5.2, it may happen nevertheless that coefficients of *system* (2.1), (5.1) written as a single equation with respect to z_t do not satisfy conditions like (5.10) and (5.11). An example of such a situation is given by the system of one–dimensional equations: $dx_t = x_t dt, dy_t = x_t y_t dt$. If we consider this system as one equation with respect to the two–dimensional process (x_t, y_t), then for instance the left–hand side of (5.11) is $2(x^1)^2 + 2x^1(x^2)^2$ and this function cannot be estimated from above by $K(1 + |x|^2)$ for all $x = (x^1, x^2) \in \mathbb{R}^2$.

Exercise 5.8 (method of new variables). Assume that on $[0, \infty) \times \mathbb{R}^d$ we are given functions $f(t, x)$ and $c(t, x)$. Assume that f, c are Borel functions with respect to t for every x and are continuous in x for every $t \geq 0$, and for every $T, R \in (0, \infty)$

$$\int_0^T \sup_{|x| \leq R} (|f(t, x)| + |c(t, x)|) \, dt < \infty. \tag{5.12}$$

This just means that f, c satisfy our basic assumptions on σ, b. Finally, assume that (2.1) is regular.

Consider the system of equations consisting of (2.1) and the following equation

$$x_t^{d+1} = x^{d+1} + \int_0^t c(s+r, x_r)\, dr. \tag{5.13}$$

Observe that by virtue of (5.12) this equation makes sense and prove that this system is regular. Also add to system (2.1), (5.13) the following equation

$$x_t^{d+2} = x^{d+2} + \int_0^t f(s+r, x_r) \exp\{-x_r^{d+1}\}\, dr. \tag{5.14}$$

Prove that system (2.1), (5.13), (5.14) is regular with respect to $(x_t^1, ..., x_t^{d+2})$.

5.3 Kolmogorov's equation

We consider the equation

$$x_t = x + \int_0^t \sigma(s+r, x_r)\, dw_r + \int_0^t b(s+r, x_r)\, dr, \tag{5.15}$$

with usual assumptions on dimensionalities of σ, b. We assume that σ and b are Borel functions on $[0, \infty) \times \mathbb{R}^d$ satisfying

$$\|\sigma(t, x)\| + |b(t, x)| \le K(1 + |x|),$$
$$\|\sigma(t, x) - \sigma(t, x)\| + |b(t, x) - b(t, y)| \le K|x - y|, \tag{5.16}$$

where K is a fixed constant. By Corollary 5.5 equation (5.15) is regular.

Assume that, on $[0, \infty) \times \mathbb{R}^d$, we are also given real–valued Borel functions $c(t, x) \ge 0$, $f(t, x)$, and $g(x)$ satisfying

$$|c(t, x)| + |f(t, x)| + |g(x)| \le K(1 + |x|^m),$$

where m is a nonnegative constant. Define

$$\varphi_t^{s,x} = \int_0^t c(s+r, x_r^{s,x})\, dr,$$

$$v(s, x) = E_{s,x}\Big[\int_0^{T-s} f(s+t, x_t)e^{-\varphi_t}\, dt + g(x_{T-s})e^{-\varphi_{T-s}} \Big],$$

where we use the standard stipulation about indices.

Remark 5.9. Clearly, φ_t and $v(t, x)$ are well defined and by Theorem 4.6

$$|v(t, x)| \le N(K, m, T)(1 + |x|^m).$$

Furthermore, if c, f, and g are continuous in x, Lemma 2.9 easily implies that the random variable

$$\int_0^{T-s} f(s+t, x_t^{s,x})e^{-\varphi_t^{s,x}}\, dt + g(x_{T-s}^{s,x})e^{-\varphi_{T-s}^{s,x}}$$

is continuous in probability as a function of (s, x). Theorem 4.6 provides locally uniform estimates of its moments and this implies that v is continuous in $[0, T] \times \mathbb{R}^d$ if c, f, and g are continuous in x. In the general case v is Borel in $[0, T] \times \mathbb{R}^d$ as follows from the above argument by well known measure-theoretic results.

Lemma 5.10. *For any $t \in [0,T]$, let the functions σ, b, c, f, and g be twice continuously differentiable with respect to x and let the absolute values of all their first and second order derivatives be less than $K(1+|x|^m)$ in $[0,T] \times \mathbb{R}^d$. Then*

(i) for any $t \in [0,T]$ the function $v(t,x)$ is twice continuously differentiable with respect to x and

$$|v_x(t,x)| \leq Ne^{NT}(1+|x|^{2m}), \quad |v_{xx}(t,x)| \leq Ne^{NT}(1+|x|^{3m})$$

in $[0,T] \times \mathbb{R}^d$, where N is independent of t, x, T;

(ii) for any $x \in \mathbb{R}^d$ the function $v(t,x)$ is Lipschitz with respect to $t \in [0,T]$;

(iii) in $(0,T) \times \mathbb{R}^d$, the generalized derivative $D_t v$ exists and

$$|D_t v| \leq N(K,T,R) \quad in \quad Q_R \quad (a.e.) \quad \forall R;$$

(iv) in $(0,T) \times \mathbb{R}^d$, the function v satisfies Kolmogorov's equation

$$Lv + f = 0 \quad (a.e.), \tag{5.17}$$

where

$$Lu := Lu(t,x) = D_t u(t,x) + a^{ij}(t,x)u_{x^i x^j}(t,x)$$
$$+ b^i(t,x)u_{x^i}(t,x) - c(t,x)u(t,x), \quad a = \tfrac{1}{2}\sigma\sigma^*.$$

Proof. Assertion (i) follows directly from Theorem 4.32. To prove other assertions, we use Exercise 5.8, by which system (2.1), (5.13), (5.14) is regular. Define $\bar{x} = (x, x^{d+1}, x^{d+2}) = (x^1, ..., x^{d+2})$ and

$$\bar{g}(\bar{x}) = x^{d+2} + g(x)\exp(-x^{d+1}), \quad \bar{v}(s,\bar{x}) = E_{s,\bar{x}}\bar{g}(\bar{x}_{T-s}),$$

where, of course, $\bar{x}_t^{s,\bar{x}}$ is a solution of (2.1), (5.13), (5.14). Obviously,

$$\bar{v}(s,\bar{x}) = x^{d+2} + v(s,x)\exp(-x^{d+1}). \tag{5.18}$$

In particular, $\bar{v}$ is a smooth function of $\bar{x}$. By Theorem 2.13, $\bar{v}$ is harmonic relative to $\bar{x}_t^{s,\bar{x}}$ in any bounded subdomain of $(0,T) \times \mathbb{R}^{d+2}$. By Theorem 3.5 and by smoothness of v in x, we get $\bar{v} \in B^{1,2}((0,T) \times \{|\bar{x}| < R\})$, which proves (ii) and (iii) again owing to (5.18). Finally, applying Theorem 3.6, we can write $\bar{L}\bar{v} = 0$ (a.e.) in $(0,T) \times \mathbb{R}^{d+2}$, where $\bar{L}$ is the operator corresponding to $\bar{x}_t^{s,\bar{x}}$. A very easy computation shows that the last equation amounts to (5.17) by virtue of (5.18). The lemma is proved.

Exercise 5.11. Let $\xi_t^{s,x,\xi} = x_{t(\xi)}^{s,x}$ and $\eta_t^{s,x,\xi} = x_{t(\xi)(\xi)}^{s,x}$. Prove that under the conditions of Lemma 5.10, the triplet $(x_t^{s,x}, \xi_t^{s,x,\xi}, \eta_t^{s,x,\xi})$ is a solution of a regular system, and from formulas (4.23) and (4.24) and Lemma 2.9 (i) derive that v_x and v_{xx} are continuous in (s,x), where v is the function from Lemma 5.10.

In the future we will need to view equation (5.17) in the sense of generalized functions.

Lemma 5.12. *For any $t \in [0, T]$, let the function σ be continuously differentiable with respect to x and for any finite $R > 0$ let σ_x satisfy Lipschitz condition with respect to x on Q_R. Also let c, f, and g be continuous in x for any $t \in [0, T]$. Then Kolmogorov's equation (5.17) holds in the sense of distributions which means that for any $\eta \in C_0^\infty((0, T) \times \mathbb{R}^d)$ we have*

$$\int_0^T \int_{\mathbb{R}^d} [vL^*\eta + f\eta] \, dx \, dt = 0, \qquad (5.19)$$

where

$$L^*\eta := (a^{ij}\eta)_{x^i x^j} - (b^i\eta)_{x^i} - c\eta - D_t\eta$$
$$= a^{ij}\eta_{x^i x^j} + (2a^{ij}_{x^j} - b^i)\eta_{x^i} + (a^{ij}_{x^i x^j} - b^i_{x^i} - c)\eta - D_t\eta.$$

Proof. First, use the fact that Lipschitz continuous functions are almost everywhere differentiable and have generalized derivatives. Then we see that the expressions $a^{ij}_{x^i x^j}$ and $b^i_{x^i}$ are well defined and $L^*\eta$ makes sense and is bounded in Q_R for any R.

Next, if the conditions of Lemma 5.10 were satisfied, we would get (5.19) multiplying (5.17) by η and integrating by parts. In the general situation we make the passage to the limit in (5.19).

Take a nonnegative function $\zeta \in C_0^\infty(\mathbb{R}^d)$ with unit integral and for $\varepsilon \neq 0$ define $\zeta_\varepsilon(x) = \varepsilon^{-d}\zeta(x/\varepsilon)$, $u^{(\varepsilon)}(t, x) = u(t, \cdot) * \zeta_\varepsilon(\cdot)(x)$. Also let $u^{(0)} := u$. For any ε, define $x_t^{s,x,\varepsilon}$ as a solution of the equation

$$x_t = x + \int_0^t \sigma^{(\varepsilon)}(s + r, x_r) \, dw_r + \int_0^t b^{(\varepsilon)}(s + r, x_r) \, dr$$

and let

$$\varphi_t^{s,x,\varepsilon} = \int_0^t c^{(\varepsilon)}(s + r, x_r^{s,x,\varepsilon}(\varepsilon)) \, dr,$$

$$v(s, x, \varepsilon) = E\Big[\int_0^{T-s} f^{(\varepsilon)}(s+t, x_t^{s,x,\varepsilon}) e^{-\varphi_t^{s,x,\varepsilon}} \, dt + g^{(\varepsilon)}(x_{T-s}^{s,x,\varepsilon}) e^{-\varphi_{T-s}^{s,x,\varepsilon}} \Big]. \quad (5.20)$$

It turns out that, $v(s, x, \varepsilon)$ is a continuous function of (s, x, ε). Indeed, the equation $d\varepsilon_t = 0$ is regular. In addition, obviously for any ε, the functions $\sigma^{(\varepsilon)}$ and $b^{(\varepsilon)}$ satisfy (5.16). Hence by Theorem 5.3, the couple $(x_t^{s,x,\varepsilon}, \varepsilon)$ is regular. Furthermore, $c^{(\varepsilon)}, f^{(\varepsilon)}$, and $g^{(\varepsilon)}$ as functions of (s, x, ε), satisfy the growth condition imposed on c, f, g. Therefore, the continuity of $v(s, x, \varepsilon)$ follows as in Remark 5.9. In particular, $v(s, x, \varepsilon) \to v(s, x, 0) = v(s, x)$ as $\varepsilon \to 0$ uniformly on $\bar{Q}_R$ for any finite R.

Next, by Lemma 5.10 (for each $\varepsilon \neq 0$) we have

$$\int_0^T \int_{\mathbb{R}^d} [v(\varepsilon)L_\varepsilon^* \eta + f^{(\varepsilon)}\eta] \, dx \, dt = 0, \tag{5.21}$$

where L_ε is the operator corresponding to $\sigma^{(\varepsilon)}, b^{(\varepsilon)}$, and $c^{(\varepsilon)}$. We let $\varepsilon \to 0$ using the convergence $v(\varepsilon) = v(t, x, \varepsilon) \to v(t, x)$, which is uniform on bounded sets. We also observe that

$$(b^{(\varepsilon)})_{x^i} = (b_{x^i})^{(\varepsilon)} \to b_{x^i} \quad \text{(a.e.)},$$

where b_{x^i} is Sobolev's derivative of b. This derivative is bounded so that $(b^{(\varepsilon)})_{x^i}$ are uniformly bounded. The same is true for other coefficients of L_ε, for instance,

$$2(a_\varepsilon)_{x^i x^j} = (\sigma^{(\varepsilon)}\sigma^{(\varepsilon)*})_{x^i x^j} = (\sigma_{x^i x^j})^{(\varepsilon)}\sigma^{(\varepsilon)*} + (\sigma_{x^i})^{(\varepsilon)}(\sigma_{x^j}^*)^{(\varepsilon)} + (\sigma_{x^j})^{(\varepsilon)}(\sigma_{x^i}^*)^{(\varepsilon)}$$

$$+\sigma^{(\varepsilon)}(\sigma_{x^i x^j}^*)^{(\varepsilon)} \to \sigma_{x^i x^j}\sigma^* + \sigma_{x^i}\sigma_{x^j}^* + \sigma_{x^j}\sigma_{x^i}^* + \sigma\sigma_{x^i x^j}^* = 2a_{x^i x^j}.$$

Now we obtain (5.19) from (5.21) by the dominated convergence theorem. The lemma is proved.

Two main results of this section follow. As always we suppose that the assumptions stated in the beginning of this subsection are satisfied.

Theorem 5.13. *Assume that*

$$|c(t, x) - c(t, y)| + |f(t, x) - f(t, y)|$$

$$+ |g(x) - g(y)| \leq K|x - y|(1 + |x|^m + |y|^m) \tag{5.22}$$

for all t, x, y. *Then*

(i) the function $v(s, x)$ *has first-order generalized derivatives in* $(0, T) \times \mathbb{R}^d$ *with respect to* x. *Furthermore, (a.e.)*

$$|v_x(s, x)| \leq Ne^{N(T-s)}(1 + |x|^{2m}), \tag{5.23}$$

where $N = N(K, m)$;

(ii) Kolmogorov's equation (5.17) holds in the sense of distributions which means that for any $\eta \in C_0^\infty((0, T) \times \mathbb{R}^d)$

$$\int_0^T \int_{\mathbb{R}^d} [vD_t\eta + (a^{ij}\eta)_{x^i}v_{x^j} + v(b^i\eta)_{x^i} + \eta cv - \eta f] \, dx \, dt = 0. \tag{5.24}$$

Proof. We introduce the same objects as in the proof of Lemma 5.12 but in (5.21) in the term containing the second–order derivatives of η we integrate by parts to find

$$\int_0^T \int_{\mathbb{R}^d} [v(\varepsilon)D_t\eta + (a_\varepsilon^{ij}\eta)_{x^i}v_{x^j}(\varepsilon) + v(\varepsilon)(b^{(\varepsilon)i}\eta)_{x^i} + \eta c^{(\varepsilon)}v(\varepsilon) - \eta f^{(\varepsilon)}] \, dx \, dt = 0. \tag{5.25}$$

Notice that, due to conditions (5.22) and (5.16), and Theorem 4.32, the functions $v(t, x, \varepsilon)$ are continuously differentiable with respect to x for any $t \in$

$[0, T]$ and $\varepsilon \neq 0$. Furthermore, $v_x(t, x, \varepsilon)$ satisfies (5.23) with $N = N(K, m)$ if $0 \neq |\varepsilon| \leq 1$. In the proof of Lemma 5.12 we have seen that $v(\varepsilon) \to v$ as $\varepsilon \to 0$. This easily proves assertion (i) and also that $v_x(\varepsilon) \to v_x$ weakly in $L_2(Q_R)$ for any finite $R > 0$.

Furthermore, as in the proof of Lemma 5.12 we have $(a_\varepsilon^{ij})_{x^i} \to a_{x^i}^{ij}$ (a.e.) and all these functions are bounded by the same constant on Q_R for any finite $R > 0$. In particular, this convergence is strong in $L_2(Q_R)$ for any finite $R > 0$. It follows that

$$\int_0^T \int_{\mathbb{R}^d} \eta (a_\varepsilon^{ij})_{x^i} v_{x^j}(\varepsilon) \, dx \, dt \to \int_0^T \int_{\mathbb{R}^d} \eta a_{x^i}^{ij} v_{x^j} \, dx, dt.$$

Similar conclusion for other terms in (5.25) is much simpler to prove, so that letting $\varepsilon \to 0$ in (5.25) we get (5.24) and the theorem is proved.

The following theorem generalizes Lemma 5.12. Remember the assumptions stated in the beginning of this subsection before Remark 5.9.

Theorem 5.14. *Let the generalized function*

$$\sum_{i,j=1}^d a_{x^i x^j}^{ij} \tag{5.26}$$

be a locally finite measure in $(0, T) \times \mathbb{R}^d$. Then Kolmogorov's equation (5.17) holds in the sense of distributions which means that for any $\eta \in C_0^\infty((0, T) \times \mathbb{R}^d)$ equation (5.19) holds.

Proof. If in addition condition (5.22) is satisfied, then we can integrate by parts in (5.24) to get (5.19). However, notice that if before our generalized derivatives were bounded, so that we were dealing with absolutely continuous functions, this time we deal with measures and perhaps some additional explanation is needed. What we actually need to do is to approximate v on the support of η by functions v_n from $C_0^\infty((0, T) \times \mathbb{R}^d)$ so that

$$\int_0^T \int_{\mathbb{R}^d} \eta a_{x^i x^j}^{ij} v \, dx \, dt$$

$$= \lim_{n \to \infty} \int_0^T \int_{\mathbb{R}^d} \eta a_{x^i x^j}^{ij} v_n \, dx \, dt = -\lim_{n \to \infty} \int_0^T \int_{\mathbb{R}^d} (\eta a_{x^i}^{ij} v_{nx^j} + \eta_{x^j} a_{x^i}^{ij} v_n) \, dx \, dt$$

$$= -\int_0^T \int_{\mathbb{R}^d} (\eta a_{x^i}^{ij} v_{x^j} + \eta_{x^j} a_{x^i}^{ij} v) \, dx \, dt.$$

One can do this by taking mollifiers of v multiplied by a cut-off function since in this case on the support of η: $v_n \to v$ at any point, these functions are uniformly bounded, and $v_{nx} \to v_x$ (a.e.) and these functions are uniformly bounded too.

Once we get (5.19) under condition (5.22), we fix η, σ, and b and consider the set $\mathfrak{B}$ of all (c, f, g) (satisfying the conditions from the beginning of the subsection with the same K, m) for which (5.19) holds. Notice that if $(c_n, f_n, g_n) \in \mathfrak{B}$ and $(c_n, f_n, g_n) \to (c, f, g)$ pointwise, then, obviously, for v_n constructed on the basis of c_n, f_n, g_n we have $v_n \to v$ pointwise and all the functions c_n, f_n, g_n, v_n are uniformly bounded on the support of η. This allows us to use the dominated convergence theorem and prove that $(c, f, g) \in \mathfrak{B}$ as well. This property of (5.19) and a standard measure-theoretic argument (see, for instance, Theorem 19, Sec. 2, Ch. 1 of [M]) finishes the proof of the theorem.

6. Some integral approximations of differential operators

Assume the conditions and notation from Subsec. 5.3. Here we start preparing an important tool which will allow us to derive Kolmogorov's equations for generally nonsmooth functions which are harmonic in domains.

Let Q be a subdomain of $(0, T) \times \mathbb{R}^d$. For Borel functions $h(t, x)$ and $g(t, x)$ which are defined in Q and ∂Q, respectively, and $\lambda \geq 0$ denote

$$R_\lambda(Q)h(s, x) = E_{s,x} \int_0^T e^{-\varphi_t - \lambda t} h(s + t, x_t) \, dt,$$

$$R(Q) = R_0(Q), \quad R_\lambda = R_\lambda((0, T) \times \mathbb{R}^d),$$

$$\pi_\lambda(Q)g(s, x) = E_{s,x} e^{-\varphi_\tau - \lambda \tau} g(s + \tau, x_\tau), \quad \pi(Q) = \pi_0(Q),$$

where

$$\tau = \tau_Q = \inf\{t \geq 0 : (s + t, x_t) \notin Q\}.$$

A useful point of view at $R_\lambda(Q)h$ and $\pi_\lambda(Q)g$ is the following. Take a random variable γ, which is exponentially distributed with parameter 1 and independent of $w.$. Then, for $\lambda > 0$, the variable γ/λ is exponentially distributed with parameter λ. By using this, one proves easily that

$$\lambda R_\lambda(Q)h(s, x) = E_{s,x} e^{-\varphi_{\gamma/\lambda}} h(s + \gamma/\lambda, x_{\gamma/\lambda}) I_{\gamma/\lambda \leq \tau},$$

$$\pi_\lambda(Q)g(s, x) = E_{s,x} e^{-\varphi_\tau} g(s + \tau, x_\tau) I_{\gamma/\lambda > \tau},$$

$$\lambda R_\lambda(Q)h(s, x) + \pi_\lambda(Q)h(s, x) = E_{s,x} e^{-\varphi_{(\gamma/\lambda) \wedge \tau}} h(s + (\gamma/\lambda) \wedge \tau, x_{(\gamma/\lambda) \wedge \tau})$$

Since $\gamma/\lambda \to 0$ as $\lambda \to \infty$, it is natural that $\lambda R_\lambda(Q)h \to h$ under reasonable assumptions on h. Also Lemma 3.3 shows that $\lambda^n \pi_\lambda(Q)g(s, x) \to 0$ for any n. Furthermore, if Itô's formula is applicable, then

$$\lambda R_\lambda(Q)h(s, x) + \pi_\lambda(Q)h(s, x) = h(s, x) + E_{s,x} \int_0^{(\gamma/\lambda) \wedge \tau} e^{-\varphi_t} Lh(s + t, x_t) dt$$

$$= h(s, x) + \lambda E_{s,x} \int_0^\infty e^{-\lambda t} \int_0^{t \wedge \tau} e^{-\varphi_r} Lh(s + r, x_r) \, dr dt$$

$$= h(s,x) + E_{s,x} \int_0^\tau e^{-\varphi_r - \lambda r} Lh(s+r, x_r)\, dr = h(s,x) + R_\lambda(Q) Lh(s,x).$$

Thus
$$\lambda(\lambda R_\lambda(Q) h - h) = \lambda R_\lambda(Q) Lh - \lambda \pi_\lambda(Q) h,$$

which makes it natural that if Lh satisfies "reasonable" assumptions so that $\lambda R_\lambda(Q) Lh \to Lh$, then

$$\lambda(\lambda R_\lambda(Q) h - h) \to Lh$$

as $\lambda \to \infty$ and leads us to "integral" approximations of the operator L.

In what follows we justify the above conclusions.

Lemma 6.1. *Let $g(t,x)$ be a bounded Borel function defined on ∂Q. Then for any $n \geq 0$ we have $\lambda^n \pi_\lambda(Q) g \to 0$ as $\lambda \to \infty$ uniformly on any closed bounded subset of Q.*

Proof. Obviously, we may assume $g \equiv 1$. Take a $\delta \in (0,1)$ and let Q_δ be the set of all points $(t,x) \in Q$ such that $|x| < \delta^{-1}$ and for the closest to (t,x) point (s,y) on the boundary of Q it holds that $|t-s| > \delta$ and $|x-y| > \delta$. Then by Lemma 3.3 for any $\varepsilon \leq \delta$

$$\pi_\lambda(Q) 1(s,x) \leq E_{s,x} e^{-\lambda \tau} I_{\tau \leq \varepsilon} + e^{-\lambda \varepsilon} \leq P_{s,x}\{\sup_{r \leq \varepsilon} |x_r - x| \geq \delta\} + e^{-\lambda \varepsilon}$$

$$\leq N\varepsilon^{3n} + e^{-\lambda \varepsilon}$$

with N depending only on δ, n and $K(\delta + \delta^{-1})$. For $\varepsilon = \lambda^{-1/2}$ and $\lambda \geq \delta^{-2}$ we find
$$\lambda^n \pi_\lambda(Q) 1(t,x) \leq N\lambda^{-n/2} + \lambda^n e^{-\lambda^{1/2}},$$

which shows that $\lambda^n \pi_\lambda(Q) 1 \to 0$ uniformly in Q_δ. The arbitrariness of δ proves that $\lambda^n \pi_\lambda(Q) 1 \to 0$ uniformly in any closed compact subset of Q. The lemma is proved.

Lemma 6.2. *Let h be a bounded Borel function defined in Q.*

(i) Let Q' be a domain, Γ be a closed bounded set and let $\Gamma \subset Q' \subset Q$. Then for any $n \geq 0$

$$\lim_{\lambda \to \infty} \lambda^n \sup_\Gamma |R_\lambda(Q')h - R_\lambda(Q)h| = 0.$$

(ii) Assume that either
(a) generalized function (5.26) is a locally finite measure on Q, or
(b) the function $h(t,x)$ is continuous in x for any t.
Then
$$\lambda R_\lambda(Q) h \to h \tag{6.1}$$

as $\lambda \to \infty$ in the sense of generalized functions on Q.

Proof. Assertion (i) follows from Lemma 6.1 and the formula

$$|R_\lambda(Q)h(s,x) - R_\lambda(Q')h(s,x)|$$

$$= |E_{s,x}e^{-\varphi_{\tau_{Q'}}-\lambda\tau_{Q'}}\int_{\tau_{Q'}}^{\tau_Q} h(s+t,x_t)\exp\{-\int_{\tau_{Q'}}^t c(s+r,x_r)\,dr - \lambda(t-\tau_{Q'})\}\,dt|$$

$$\leq T\sup|h|E_{s,x}e^{-\varphi_{\tau_{Q'}}-\lambda\tau_{Q'}} = T\sup|h|\pi_\lambda(Q')1(s,x).$$

To prove assertion (ii) first of all notice that proving that $\lambda R_\lambda(Q)h \to h$ in the sense of generalized functions on Q is equivalent to proving that this convergence holds in the sense of generalized functions on any subdomain of Q. Therefore, we need only to prove that (6.1) holds in the sense of generalized functions on Q_1 for any cylinder $Q_1 \subset \bar{Q}_1 \subset Q$. By assertion (i) it suffices to show that

$$\lambda R_\lambda(Q_1)h \to h \tag{6.2}$$

in the sense of generalized functions on Q_1.

To proceed further assume condition (a) and notice that $R_\lambda(Q_1)h$ is uniquely determined by the values of σ, b, c, and h in Q_1 and condition (a) bears on Q. It follows that by multiplying σ, b, c, and h by a cut-off function we will not change $R_\lambda(Q_1)h$ and will have condition (a) satisfied in $(0,T)\times\mathbb{R}^d$. Now we make another twist in the above argument concluding that instead of proving (6.2) under condition (a) it suffices to prove (6.1) in the case $Q = (0,T)\times\mathbb{R}^d$ assuming (a) for this domain Q. In this situation by Theorem 5.14 we have $\lambda R_\lambda h - h = LR_\lambda h$ and obviously $|R_\lambda h| \leq \lambda^{-1}\sup|h| \to 0$ uniformly on Q. The latter implies that $LR_\lambda h \to 0$ in the sense of generalized functions on Q. We have proved (6.1) under condition (a).

We will prove that under condition (b) the convergence in (6.1) is even stronger, namely, in L_p for any p on any closed bounded subset of Q. In the same way as above we see that while proving this we may assume that $Q = (0,T)\times\mathbb{R}^d$ and c is bounded. Also, the functions $\lambda R_\lambda h$ are uniformly bounded, so that to prove the convergence in L_p on any closed bounded subset of Q it suffices to prove that (6.1) holds almost everywhere, for instance, at any point (s,x) such that s is a Lebesgue point of the functions $h(t,x)$ and

$$\sup_{y:|x-y|\leq 1/n} |h(t,x) - h(t,y)|$$

as functions of t, where $n = 1,2,\dots.$ Denote Θ the set of all such points (s,x).

Observe, that if s is a Lebesgue point of an integrable function $g(t)$ given on $(0,T)$, then

$$f(t) := t^{-1}\int_0^t |g(s+r) - g(s)|\,dr$$

is a bounded continuous function on $(0, T-s)$ which tends to zero as $t \downarrow 0$ so that by the dominated convergence theorem

$$\lambda \int_0^{T-s} |g(s+t) - g(s)| e^{-\lambda t} \, dt = \lambda^2 \int_0^{T-s} \left(\int_0^t |g(s+r) - g(s)| \, dr \right) e^{-\lambda t} \, dt$$

$$= \int_0^\infty I_{t/\lambda < T-s} f(t/\lambda) t e^{-t} \, dt \to 0$$

as $\lambda \to \infty$.

Bearing in mind this observation, take $(s, x) \in \Theta$ and an integer $n \geq 1$ and write

$$R_\lambda h(s, x) = E_{s,x} \int_0^{T-s} (e^{-\varphi_t} - 1) e^{-\lambda t} h(s + t, x_t) \, dt$$

$$+ E_{s,x} \int_0^{T-s} e^{-\lambda t} [h(s+t, x_t) - h(s+t, x)] \, dt + \int_0^{T-s} h(s+t, x) e^{-\lambda t} \, dt = I_1 + I_2 + I_3.$$

Here

$$|I_1| \leq N \int_0^{T-s} t e^{-\lambda t} \, dt \leq N\lambda^{-2},$$

$$\lim_{\lambda \to \infty} \lambda I_3 = h(s, x) \lim_{\lambda \to \infty} \lambda \int_0^{T-s} e^{-\lambda t} \, dt = h(s, x),$$

$$|I_2| \leq \int_0^{T-s} e^{-\lambda t} P_{s,x} \{ |x_t - x| \geq 1/n \} \, dt$$

$$+ \int_0^{T-s} e^{-\lambda t} \sup_{y: |x-y| \leq 1/n} |h(s + t, y) - h(s + t, x)| \, dt.$$

Furthermore, $P_{s,x}\{|x_t - x| \geq 1/n\} \leq Nt$, where N is independent of λ. Hence

$$\overline{\lim_{\lambda \to \infty}} \lambda |I_2| \leq \lim_{\lambda \to \infty} \lambda \int_0^{T-s} e^{-\lambda t} \sup_{y: |x-y| \leq 1/n} |h(s + t, y) - h(s + t, x)| \, dt$$

$$= \sup_{y: |x-y| \leq 1/n} |h(s, y) - h(s, x)|.$$

The last expression tends to zero as $n \to \infty$. This along with the above equations implies that $\lim_{\lambda \to \infty} \lambda R_\lambda h(s, x) = h(s, x)$ and the lemma is proved.

The following theorem is the main result of this section.

Theorem 6.3. *Let $h(t, x)$ be a bounded Borel functions defined in Q. Assume that either*

(i) the generalized function (5.26) is a locally integrable function on Q,

or

(ii) for any finite $R > 0$ there exists a constant N such that

$$|h(t, x) - h(t, y)| \leq N|x - y|, \quad |c(t, x) - c(t, y)| \leq N|x - y| \tag{6.3}$$

for all $(t, x), (t, y) \in Q \cap Q_R$. Then

$$\lambda(\lambda R_\lambda(Q)h - h) \to Lh \tag{6.4}$$

as $\lambda \to \infty$ in the sense of generalized functions on Q.

Proof. First, assume condition (i). Similarly to our arguments before and after (6.2) it suffices to consider the case $Q = (0, T) \times \mathbb{R}^d$. In this case by Theorem 5.14 we have

$$\lambda(\lambda R_\lambda h - h) = L(\lambda R_\lambda h). \tag{6.5}$$

By definition for any $\eta \in C_0^\infty(Q)$

$$\int_Q \eta L(\lambda R_\lambda h) \, dx \, dt = \int_Q \zeta \lambda R_\lambda h \, dx \, dt, \tag{6.6}$$

where $\zeta = L^* \eta$ is an integrable function with compact support lying in Q. Observe again that $\lambda R_\lambda h$ are bounded uniformly in Q by $2 \sup |h|$ and for any $\tilde{\zeta} \in C_0^\infty(Q)$ write

$$\varlimsup_{\lambda \to \infty} \left| \int_Q \zeta(\lambda R_\lambda h - h) \, dx \, dt \right|$$

$$\leq 2 \sup |h| \int_Q |\zeta - \tilde{\zeta}| \, dx \, dt + \lim_{\lambda \to \infty} \int_Q \tilde{\zeta}(\lambda R_\lambda h - h) \, dx \, dt. \tag{6.7}$$

The latter limit is zero by Lemma 6.2. The first term on the right can be made arbitrary small by choice of $\tilde{\zeta}$. Hence,

$$\lim_{\lambda \to \infty} \int_Q \zeta \lambda R_\lambda h \, dx \, dt = \int_Q \zeta h \, dx \, dt = \int_Q \eta L h \, dx \, dt,$$

which means that $L(\lambda R_\lambda h) \to Lh$. Coming back to (6.5) we get our assertion under condition (i) and the assumption that $Q = (0, T) \times \mathbb{R}^d$.

Under condition (ii) we follow the same scheme which shows that we only need to consider the case $Q = (0, T) \times \mathbb{R}^d$ and, in addition, we may assume that (6.3) holds with the same constant N for all $t \in [0, T]$ and $x, y \in \mathbb{R}^d$. In this case we can start as above. We have (6.5) again, this time by Theorem 5.13. However, now instead of (6.6) we have

$$\int_Q \eta L(\lambda R_\lambda h) \, dx \, dt = \int_Q \zeta \lambda R_\lambda h \, dx \, dt - \int_Q \lambda (R_\lambda h)_{x^i} \zeta_i \, dx \, dt,$$

where

$$\zeta = -D_t \eta - (b^i \eta)_{x^i} - c\eta, \quad \zeta_i = (a^{ij} \eta)_{x^j}$$

are integrable functions with compact support lying in Q. Now the estimate $\lambda |R_\lambda h| \leq \sup |h|$, Lemma 6.2, and formulas like (6.7) imply that to finish the proof it suffices to show that for any $R > 0$ there is a constant N such that for all $\lambda \geq 1$ on Q_R (a.e.)

$$\lambda |(R_\lambda h)_x| \leq N.$$

In the same way as in the proof of Theorem 5.13 it suffices to prove that for any $R > 0$ there is a constant N such that for all $\lambda \geq 1$ and $0 < |\varepsilon| \leq 1$ on Q_R

$$\lambda |v_x(s,x,\varepsilon)| \le N, \tag{6.8}$$

where v is taken from (5.20) with $g \equiv 0$ and $c + \lambda, h$ in place of c, f, respectively.

To prove (6.8) we repeat some arguments from the proof of Theorem 4.32. By (4.23) we have

$$
\begin{aligned}
v_{(\xi)}(s,x,\varepsilon) = E_{s,x} \int_0^{T-s} [h_{(\xi_t)}^{(\varepsilon)}(s+t,x_t(\varepsilon)) \\
- h^{(\varepsilon)}(s+t,x_t(\varepsilon)) \int_0^t c_{(\xi_r)}^{(\varepsilon)}(s+r,x_r(\varepsilon)) \, dr] e^{-\int_0^t c^{(\varepsilon)}(s+r,x_r(\varepsilon)) \, dr - \lambda t} \, dt \\
\le N E_{s,x} \sup_{t \le T-s} |\xi_t| \int_0^{T-s} (1+t) e^{-\lambda t} \, dt \le 2N\lambda^{-1} E_{s,x} \sup_{t \le T-s} |\xi_t|,
\end{aligned}
$$

where the constant N depends only on h and c and ξ_t is a solution of the equation

$$\xi_t = \xi + \int_0^t \sigma_{(\xi_r)}^{(\varepsilon)}(s+r,x_r(\varepsilon)) \, dw_r + \int_0^t b_{(\xi_r)}^{(\varepsilon)}(s+r,x_r(\varepsilon)) \, dr.$$

To finish the proof of (6.8) it only remains to use the same argument as in the proof of Theorem 4.32 after (4.23). The theorem is proved.

Remark 6.4. Observe that whenever (6.4) holds in the sense of generalized functions on Q, $\lambda R_\lambda(Q)h \to h$ in the sense of generalized functions on Q.

Remark 6.5. We need condition (5.16) satisfied only in Q.

7. Kolmogorov's equations in domains

We continue investigating the objects from Subsec. 5.3 and Sec. 6.. Thus we consider the equation

$$x_t = x + \int_0^t \sigma(s+r,x_r) \, dw_r + \int_0^t b(s+r,x_r) \, dr,$$

with usual assumptions on dimensionalities of σ, b. We assume that σ and b are Borel functions on $[0,\infty) \times \mathbb{R}^d$ satisfying

$$
\begin{aligned}
\|\sigma(t,x)\| + |b(t,x)| &\le K(1+|x|), \\
\|\sigma(t,x) - \sigma(t,x)\| + |b(t,x) - b(t,y)| &\le K|x-y|,
\end{aligned}
$$

where K is a fixed constant.

Theorem 7.1. *Assume that generalized function (5.26) is a locally finite measure on Q. Assume that we are given bounded Borel functions $h(t,x)$ and $g(t,x)$ defined in Q and ∂Q, respectively. Then the functions $u_1 := R(Q)h$ and $u_2 := \pi(Q)g$ satisfy*

$$Lu_1 + h = 0, \quad Lu_2 = 0 \tag{7.1}$$

in the sense of generalized functions on Q.

Proof. If (5.26) were a locally integrable function in Q, one could give quite a short proof of the theorem. To show this observe that if $\lambda > 0$ and γ is a random variable exponentially distributed with parameter λ and independent of w., then by the Markov property, we have for $(s, x) \in Q$ and $r < t$ that

$$E\{I_{t<\tau} e^{-\varphi_t} h(s+t, x_t)\, dt | \mathcal{F}_r\} = e^{-\varphi_r} E_{s+r, x_r} I_{t-r<\tau} e^{-\varphi_{t-r}} h(s+t, x_{t-r}) \quad \text{(a.s.)},$$

which implies

$$u_1(s, x) = E_{s,x} \int_0^{\gamma \wedge \tau} e^{-\varphi_t} h(s+t, x_t)\, dt + E_{s,x} I_{\tau > \gamma} \int_\gamma^\tau e^{-\varphi_t} h(s+t, x_t)\, dt$$

$$= E_{s,x} \int_0^\infty \lambda e^{-\lambda r} \int_0^{r \wedge \tau} e^{-\varphi_t} h(s+t, x_t)\, dt dr$$

$$+ \int_0^\infty \lambda e^{-\lambda r} \int_r^\infty E_{s,x} I_{\tau > r} E\{I_{t<\tau} e^{-\varphi_t} h(s+t, x_t) | \mathcal{F}_r\}\, dt dr$$

$$= E_{s,x} \int_0^\tau e^{-\varphi_t - \lambda t} h(s+t, x_t)\, dt$$

$$+ \int_0^\infty \lambda e^{-\lambda r} \int_r^\infty E_{s,x} I_{\tau > r} e^{-\varphi_r} E_{s+r, x_r} I_{t-r<\tau} e^{-\varphi_{t-r}} h(s+t, x_{t-r})\, dt dr$$

$$= R_\lambda(Q)h(s, x) + E_{s,x} \int_0^\tau \lambda e^{-\varphi_r - \lambda r} \int_0^\infty E_{s+r, x_r} I_{p<\tau} e^{-\varphi_p} h(s+r+p, x_p)\, dp dr$$

$$= R_\lambda(Q)h(s, x) + \lambda R_\lambda(Q) R(Q) h(s, x).$$

In this way we arrive at the Hilbert identity

$$R(Q) = R_\lambda(Q) + \lambda R_\lambda(Q) R(Q).$$

Similarly, $\pi(Q) = \pi_\lambda(Q) + \lambda R_\lambda(Q)\pi(Q)$, so that

$$\lambda(\lambda R_\lambda(Q)u_1 - u_1) = -\lambda R_\lambda(Q)h, \quad \lambda(\lambda R_\lambda(Q)u_2 - u_2) = -\lambda \pi_\lambda(Q)u_2.$$

Hence (7.1) follows from Theorem 6.3 and Lemmas 6.1 and 6.2.

In the general case take an increasing sequence of bounded domains $Q_n \subset \bar{Q}_n \subset Q$ such that $Q = \cup_n Q_n$ and take functions $\zeta_n \in C^\infty((0, T) \times \mathbb{R}^d)$ so that $\zeta_n = 1$ outside Q_{n+1}, $0 \le \zeta_n \le 1$, and $\zeta_n = 0$ on Q_n. Also continue h outside $\bar{Q}$ as zero and for $k \ge 1$ define

$$u_{kn}(s,x) = E_{s,x} \int_0^{T-s} e^{-\varphi_t - k\varphi_{nt}} h(s+t, x_t)\, dt,$$

where

$$\varphi_{nt} := \int_0^t \zeta_n(s+r, x_r)\, dr.$$

By Theorem 5.14

$$Lu_{kn} - k\zeta_n u_{kn} + h = 0$$

in the sense of generalized functions on $(0,T) \times \mathbb{R}^d$. In particular,

$$Lu_{kn} + h = 0$$

in Q_n. Also, obviously the functions u_{kn} are uniformly bounded on any compact subset of Q. It follows that to prove the first equality in (7.1) we only need to prove that $u_{kn} \to u$ at any point $(s,x) \in Q$ as first $k \to \infty$ and then $n \to \infty$.

While proving that $u_{kn} \to u$ we may an will assume that $h \geq 0$. In this case, obviously, u_{kn} decrease as k increase and $u_{kn} \geq R(Q_n)h$. In addition observe that $\tau_n := \tau_{Q_n} \uparrow \tau_Q$. Hence,

$$R(Q_n)h \uparrow u, \quad \lim_{n\to\infty} \lim_{k\to\infty} u_{kn} \geq \lim_{n\to\infty} R(Q_n)h = u. \tag{7.2}$$

On the other hand, by general form of the strong Markov property

$$u_{kn}(s,x) = E_{s,x}\Big\{ \int_0^{\tau_{n+1}} e^{-\varphi_t - k\varphi_{nt}} h(s+t, x_t)\, dt$$
$$+ u_{kn}(s + \tau_{n+1}, x_{\tau_{n+1}}) \exp(-\varphi_{\tau_{n+1}} - k\varphi_{n\tau_{n+1}})\Big\}$$
$$\leq R(Q_{n+1})h(s,x) + N E_{s,x} \exp(-k\varphi_{n\tau_{n+1}}),$$

where $N = T \sup h \geq \sup u_{kn}$. Here

$$\lim_{k\to\infty} E_{s,x} \exp(-k\varphi_{n\tau_{n+1}}) = P_{s,x}\{\varphi_{n\tau_{n+1}} = 0\}.$$

The last probability is zero if $(s,x) \in Q_{n+1}$, since $0 < \tau_{n+1} \leq T - s$ for any ω and for r close to τ_{n+1} we have $\zeta(s+r, x_r) > 0$. This and (7.2) prove that $u_{kn} \to u$ and finish the proof of the first equality in (7.1).

A standard measure-theoretic argument shows that in the proof of the second equality in (7.1) we may confine ourselves to the case of g which are twice continuously differentiable in (t,x) and have bounded derivatives of the first and second order. In that case by Itô's formula $u_2 = \pi(Q)g = g + R(Q)(D_t g + Lg)$, so that by the above result

$$L(u_2 - g) + Lg = 0,$$

which coincides with the second formula in (7.1). The theorem is proved.

For elliptic equations or more generally for the case $T = \infty$ one has the following result.

Theorem 7.2. *Assume that generalized function (5.26) is a locally finite measure on* $(0, \infty) \times D$, *where* D *is a domain in* $\mathbb{R}^d$. *Assume that we are given bounded Borel functions* $h(t, x)$ *and* $g(t, x)$ *defined in* $(0, \infty) \times D$ *and* ∂D, *respectively. Also assume that the function*

$$u_1(s, x) := E_{s,x} \int_0^\tau e^{-\varphi_t} h(s + t, x_t) \, dt,$$

where $\tau = \inf\{t \geq 0 : x_t \notin D\}$, *is well defined and locally bounded in* $(0, \infty) \times D$ *and let*

$$u_2(s, x) := E_{s,x} e^{-\varphi_\tau} g(s + \tau, x_\tau).$$

Then u_1 *and* u_2 *satisfy again (7.1) in the sense of generalized functions on* $(0, \infty) \times D$.

To derive this theorem from Theorem 7.1 it suffices to notice that by the strong Markov property, for any $T \in (0, \infty)$, bounded domain $G \subset \bar{G} \subset D$, and $(s, x) \in Q := (0, T) \times G$ we have

$$u_1 = R(Q)h + \pi(Q)u_1, \quad u_2 = \pi(Q)u_2.$$

In general one cannot expect much more than stated in Theorems 7.1 or 7.2.

Exercise 7.3. Let $d = 2$, $d_1 = 1$, $D = B_2 \setminus \bar{B}_1$, $dx_t^1 = dw_t$, $dx_t^2 = 0$,

$$u(s, x) = u(x) = P\{|x_{\gamma(x)}| = 2\} = E_x g(x_{\tau(x)}),$$

where $\tau(x)$ is the first exit time from D of the time–homogeneous process x_t starting at x and $g(x) = |x| - 1$. By Theorem 7.2

$$u_{x^1 x^1} = 0$$

in $(0, \infty) \times D$ or in D in the sense of generalized functions. Show that, for any x^2 satisfying $|x^2| < 2$, the function $u(x^1, x^2)$ is indeed (piecewise) linear with respect to x^1 but is discontinuous as a function of x on the lines $x^2 = \pm 1$.

Finally, we give an extremely important extension of Theorem 7.1 to a situation occurring in applications to controlled diffusion processes where we are more interested in cases when Kolmogorov's equation becomes an inequality. One uses inequalities like (7.4) when it is known that the second order derivatives of u are bounded from below in the sense of generalized functions and one wants to estimate them from above. The idea is as follows. If for a function u of two variables (x, y) we have $u_{xx}, u_{yy} \geq 0$, then from the inequality like $u_{xx} + u_{yy} \leq 1$ one gets that $u_{xx} \leq 1, u_{yy} \leq 1$, so that $|u_{xx}| \leq 1, |u_{yy}| \leq 1$. For more details see [KR89].

Theorem 7.4. *Let $u(t, x)$ be a bounded Borel functions defined in a bounded domain $Q \subset (0, T) \times \mathbb{R}^d$. Assume that either*

(i) the generalized function (5.26) is a locally integrable function on Q,

or

(ii) the function $f(t, x)$ is continuous in x and there exists a constant N such that

$$|u(t, x) - u(t, y)| \leq N|x - y|, \quad |c(t, x) - c(t, y)| \leq N|x - y|$$

for all $(t, x), (t, y) \in Q$. Finally, assume that for any $(s, x) \in Q$

$$u(s, x) \geq E_{s,x}\Big[\int_0^{\tau \wedge t} f(s+r, x_r)e^{-\varphi_r}\, dr + u(s+\tau \wedge t, x_{\tau \wedge t})e^{-\varphi_{\tau \wedge t}}\Big] + o(t) \quad (7.3)$$

as $t \downarrow 0$, where $o(t)$ is uniform for (s, x) in any closed subset of Q. Then

$$Lu + f \leq 0 \qquad (7.4)$$

in the sense of generalized functions on Q.

Proof. Define $\varepsilon(s, x, t) = o(t)/t$. It is convenient to assume that $\varepsilon(s, x, t)$ is defined for all $t > 0$. Furthermore, without loss of generality we assume that $\varepsilon(s, x, t) = \varepsilon(s, x, T - s)$ for $t \geq T - s$. This is possible since $\tau \leq T - s$ so that the first term on the right in (7.3) does not change for $t \geq T - s$ and the left–hand side does not depend on t at all. Also $|o(t)|$ need not be larger than sup over $(s, x) \in Q$ and $t \leq T - s$ of the absolute value of the difference of the left–hand side and the first term on the right in (7.3). Hence, we may assume that $|o(t)|$ is bounded, and since $\varepsilon(s, x, t) \to 0$ as $t \downarrow 0$ uniformly on any closed subset of Q, we may also assume that for (s, x) in such a set, $\varepsilon(s, x, t)$ is bounded by a constant independent of t. Then by the dominated convergence theorem

$$I_\lambda(s, x) := \lambda^2 \int_0^\infty t\varepsilon(s, x, t)e^{-\lambda t}\, dt = \int_0^\infty \varepsilon(s, x, t/\lambda)te^{-t}\, dt \to 0$$

as $\lambda \to \infty$ uniformly on any closed subset of Q or in the sense of generalized functions on Q.

After this preparations multiply (7.3) by $\lambda e^{-\lambda t}$ and integrate over $(0, \infty)$. Then by Fubini's theorem we easily get

$$u(s, x) \geq R_\lambda(Q)(f + \lambda u)(s, x) + \pi_\lambda(Q)u(s, x) + \lambda^{-1}I_\lambda(s, x).$$

It follows that in the sense of generalized functions on Q

$$\lim_{\lambda \to \infty} \lambda\big(\lambda R_\lambda(Q)u - u\big) \leq -\lim_{\lambda \to \infty} \lambda R_\lambda(Q)f - \lim_{\lambda \to \infty} \lambda\pi_\lambda(Q)u$$

and it only remains to use Theorem 6.3 and Lemmas 6.2 and 6.1. The theorem is proved.

Remark 7.5. Take a bounded Borel function $g(t, x)$ and define $u = R(Q)f + \pi(Q)g$. Then under assumption (ii) of Theorem 7.4 we have $Lu + f = 0$ in Q since by the strong Markov property we have equality in (7.3) with $o(t) \equiv 0$.

The same is true under condition (i), however this is a weaker statement than Theorem 7.1.

References

[KO] A.N. Kolmogorov, *Über die analytischen Methoden in der Wahrscheinlichkeit- srechnung*, Math. Ann., Vol. 104 (1931), 415–458.

[KR89] N.V. Krylov, *Smoothness of the value function for a controlled diffusion process in a domain*, Izvestija Akademii Nauk SSSR, serija matematicheskaja, Vol. 53, No. 1 (1989), 66–96 in Russian; English translation: Russian Acad. Sci. Izv. Math., Vol. 34, No. 1 (1990), 65–96.

[KR92] N.V. Krylov, *On first quasiderivatives of solutions of Itô's stochastic equa- tions*, Izvestija Akademii Nauk SSSR, serija matematicheskaja, Vol. 56, No. 2 (1992), 398-426 in Russian; English translation: Russian Acad. Sci. Izv. Math., Vol. 40, No. 2 (1992), 377-403.

[KR] N.V. Krylov "Nonlinear elliptic and parabolic equations of second order", Nauka, Moscow, 1985 in Russian; English translation: Reidel, Dordrecht, 1987.

[KR1] N.V. Krylov and B.L. Rozovsky, *Stochastic evolution equations*, "Itogy nauki i tekhniki", Vol. 14, VINITI, Moscow, 1979, 71-146 in Russian; English transla- tion in J. Soviet Math., Vol. 16, No. 4 (1981), 1233–1277.

[M] P. A. Meyer, "Probability and potentials", Blaisdell Publishing Company, A Division of Ginn and Company, Waltham, Massachusetts, Toronto, London, 1966.

L^p–analysis of finite and infinite dimensional diffusion operators

Michael Röckner

Fakultät für Mathematik Universität Bielefeld Postfach 10 01 31 D-33501 Bielefeld Germany

1. Introduction

The purpose of these lectures is to present an approach to Kolmogorov equations in infinite dimensions which is based on an $L^p(\mu)$–analysis of the corresponding diffusion operators w.r.t. suitably chosen measures μ. More precisely, we solve

$$\frac{du}{dt} = Lu \ , \ u(0, \cdot) = f \tag{1.1}$$

for functions f, $u(t, \cdot)$ on an infinite dimensional (vector) space E and diffusion operators L (cf. Definition 3.1 below) by a C^0–semigroup

$$u(t, \cdot) = T_t f$$

generated by some closed extension $\hat{L}$ of L on $L^p(E; \mu)$ for some appropriate measure μ. Here we think of L as a–priori *only* given explicitly on a set D of nice (smooth) functions. In principle there are at least two types of measures that appear appropriate in applications:

(1) A given reference (e.g. Gaussian) measure μ so that L can be written as a sum of an operator L_0, which is symmetric on $L^2(E; \mu)$, and a small perturbation in the sense of sectorial forms. (This case is refered to in the text below as the "sectorial case").

(2) μ is an invariant measure for L (or more generally for $\alpha - L$, $\alpha \geq 0$) in the sense that

$$\int Lu \, d\mu = 0 \text{ for all } u \in D \ .$$

(This situation we refer to below as the "non–sectorial case").

So, in case (2) the first problem is to find such a measure μ. The second problem is then to construct a closed extension $\hat{L}$ of L on $L^p(E; \mu)$ that generates a C^0–semigroup on $L^p(E; \mu)$ ("existence problem"), and the third is to prove that it is the only such extension ("uniqueness problem").

AMS Subject Classification (1991) Primary: 60 H 30, 35 K 15 Secondary: 31 C 25, 35 K 10, 60 J 60

Key words and phrases. Kolmogorov equations, diffusion operators, sectorial forms, Dirichlet forms, strongly continuous semi–groups, martingale problem

In these notes we address all three problems presenting results obtained over the last few years in implementing this program. For the precise structure of the corresponding parts of the text, i.e. Sections 2. – 5. we refer to the list of contents. We would only like to point out here that we also include results on properties as "positivity preserving", "(sub-)Markovian" of the C^0–semigroups solving (1.1), in regard to their probabilistic relevance as transition functions of Markov processes. The latter is made precise in Section 6.. In Section 7. we present a case study where E is an infinite dimensional manifold rather than a vector space. In particular, we discuss ergodic properties of the above C^0–semigroups in this situation.

The main purpose of these lectures is to give an updated and an as complete as possible presentation of what has been achieved so far and what the aims are, rather than to repeat detailed proofs. We complement instead arguments needed to derive the statements given here from those recent results, giving precise reference for the latter.

The material presented in these lectures is taken from a number of joint papers with various co–authors as well as the recent work of Andreas Eberle and Wilhelm Stannat. I would like to thank the latter two as well as all my co–authors, in particular Sergio Albeverio, Vladimir Bogachev, Yuri Kondratiev, Vitali Liskevich, Tatjana Tsikalenko and Tu–Sheng Zhang for permission to include very recent, so far unpublished joint results.

Finally, it is a great pleasure to thank C.I.M.E. and, in particular, the organizer of this session, Giuseppe Da Prato, for a very stimulating summer school and a fantastic time in Cetraro. I would also like to thank my two colleague lecturers Nick Krylov and Jerzy Zabczyk for teaching me so much through their lectures and intensive discussion. I also would like to express my deep thanks to all other participants for their steady interest in the subject and for creating a really marvellous atmosphere of a type I had never experienced before.

2. Solution of Kolmogorov equations via sectorial forms

2.1 Preliminaries

In this subsection we recall some well–known classical results. We assume the reader to be familiar with the notion of a C^0–semigroup $(T_t)_{t\geq0}$ resp. its generator $(L, D(L))$ on a Banach space (cf. e.g. [Pa 85]). We only recall here that if $u \in D(L)$ then $T_t u \in D(L)$ for all $t \geq 0$ and

$$\frac{d}{dt} T_t u = LT_t u = T_t Lu \text{ for all } t \geq 0 . \tag{2.1}$$

Below let $(B, \|\cdot\|)$ be a real Banach space with dual B^* and corresponding dualization $\langle\,,\,\rangle$. Let L with domain D be a densely defined linear operator on B.

Definition 2.1. (L, D) is called *dissipative*, if for all $u \in D$ there exists $u^* \in B^*$ such that

$$\langle u^*, u \rangle = \|u\|_B^2 = \|u^*\|_B^2 \text{ and } \langle u^*, Lu \rangle \leq 0 \qquad (2.2)$$

or, equivalently, (cf. [Pa 85, Ch. I, Thm. 4.2])

$$\|\alpha u - Lu\| \geq \alpha \|u\| \text{ for all } \alpha > 0, \, u \in D .$$

Remark 2.2. If (L, D) is dissipative, then it is closable. Below we shall denote its closure by $(\overline{L}, \overline{D})$ or shortly $\overline{L}$.

Theorem 2.3. *(Lumer/Phillips) Suppose (L, D) is dissipative. Then $\overline{L}$ is the generator of a C^0-semigroup of contractions (i.e., the operator norms are bounded by 1) on B if and only if $(\alpha - L)(D)$ is dense in B for some (or equivalently all) $\alpha > 0$.*

Let $L^* : D(L^*) \subseteq B^* \to B^*$ denote the adjoint of (L, D). Then as a consequence of Theorem 2.3 and the Hahn–Banach theorem one immediately obtains.

Corollary 2.4. *(1st adjoint criterion) Let (L, D) be dissipative. Then $\overline{L}$ generates a C^0-semigroup of contractions on B if and only if $\ker(\alpha - L^*) = \{0\}$ for some (or equivalently all) $\alpha > 0$.*

This in turn obviously implies:

Corollary 2.5. *(2nd adjoint criterion) Let both (L, D) and $(L^*, D(L^*))$ be dissipative. Then $\overline{L}$ generates a C^0-semigroup of contractions on B.*

Later we shall also use the following "uniqueness result":

Theorem 2.6. *Suppose there exists a C^0-semigroup $(T_t)_{t \geq 0}$ on B such that its generator extends (L, D). Then the following assertions are equivalent:*

  (i) D is a core for the generator of $(T_t)_{t \geq 0}$ (i.e., D is dense in the domain of the generator w.r.t. the norm induced by its graph norm).
  (ii) The closure $\overline{L}$ of (L, D) generates a C^0-semigroup.
  (iii) $(T_t)_{t \geq 0}$ is the only C^0-semigroup on B whose generator extends (L, D).

If $(T_t)_{t \geq 0}$ consists of contractions and if $\varepsilon > 0$, then (i) - (iii) is equivalent to

  (iii') $(T_t)_{t \geq 0}$ is the only C^0-semigroup on B such that its generator extends (L, D) and $\|T_t\| \leq e^{\varepsilon t}$ for all $t \geq 0$.

Proof. See [Eb 98, Ch. I, Thm. 1.2] and [Ar 86, A - II, Thm. 1.33]. □

A sufficient condition for uniqueness (which will, however, turn out to be typically *not* applicable to cases of interest in these lectures) is the following:

Proposition 2.7. *Let $(T_t)_{t\geq 0}$ be a C^0-semigroup on B whose generator extends (L, D). If there exists a dense subspace $B_0 \subset B$ such that $T_t(B_0) \subset D$ for all $t \geq 0$, then $(T_t)_{t\geq 0}$ is the only C^0-semigroup whose generator extends (L, D).*

Proof. See e.g. [EthKur 86, Ch. 1, Prop. 3.3].   □

2.2 Sectorial forms.

The terminology w.r.t. "sectorial forms" used in these lectures is according to [MR 92]. Below we fix a Hilbert space $(\mathcal{H}, (,))$ and set $\|\cdot\| := (,)^{1/2}$.

Definition 2.8. A bilinear form $\mathcal{E} : D(\mathcal{E}) \times D(\mathcal{E}) \to \mathbb{R}$ with dense domain $D(\mathcal{E}) \subset \mathcal{H}$ is called a *(closed) coercive form* if it satisfies the following conditions:

(i) Its symmetric part $(\tilde{\mathcal{E}}, D(\tilde{\mathcal{E}}))$ defined by $\tilde{\mathcal{E}}(u, v) := \frac{1}{2}(\mathcal{E}(u, v) + \mathcal{E}(v, u))$; $u, v \in D(\tilde{\mathcal{E}}) := D(\mathcal{E})$, is positive definite and *closed* (i.e., $D(\mathcal{E})$ is complete with respect to the norm $\tilde{\mathcal{E}}_1^{1/2}$ coming from the symmetric part of $\mathcal{E}_\alpha(u, v) := \mathcal{E}(u, v) + \alpha(u, v)$; $\alpha > 0$; $u, v \in D(\mathcal{E})$ for $\alpha := 1$).

(ii) $(\mathcal{E}, D(\mathcal{E}))$ satisfies the *weak sector condition*, i.e., for one (and hence all) $\alpha \in (0, \infty)$ there exists $K_\alpha \in (0, \infty)$ such that

$$|\mathcal{E}_\alpha(u, v)| \leq K_\alpha \, \mathcal{E}_\alpha(u, u)^{1/2} \, \mathcal{E}_\alpha(v, v)^{1/2} \text{ for all } u, v \in D(\mathcal{E}) . \qquad (2.3)$$

We shall use (2.3) below almost always with $\alpha := 1$ and set $K := K_1$.

Remark 2.9.   (i) If $(\mathcal{E}, D(\mathcal{E}))$ is a coercive form on $\mathcal{H}$, then so is $(\mathcal{E}^*, D(\mathcal{E}^*))$ defined by

$$\mathcal{E}^*(u, v) := \mathcal{E}(v, u) ; \ u, v \in D(\mathcal{E}^*) := D(\mathcal{E}) . \qquad (2.4)$$

The same holds for $(\tilde{\mathcal{E}}, D(\tilde{\mathcal{E}}))$.

(ii) The *antisymmetric part* $(\check{\mathcal{E}}, D\check{\mathcal{E}})$ of a bilinear form $(\mathcal{E}, D(\mathcal{E}))$ on $\mathcal{H}$ is defined by

$$\check{\mathcal{E}}(u, v) := \frac{1}{2}(\mathcal{E}(u, v) - \mathcal{E}(v, u)) ; \ u, v \in D(\check{\mathcal{E}}) := D(\mathcal{E}) .$$

It is easy to see that if $(\mathcal{E}, D(\mathcal{E}))$ is positive definite, then $(\mathcal{E}, D(\mathcal{E}))$ satisfies the weak sector condition if and only if for some (hence all) $\alpha \in (0, \infty)$ there exists $C_\alpha \in (0, \infty)$ such that

$$\check{\mathcal{E}}(u, v) \leq C_\alpha \, \mathcal{E}_\alpha(u, u)^{1/2} \, \mathcal{E}_\alpha(v, v)^{1/2} \text{ for all } u, v \in D(\mathcal{E}) .$$

A coercive form is always associated with a linear operator which generates a C^0-semigroup of contractions on $\mathcal{H}$. We restate and reprove this fact here (cf. e.g. [MR 92, Proposition 2.16]), which is an easy consequence of the Lax–Milgram Theorem.

Proposition 2.10. *Let $(\mathcal{E}, D(\mathcal{E}))$ be a coercive form on $\mathcal{H}$. Define*

$$D(L) := \{u \in D(\mathcal{E}) | v \mapsto \mathcal{E}(u, v) \text{ is continuous w.r.t. } \| \cdot \| \text{ on } D(\mathcal{E})\} , (2.5)$$

and for $u \in D(L)$ let Lu denote the unique element in $\mathcal{H}$ such that

$$(-Lu, v) = \mathcal{E}(u, v) \text{ for all } v \in D(\mathcal{E}) . (2.6)$$

Then $(L, D(L))$ is a linear operator on $\mathcal{H}$ that generates a C^0–semigroup $(T_t)_{t \geq 0}$ of contractions on $\mathcal{H}$.

Proof. Clearly $(L, D(L))$ is a well–defined linear operator on $\mathcal{H}$ and applying the above definition to the coercive form $(\mathcal{E}^*, D(\mathcal{E}^*))$ we likewise obtain an operator $(L^*, D(L^*))$ which obviously is exactly the adjoint operator of $(L, D(L))$ on $\mathcal{H}$.

Claim 1. $(1 - L)(D(L)) = \mathcal{H}$ and $D(L)$ is dense in $\mathcal{H}$.

Indeed, applying for $f \in \mathcal{H}$ the Lax–Milgram Theorem to the linear functional $v \mapsto (v, f)$ on $D(\mathcal{E})$ (which is continuous w.r.t. $\tilde{\mathcal{E}}_1^{1/2}$) we obtain that there are $G_1 f, G_1^* f \in D(\mathcal{E})$ such that for $v \in D(\mathcal{E})$

$$(f, v) = \mathcal{E}_1(G_1 f, v) = \mathcal{E}_1^*(G_1^* f, v) = \mathcal{E}_1(v, G_1^* f).$$

This in turn implies that $G_1 f \in D(L), G_1^* f \in D(L^*)$ and that $(1-L)G_1 f = f$ and $(1 - L^*)G_1^* f = f$. Consequently, $(1 - L)(D(L)) = \mathcal{H}$ and if $g \in \mathcal{H}$ such that $(g, u) = 0$ for all $u \in D(L)$, then for all $f \in \mathcal{H}$

$$0 = (g, G_1 f) = \mathcal{E}_1(G_1 f, G_1^* g) = (f, G_1^* g) .$$

Hence $G_1^* g = 0$, therefore, $g = (1 - L^*)G_1^* g = 0$. So, $D(L)$ is dense in $\mathcal{H}$.

Claim 2. L is dissipative and closed.

Clearly, for all $u \in D(L), \alpha > 0$, we have

$$\alpha \|u\|^2 \leq \mathcal{E}_\alpha(u, u) = ((\alpha - L)u, u) \leq \|\alpha u - Lu\| \|u\| .$$

In particular, $1 - L$ is invertible with bounded inverse $(1 - L)^{-1}$ which by Claim 1 is defined on all of $\mathcal{H}$. Hence $(1 - L)^{-1}$ and thus also L is closed.

Claims 1 and 2 prove the assertion.   □

It follows by (2.3) that the operator L in the previous proposition is *weakly sectorial*, i.e., for all $\alpha \in (0, \infty)$ there exists $K_\alpha \in (0, \infty)$ such that

$$|((\alpha - L)u, v)| \leq ((\alpha - L)u, u)^{1/2}((\alpha - L)v, v)^{1/2} \text{ for all } u, v \in D(L) . (2.7)$$

In fact Proposition 2.10 establishes a one–to–one correspondence between sectorial forms and such operators resp. C^0–semigroups of a certain type.

Theorem 2.11. *The association of L and hence $(T_t)_{t \geq 0}$ to $(\mathcal{E}, D(\mathcal{E}))$ in Proposition 2.10 defines bijective maps between the following three sets of mathematical objects:*

(a) *Sectorial forms* $(\mathcal{E}, D(\mathcal{E}))$ *on* $\mathcal{H}$.

(b) *Weakly sectorial, closed, dissipative operators* $(L, D(L))$ *on* $\mathcal{H}$ *for which* $(1 - L)(D(L))$ *is dense in* $\mathcal{H}$.

(c) C^0-*semigroups* $(T_t)_{t \geq 0}$ *of contractions on* $\mathcal{H}$ *having the property that for their natural extensions* $(T_t^{\mathbb{C}})_{t \geq 0}$ *to the complexification* $\mathcal{H}_{\mathbb{C}}$ *of* $\mathcal{H}$ *the semigroups* $(e^t T_t^{\mathbb{C}})_{t \geq 0}$ *are restrictions of holomorphic contraction semigroups* $(e^z T_z^{\mathbb{C}})_{z \in K}$ *where* K *is a sector in* $\mathbb{C}$ *containing* $[0, \infty)$.

For the proof of Theorem 2.11 we refer to [MR 92, Ch. I, Sect. 2] (in particular, cf. the diagram on p. 27). We only note here that $(T_t)_{t \geq 0}$ in Proposition 2.10 is, therefore, analytic. In particular,

$$T_t(\mathcal{H}) \subset D(L) \text{ for all } t > 0$$

(cf. e.g. [ReS 75, Corollary 2 of Theorem X.52]) and thus for all $f \in \mathcal{H}$

$$\frac{d}{dt} T_t f = L T_t f \text{ for all } t > 0$$

$$T_0 f = f .$$

2.3 Sectorial forms on $L^2(E; m)$

Let $(E; \mathcal{B}; m)$ be a measure space and $\mathcal{H} := L^2(E; m) := L^2(E; \mathcal{B}; m)$ the corresponding (real) L^2-space with its usual inner product. In particular, $\mathcal{H}$ now has a vector lattice structure w.r.t. the usual order "$\leq$" (defined representativewise m-a.e.). As usual for $f \in L^2(E; m)$ we set $f^+ := \sup(f, 0)$ and $f^- := -\inf(f, 0)$.

In this subsection we recall a few results giving conditions on a generator (resp. the associated coercive form, if it exists) ensuring that the C^0-semigroup respects this order. These results (at least for symmetric semigroups) are known under the key-word "Beurling–Deny criteria". We first recall

Definition 2.12. A C^0-semigroup $(T_t)_{t \geq 0}$ on $L^2(E; m)$ is called *positivity preserving*, if

$$f \in L^2(E; m) , \ f \geq 0 \Rightarrow T_t f \geq 0 \text{ for all } t \geq 0 ,$$

and *sub-Markovian* if

$$f \in L^2(E; m) , \ 0 \leq f \leq 1 \Rightarrow 0 \leq T_t f \leq 1 \text{ for all } t \geq 0 .$$

If $m(E) < \infty$, a positivity preserving C^0-semigroup is called *Markovian* if $T_t 1 = 1$ for all $t > 0$.

It is easy to see that "sub-Markovian" implies "positivity preserving".

Theorem 2.13. *Let* $(T_t)_{t\geq 0}$ *be a* C^0*–semigroup of contractions on* $L^2(E;m)$ *with generator* $(L, D(L))$*. Then*

(i) $(T_t)_{t\geq 0}$ *is positivity preserving if and only if*

$$(Lu, u^+) \leq 0 \text{ for all } u \in D(L) \ .$$

(ii) $(T_t)_{t\geq 0}$ *is sub–Markovian if and only if*

$$(Lu, (u - 1)^+) \leq 0 \text{ for all } u \in D(L) \ .$$

Proof. For (i) see e.g. [MR 95, Theorem 1.7] and for (ii) see e.g. [MR 92, Ch. I, Proposition 4.3]. □

Unfortunately, the conditions on L in the previous theorem are often hard to check. As we shall see below, this is different if $(T_t)_{t\geq 0}$ is associated with a sectorial form, since then there are respective easily checkable conditions on the latter.

Theorem 2.14. *Let* $(\mathcal{E}, D(\mathcal{E}))$ *be a sectorial form on* $L^2(E;m)$ *with corresponding (generator L and) C^0–semigroup* $(T_t)_{t\geq 0}$ *of contractions. Then:*

(i) $(T_t)_{t\geq 0}$ *is positivity preserving if and only if for all* $u \in D(\mathcal{E})$ *also* $u^+ \in D(\mathcal{E})$ *and*

$$\mathcal{E}(u, u^+) \geq 0 \ .$$

(ii) $(T_t)_{t\geq 0}$ *is sub–Markovian if and only if for all* $u \in D(\mathcal{E})$ *also* $u^\# := \inf(u^+, 1) \in D(\mathcal{E})$ *and*

$$\mathcal{E}(u, u - u^\#) \geq 0 \ .$$

Proof. (i): [MR 95, Theorem 1.5 and Proposition 1.3].
(ii): [MR 92, Ch. I, Theorem 4.4] and its proof. □

Remark 2.15. (i) In case (i) of Theorem 2.14 the coercive form $(\mathcal{E}, D(\mathcal{E}))$ is also called *positivity preserving* while in case (ii) it is called a *Semi–Dirichlet form*. If also $(\mathcal{E}^*, D(\mathcal{E}^*))$ is a Semi–Dirichlet form, then $(\mathcal{E}, D(\mathcal{E}))$ is called a *Dirichlet form*.
(ii) We note that if $(\mathcal{E}, D(\mathcal{E}))$ is positivity preserving, then so is $(\mathcal{E}^*, D(\mathcal{E}^*))$ and therefore so is $(\bar{\mathcal{E}}, D(\bar{\mathcal{E}}))$. (Indeed, setting $u_n := u^+ - \frac{1}{n}u^-$ we obtain for all $n \in \mathbb{N}$

$$\mathcal{E}(u^+, u) = \mathcal{E}(u_n, u) + \frac{1}{n}\mathcal{E}(u^-, u)$$
$$= \mathcal{E}(u_n, (u_n)^+) + n\mathcal{E}(-u_n, (-u_n)^+) + \frac{1}{n}\mathcal{E}(u^-, u) \geq \frac{1}{n}\mathcal{E}(u^-, u) \ .)$$

But if $(\mathcal{E}, D(\mathcal{E}))$ is a Semi–Dirichlet form, $(\mathcal{E}^*, D(\mathcal{E}^*))$ might be not (cf. [MOR 95, Remark 2.2 (ii)]). So, not every Semi–Dirichlet form is a Dirichlet form.

Since in many applications a coercive form $(\mathcal{E}, D(\mathcal{E}))$ is explicitly given only on a core $D \subset D(\mathcal{E})$, we recall the following useful result to check whether $(\mathcal{E}, D(\mathcal{E}))$ is positivity preserving. Analogous results hold for (Semi–)Dirichlet forms (see e.g. [MR 92, Prop. 4.7 and Ex. 4.8]).

Lemma 2.16. *Let $(\mathcal{E}, D(\mathcal{E}))$ be a coercive form on $L^2(E; m)$. Let D be a subset of $D(\mathcal{E})$, dense w.r.t. $\tilde{\mathcal{E}}_1^{1/2}$ such that for all $u \in D$ the following holds: for all $\varepsilon > 0$ there exists a function $\varphi_\varepsilon : \mathbb{R} \to [-\varepsilon, \infty)$ such that $\varphi_\varepsilon(t) = t$ for all $t \in [0, \infty)$, $0 \le \varphi_\varepsilon(t_2) - \varphi_\varepsilon(t_1) \le t_2 - t_1$ if $t_1 \le t_2$, $\varphi_\varepsilon \circ u \in D(\mathcal{E})$ and*

$$\sup_{\varepsilon > 0} \mathcal{E}(\varphi_\varepsilon \circ u, \varphi_\varepsilon \circ u) \le \text{ const. } \mathcal{E}(u, u)$$

as well as

$$\limsup_{\varepsilon \to 0} \mathcal{E}(u, \varphi_\varepsilon \circ u) \ge 0 \ .$$

Then $(\mathcal{E}, D(\mathcal{E}))$ is positivity preserving.

Proof. [MR 95, Proposition 1.2]. □

2.4 Examples and Applications

2.4.1 $E :=$ open subset of $\mathbb{R}^d$.

Let $E := U \subset \mathbb{R}^d$, U open, $\mathcal{B} := \mathcal{B}(U)$ the Borel σ–algebra on U, and let dx denote Lebesgue measure on U. For $1 \le i, j \le d$ let $a_{ij} \in L^1_{\text{loc}}(U; dx)$ such that $\check{a}_{ij} := \frac{1}{2}(a_{ij} - a_{ji}) \in L^\infty(U; dx)$, $b_i, \hat{b}_i \in L^\infty(U; dx) + L^d(U; dx)$, $c \in L^1_{\text{loc}}(U; dx)$ such that $c^- \in L^\infty(U; dx) + L^{d/2}(U; dx)$. Define $A := (a_{ij})_{1 \le i, j \le d}$, $b := (b_1, \ldots, b_d)$, $\hat{b} = (\hat{b}_1, \ldots, \hat{b}_d)$ and let $\langle \ , \ \rangle_{\mathbb{R}^d}$ denote the Euclidean inner product on $\mathbb{R}^d$. Assume that there exists $\delta > 0$ such that for all $\xi \in \mathbb{R}^d$

$$\langle A(x)\xi, \xi \rangle_{\mathbb{R}^d} \ge \delta \langle \xi, \xi \rangle_{\mathbb{R}^d} \text{ for } dx\text{–a.e. } x \in U \ .$$

Define for $u, v \in C_0^\infty(U)$ (: = the set of all infinitely differentiable real–valued functions on U with compact support).

$$
\begin{aligned}
\mathcal{E}(u, v) \ := \ & \int_U \langle A(x)\nabla u(x), \nabla v(x) \rangle_{\mathbb{R}^d} \ dx + \int_U \langle b(x), \nabla u(x) \rangle_{\mathbb{R}^d} \ v(x) \ dx \\
& + \int_U u(x) \langle \hat{b}(x), \nabla v(x) \rangle_{\mathbb{R}^d} \ dx + \int_U u(x) \ v(x) \ c(x) \ dx \ . \quad (2.8)
\end{aligned}
$$

Then there exists an $\alpha \in (0, \infty)$ such that $(\tilde{\mathcal{E}}_\alpha, C_0^\infty(U))$ is a closable, positive definite, symmetric bilinear form on $L^2(U; dx)$. Let $(\tilde{\mathcal{E}}_\alpha, D(\tilde{\mathcal{E}}_\alpha))$ denote its closure. $(\mathcal{E}_\alpha, C_0^\infty(U))$ satisfies the weak sector condition, and hence has a unique (w.r.t. $\tilde{\mathcal{E}}_\alpha^{1/2}$) continuous bilinear extension to the domain $D(\tilde{\mathcal{E}}_\alpha)$, and this extension, denoted by $(\mathcal{E}_\alpha, D(\mathcal{E}_\alpha))$ (with $D(\mathcal{E}_\alpha) := D(\tilde{\mathcal{E}}_\alpha)$), is a coercive form on $L^2(U; dx)$. Furthermore, $(\mathcal{E}_\alpha, D(\mathcal{E}_\alpha))$ is positivity preserving.

The proofs for all these statements are contained in [MR 95, Subsect. 2.1] (and they will be essentially repeated in the infinite dimensional case in the next subsection).

Let $L^{(\alpha)}$ denote the generator of $(\mathcal{E}_\alpha, D(\mathcal{E}_\alpha))$ and $(T_t^{(\alpha)})_{t\geq 0}$ the corresponding positivity preserving C^0–semigroup of contractions on $L^2(U; dx)$. Define

$$L := L^{(\alpha)} + \alpha \,, \ D(L) := D(L^{(\alpha)}) \,,$$
$$T_t := e^{\alpha t} \, T_t^{(\alpha)} \,, \ t \geq 0 \,.$$

Then for all $f \in L^2(U; dx)$

$$\tfrac{d}{dt} T_t f = L T_t f \text{ for all } t > 0 \text{ (and all } t \geq 0, \text{ if } f \in D(L))$$
$$T_0 f = f \,,$$

where L (heuristically) is given by

$$Lu = \text{div } (A\nabla u + \hat{b} u) - \langle b, \nabla u \rangle - c\, u \,. \tag{2.9}$$

$(T_t)_{t\geq 0}$ is positivity preserving, since so is $(T_t^{(\alpha)})_{t\geq 0}$.

2.4.2 $E :=$ topological vector space.

Let E be a locally convex topological real vector space which is a (topological) Souslin space, that is, the continuous image of a separable complete metric space (e.g. E is a separable real Banach space). Let $\mathcal{B} := \mathcal{B}(E)$ denote its Borel σ–algebra and E' its dual space. Let $(H, \langle \,, \, \rangle_H)$ be a separable real Hilbert space, densely and continuously embedded into E. Identifying H with its dual H' we have that

$$E' \subset H \subset E \text{ densely and continuously} \tag{2.10}$$

and the dualization $_{E'}\langle \,, \, \rangle_E$ between E' and E restricted to $E' \times H$ coincides with $\langle \,, \, \rangle_H$. Here the dual E' of E is endowed with the strong topology. H should be thought of as a *tangent space* to E at each point.

For a dense linear subspace $K \subset E'$ we introduce the usual test function space $\mathcal{F}C_b^\infty(K)$ on E consisting of finitely based smooth bounded functions, i.e.,

$$\mathcal{F}C_b^\infty := \mathcal{F}C_b^\infty(K) := \{f(\ell_1, \ldots, \ell_m) \mid m \in \mathbb{N}, \ell_1, \ldots, \ell_m \in E', f \in C_b^\infty(\mathbb{R}^m)\}, \tag{2.11}$$

where $C_b^\infty(\mathbb{R}^m)$ denotes the set of all infinitely differentiable functions on $\mathbb{R}^m$ with bounded partial derivatives of all order. Note that since $\mathcal{B}(E) = \sigma(E')$, $\mathcal{F}C_b^\infty$ is dense in $L^2(E; m)$ for any finite positive measure m on $(E, \mathcal{B}(E))$. Let $u \in \mathcal{F}C_b^\infty$. For $k \in E$ as usual we set

$$\frac{\partial u}{\partial k}(z): \quad = \quad \frac{d}{dt}\, u(z + sk)_{|s=0}$$

$$\tag{2.12}$$

$$= \quad \sum_{i=1}^{m} \partial_i f(\ell_1(z), \ldots, \ell_m(z)) \ {}_{E'}\langle \ell_i, k \rangle_E, \ z \in E \,,$$

where ∂_i denotes derivative w.r.t. the i-th coordinate. The latter equality follows by the chain rule if u has the representation $u = f(\ell_1, \ldots, \ell_m)$ as in (2.11). For $z \in E$ the *H-gradient* $\nabla_H u(z) \in H$ is defined as the unique element in H such that

$$\langle \nabla_H u(z), h \rangle_H = \frac{\partial u}{\partial h}(z) \tag{2.13}$$

(where we note that by (2.10), (2.12) the linear map $h \mapsto \frac{\partial u}{\partial h}(z)$ is continuous on $(H, \langle\ ,\ \rangle_H)$).

We now fix a finite positive measure μ on $(E, \mathcal{B}(E))$ such that $\mu(U) > 0$ for all non-empty open subsets $U \subset E$. Define

$$\mathcal{E}_\mu(u,v) := \int_E \langle \nabla_H u, \nabla_H v \rangle_H \, d\mu \ ; \quad u, v \in \mathcal{F}C_b^\infty \,. \tag{2.14}$$

Then $(\mathcal{E}_\mu, \mathcal{F}C_b^\infty)$ is a positive definite symmetric bilinear form on $L^2(E; \mu)$ with dense domain. We assume

(C) $\hspace{3cm}$ $(\mathcal{E}_\mu, \mathcal{F}C_b^\infty)$ is closable on $L^2(E; \mu)$.

For sufficient (and essentially necessary) conditions implying (C) we refer to [AR 90, Sect. 3]. Next we are going to "build" coercive forms on $L^2(E; \mu)$ of type (1.7) starting with $(\mathcal{E}_\mu, \mathcal{F}C_b^\infty)$.

Let $\mathcal{L}_\infty(H)$ denote the set of all bounded linear operators on H with operator norm $\|\ \|$. Suppose $z \mapsto A(z)$, $z \in E$, is a map from E to $\mathcal{L}_\infty(H)$ such that $z \mapsto \langle A(z)h_1, h_2 \rangle_H$ is $\mathcal{B}(E)$-measurable for all $h_1, h_2 \in H$ and $\|\tilde{A}\| \in L^1(E; \mu)$, $\|\check{A}\| \in L^\infty(E; \mu)$, where $\tilde{A} := \frac{1}{2}(A + A')$, $\check{A} := \frac{1}{2}(A - A')$ and $A'(z)$ denotes the adjoint of $A(z)$, $z \in E$, on H.

Furthermore, assume that the following "ellipticity" condition holds:
(E) There exists $\delta \in (0, \infty)$ such that

$$\langle \tilde{A}(z)h, h \rangle_H \geq \delta \, \langle h, h \rangle_H \text{ for all } h \in H \,.$$

Let $c \in L^1(E; \mu)$ and define

$$Q(u,v) := \int_E \langle \tilde{A}(x) \nabla_H u(x), \nabla_H v(x) \rangle_H \, \mu(dx) +$$

$$+ \int_E u(x)v(x)c^+(x) \, \mu(dx); \qquad u, v \in \mathcal{F}C_b^\infty \,.$$

Then by (E) also $(Q, \mathcal{F}C_b^\infty)$ is closable on $L^2(E; \mu)$ (see [MR 92, Ch. II, Subsect. 3 e)]). Let $(Q, D(Q))$ denote its closure. Let $b, \hat{b} \in L^2(E \to H; \mu)$, $c \in L^1(E; \mu)$ such that they define a "small form perturbation" of $(Q, D(Q))$, i.e., they satisfy

(SFP) (i) there exist $\varepsilon \in (0,1)$ and $\alpha \in (0,\infty)$ such that

$$\int \left(\langle b + \hat{b}, \nabla_H u \rangle_H - c^- u \right) u \, d\mu \geq -\varepsilon \ Q(u,u) - \alpha \int u^2 \, d\mu$$

for all $u \in \mathcal{F}C_b^\infty$;
(ii) there exists $C \in (0,\infty)$ such that

$$\left| \int \langle b - \hat{b}, \nabla_H u \rangle_H \ v \, d\mu \right| \leq C \ Q_1(u,u)^{1/2} \ Q_1(v,v)^{1/2}$$

for all $u, v \in \mathcal{F}C_b^\infty$ (where for $\alpha > 0$ again $Q_\alpha(u,v) := Q(u,v) + \alpha \int u \, v \, d\mu$).

Then define

$$\mathcal{E}(u,v) := \int_E \langle A(x) \nabla_H u(x), \nabla_H v(x) \rangle_H \ \mu(dx)$$

$$+ \int_E \langle b(x), \nabla_H u(x) \rangle_H \ v(x) \ \mu(dx) + \int_E u(x) \ \langle \hat{b}(x), \nabla_H v(x) \rangle_H \ \mu(dx)$$

$$+ \int_E u(x) v(x) c(x) \ \mu(dx); \ u, v \in \mathcal{F}C_b^\infty . \tag{2.15}$$

It follows by (SFP) (i) that for all $u \in \mathcal{F}C_b^\infty$

$$\mathcal{E}_{\alpha+1}(u,u) \left(:= \mathcal{E}(u,u) + (\alpha + 1) \int u^2 \, d\mu \right)$$

$$\geq (1 - \varepsilon) \ Q_1(u,u) + \varepsilon \ Q(u,u) + \int (\langle b + \hat{b}, \nabla_H u \rangle_H + (\alpha - c^-)u) \ u \, d\mu$$

$$\geq (1 - \varepsilon) \ Q_1(u,u) \tag{2.16}$$

and by (SFP) (ii) that for all $u, v \in \mathcal{F}C_b^\infty$

$$|\check{\mathcal{E}}(u,v)| \leq (M + C) \ Q_1(u,u)^{1/2} \ Q_1(v,v)^{1/2}$$

where $M := \mu-$ ess sup$_{x \in E} \|\check{A}(x)\|$. Hence by Remark 2.9 (ii) there exists $K_{\alpha+1} \in (0,\infty)$ such that for all $u, v \in \mathcal{F}C_b^\infty$

$$|\mathcal{E}_{\alpha+1}(u,v)| \leq K_{\alpha+1} \ Q_{\alpha+1}(u,u)^{1/2} \ Q_{\alpha+1}(v,v)^{1/2} . \tag{2.17}$$

(2.16) and (2.17) imply that $(\tilde{\mathcal{E}}_\alpha, \mathcal{F}C_b^\infty)$ is a closable, positive definite, symmetric bilinear form on $L^2(E; \mu)$, and for its closure $(\tilde{\mathcal{E}}_\alpha, D(\tilde{\mathcal{E}}_\alpha))$ we have that $D(\tilde{\mathcal{E}}_\alpha) = D(Q)$. By (2.17) $\mathcal{E}_\alpha$ uniquely extends to a (w.r.t. $Q_{\alpha+1}^{1/2}$) continuous bilinear form with domain $D(Q)$, and if $D(\mathcal{E}_\alpha) := D(Q)$, then (2.16), (2.17) imply that $(\mathcal{E}_\alpha, D(\mathcal{E}_\alpha))$ is a coercive form on $L^2(E; \mu)$.

Now we want to show that $(\mathcal{E}_\alpha, D(\mathcal{E}_\alpha))$ is positivity preserving. We shall use Lemma 2.16. So, for $\varepsilon > 0$ let $\varphi_\varepsilon \in C^\infty(\mathbb{R})$ such that $\varphi(\mathbb{R}) \subset [-\varepsilon, \infty)$, $\varphi_\varepsilon(t) = t$ for all $t \in [0, \infty)$, $0 < \varphi_\varepsilon' \leq 1$, and $\varphi_\varepsilon(t) = -\varepsilon$ for $t \in (-\infty, -2\varepsilon]$.

Then the assumptions in Lemma 2.16 with $D := \mathcal{F}C_b^\infty$ are fulfilled. Indeed, obviously for all $u \in \mathcal{F}C_b^\infty$, $\varepsilon > 0$, $\varphi_\varepsilon \circ u \in \mathcal{F}C_b^\infty \subset D(\mathcal{E}_\alpha) = D(Q)$ and

$$Q_{\alpha+1}(\varphi_\varepsilon \circ u, \varphi_\varepsilon \circ u) \le Q_{\alpha+1}(u,u) ,$$

hence by (2.16), (2.17)

$$\sup_{\varepsilon > 0} \mathcal{E}_\alpha(\varphi_\varepsilon \circ u, \varphi_\varepsilon \circ u) \le \text{const. } \mathcal{E}_{\alpha+1}(u,u) .$$

Furthermore, since $|\varphi_\varepsilon(u)| \le |u|$, $\varphi_\varepsilon(u) \underset{\varepsilon \searrow 0}{\longrightarrow} 1_{\{u \ge 0\}} u$, $|\varphi'_\varepsilon(u)| \le 1$, $\varphi'_\varepsilon(u) \underset{\varepsilon \searrow 0}{\longrightarrow}$ $1_{\{u \ge 0\}}$, we obtain for $\varepsilon_n := \frac{1}{n}$, $n \in \mathbb{N}$, by Lebesgue's dominated convergence theorem

$$\lim_{n \to \infty} \mathcal{E}_\alpha(u, \varphi_{\varepsilon_n} \circ u)$$

$$= \int 1_{\{u \ge 0\}} \left(\langle A\nabla_H u, \nabla_H u \rangle_H + \langle b + \hat{b}, \nabla_H u \rangle_H\, u + u^2(c + \alpha) \right)\, d\mu$$

$$= \lim_{n \to \infty} \left(\int \varphi'_{\varepsilon_n}(u)^2\, \langle A\nabla_H u, \nabla_H u \rangle_H\, d\mu \right.$$

$$+ \left. \int \varphi'_{\varepsilon_n}(u)\, \langle b + \hat{b}, \nabla_H u \rangle_H\, \varphi_{\varepsilon_n}(u)\, d\mu + \int \varphi_{\varepsilon_n}(u)^2(c + \alpha)\, d\mu \right)$$

$$= \lim_{n \to \infty} \mathcal{E}_\alpha(\varphi_{\varepsilon_n} \circ u, \varphi_{\varepsilon_n} \circ u) \ge 0 .$$

So, Lemma 1.17 implies that $(\mathcal{E}_\alpha, D(\mathcal{E}_\alpha))$ is positivity preserving.

Let $L^{(\alpha)}$ denote the generator of $(\mathcal{E}_\alpha, D(\mathcal{E}_\alpha))$ and $(T_t^{(\alpha)})_{t \ge 0}$ the corresponding positivity preserving C^0–semigroup of contractions on $L^2(E; \mu)$. Define

$$L := L^{(\alpha)} + \alpha , \quad D(L) := D(L^{(\alpha)}) ,$$
$$T_t := e^{\alpha t}\, T_t^{(\alpha)} , \qquad t \ge 0 .$$

Then for all $f \in L^2(E; \mu)$

$$\frac{d}{dt}\, T_t f = LT_t f \text{ for all } t > 0 \text{ (and all } t \ge 0, \text{ if } f \in D(L))$$

$$T_0 f = f,$$

where L (heuristically) is given by

$$Lu = \text{div}_\mu(A\nabla_H u + \hat{b}u) - \langle b, \nabla_H u \rangle_H - cu . \tag{2.18}$$

Here div_μ denotes the divergence w.r.t. to μ, i.e., (-1) times the adjoint $\nabla_H^{*,\mu}$ of ∇_H on $L^2(\mu)$. $(T_t)_{t \ge 0}$ is positivity preserving, since so is $(T_t^{(\alpha)})_{t \ge 0}$.

To conclude this subsection we discuss conditions on $b, \hat{b}, c^-$ that ensure that (SFP) holds. This is e.g. in fact the case if $b, \hat{b}, c^-$ are bounded. This is easy to see and the details can be found in [MR 92, Ch. II, Subsect. 3 e)]. The unbounded case is more delicate. In finite dimensions (see Subsect. 2.4.1)

(SFP) follows under the "$(L^d + L^\infty)$-condition" from Sobolev–embedding theorems, more precisely from the fact that

$$\left(\int |u|^{\frac{2d}{d-2}} \, dx \right)^{\frac{d-2}{2d}} \leq \text{const.} \left(\int |\nabla u|^2_{\mathbb{R}^d} \, dx \right)^{1/2}$$

(cf. [MR 92, Ch. II, Subsect. 2 d)]). Such Sobolev embeddings do not hold in infinite dimensions even for very simple (e.g. Gaussian) measures μ. However, as a substitute in some cases the following, so–called *defective log–Sobolev inequality* holds for our triple (E, H, μ):

(DLS)   There exists $\lambda_1, \lambda_2 \in [0, \infty)$ such that

$$\int_E u^2 \log |u| \, d\mu \leq \lambda_1 \int |\nabla_H u|^2_H \, d\mu + \|u\|^2_{L^2(\mu)} \log \|u\|_{L^2(\mu)}$$
$$+ \lambda_2 \|u\|^2_{L^2(\mu)} \text{ for all } f \in \mathcal{F}C_b^\infty .$$

Here $\|u\|_{L^p(\mu)} := (\int |u|^p \, d\mu)^{1/p}$. (DLS) e.g. holds even with $\lambda_2 = 0$ if (E, H, μ) forms an abstract Wiener space (cf. [Gr 93]). We shall see other examples in the next subsection.

Lemma 2.17. *Suppose that in addition to (C), (E) above, (DLS) holds for (E, H, μ), that $\|\tilde{A}\| \in L^\infty(E; \mu)$, and that*

$$e^{|b-\hat{b}|^2_H} \in \bigcup_{p>0} L^p(E; \mu) ,$$
$$e^{|b+\hat{b}|^2_H} \in L^{p_1}(E; \mu) , \quad e^{c^-} \in L^{p_2}(E; \mu)$$

with $p_1, p_2 \in [1, \infty)$ satisfying $\frac{1}{2p_1} + \frac{1}{p_2} < \delta/4\lambda_1$ (where δ, λ_1 are as in (E) resp. (DLS)). Then (SFP) holds and all the above applies.

Proof. The proof is essentially a direct consequence of the following Young inequality, i.e.,

$$st \leq s \log s - s + e^t \text{ for all } t \in \mathbb{R}, \ s \in [0, \infty) . \tag{2.19}$$

Let us first check (SFP) (i): Let $u \in \mathcal{F}C_b^\infty$. We may assume that $\|u\|_{L^2(\mu)} = 1$. Then by (2.19) and (DLS)

$$\int \langle b + \hat{b}, \nabla_H u \rangle_H \, u \, d\mu - \int u^2 c^- \, d\mu$$

$$\geq - \left(\int |b + \hat{b}|^2_H \, u^2 \, d\mu \right)^{1/2} \left(\int |\nabla_H u|^2_H \, d\mu \right)^{1/2} - \int u^2 \, c^- \, d\mu$$

$$\geq -\frac{1}{2p_1} \int u^2 \left(p_1 |b + \hat{b}|^2_H \right) \, d\mu - \frac{1}{p_2} \int u^2 \left(p_2 c^- \right) \, d\mu - \frac{1}{2} \int |\nabla_H u|^2_H \, d\mu$$

$$\underset{(2.18)}{\geq} -\frac{1}{2p_1} \left(\int u^2 \log u^2 - 1 + \int e^{p_1 |b + \hat{b}|^2_H} \, d\mu \right)$$

$$- \frac{1}{p_2} \left(\int u^2 \log u^2 - 1 + \int e^{p_2 c^-} \, d\mu \right) - \frac{1}{2} \int |\nabla_H u|^2_H \, d\mu$$

$$\underset{(DLS)}{\geq} -\varepsilon \, Q(u, u) - \alpha$$

where $\varepsilon := \left(\frac{1}{2p_1} + \frac{1}{p_2}\right)\frac{2\lambda_1}{\delta} + \frac{1}{2} < 1$ by assumption and

$$\alpha := \left(\frac{1}{2p_1} + \frac{1}{p_2}\right) 2\lambda_2 + \frac{1}{2p_1}\left(\|e^{|b+\hat{b}|_H^2}\|_{L^{p_1}(\mu)}^{p_1} - 1\right)$$
$$+ \frac{1}{p_2}\left(\|e^{c^-}\|_{L^{p_2}(\mu)}^{p_2} - 1\right).$$

So, (SFP) (i) holds. To show (SFP) (ii) let $u, v \in \mathcal{F}C_b^\infty$. We may again assume that $\|u\|_{L^2(\mu)} = \|v\|_{L^2(\mu)} = 1$. Then by (2.19) and (DLS) for any $p > 0$

$$\left|\int \langle b - \hat{b}, \nabla_H u\rangle_H \, v \, d\mu\right|$$
$$\leq \frac{1}{\sqrt{p}}\left(\int v^2\left(p|b - \hat{b}|_H^2\right) d\mu\right)^{1/2}\left(\int |\nabla_H u|_H^2 \, d\mu\right)^{1/2}$$
$$\underset{(2.18)}{\leq} \frac{1}{\sqrt{p}}\left(\int v^2 \log v^2 \, d\mu - 1 + \int e^{p|b-\hat{b}|_H^2} \, d\mu\right)^{1/2} \frac{1}{\sqrt{\delta}} Q(u,u)^{1/2}$$
$$\leq \frac{1}{\sqrt{p}}\left(\frac{2\lambda_1}{\delta} Q(v,v) + 2\lambda_2 + \|e^{|b-\hat{b}|_H^2}\|_{L^p(\mu)}^p - 1\right)^{1/2} \frac{1}{\sqrt{\delta}} Q(u,u)^{1/2}$$
$$\leq K \, Q_1(u,u)^{1/2} \, Q_1(v,v) \,,$$

where $K := \frac{1}{\sqrt{p\delta}} \max\left(1, \frac{2\lambda_1}{\delta}, 2\lambda_2 + \|e^{|b-\hat{b}|_H^2}\|_{L^p(\mu)}^p - 1\right)$ and we used that $\|u\|_{L^2(\mu)} = \|v\|_{L^2(\mu)} = 1$. $\square$

In order to apply all this to a *given* Kolmogorov equation

$$\frac{du}{dt} = Lu \,, \quad u(0, \cdot) = f$$

we have to find H and μ as above so that L is of type (2.18). Indeed, this is possible in a number of cases. We start investigating this in the next section by first looking at symmetric cases (i.e., where $\hat{b} = b \equiv 0$).

3. Symmetrizing measures

3.1 The classical finite dimensional case

Consider the following operator on an open set $U \subset \mathbb{R}^d$

$$Lu := \Delta u - \langle b, \nabla u\rangle_{\mathbb{R}^d} \,, \quad u \in D \,, \tag{3.1}$$

with domain $D := C_0^\infty(U)$. Here Δ is the usual Laplacian on $\mathbb{R}^d$ and $b : U \to \mathbb{R}^d$ a vector field. So, we are in a special case of the situation considered in Subsection 2.4.1 (more precisely: $A \equiv Id_{\mathbb{R}^d}$, $\hat{b} \equiv 0$, $c \equiv 0$).

However, we do not assume that the components of b are in $L^\infty(U; dx) +$ $L^d(U; dx)$, so cannot apply the results of Subsection 2.4.1. Instead we assume that

$$b = \nabla W$$

for some $W : U \to \mathbb{R}$, say C^1. Setting $\rho := e^{-W}$ we can rewrite (3.1) as

$$Lu = \Delta u + \langle \frac{\nabla \rho}{\rho}, \nabla u \rangle_{\mathbb{R}^d} ; \ u \in D ,$$

and see that for $\mu := \rho \, dx$ we have

$$- \int_U Lu \ v \ d\mu = \int_U \langle \nabla u, \nabla v \rangle_{\mathbb{R}^d} \ d\mu = - \int u \ Lv \ d\mu , \tag{3.2}$$

i.e., L is symmetric on $L^2(U; \mu)$. (This observation goes back to Kolmogorov, c.f. [Ko 37]). If we define for a vector field V on U

$$\begin{aligned} \mathrm{div}_\mu \ V &:= \ \mathrm{div} \ V + \langle \frac{\nabla \rho}{\rho}, V \rangle_{\mathbb{R}^d} \\ &(= -\nabla^{*,\mu} V \ \text{(by (3.2))} , \end{aligned} \tag{3.3}$$

(where $\nabla^{*,\mu}$ is the adjoint of ∇ on $L^2(\mu)$) we have

$$L = \ \mathrm{div}_\mu \ \nabla \ \text{on} \ D = C_0^\infty(U) , \tag{3.4}$$

i.e, L is a so–called μ-Laplacian. So, indeed L fits in our framework, more precisely the one described in Subsection 2.4.2 (with $A \equiv Id_\mathbb{R}$, $b = \hat{b} \equiv 0$, $c = 0$, cf. (2.18)) at least if $U = \mathbb{R}^d$ and $\mu(\mathbb{R}^d) < \infty$. However, also the general case above with even less restrictions on ρ (resp. W) can be handled using coercive forms. We refer to [MR 92, Ch. II, Subsect. 2 a)].

The above case resp. its "manifold version" is, of course, quite easy since we are more or less given $H = \mathbb{R}^d$ and A in (2.18), i.e., we basically know in advance the underlying differential geometry and only have to find a measure μ that symmetrizes L, i.e., satisfies (3.2). So, this technique seems not suitable for infinite dimensions. It is, however, a remarkable fact, that for a large class of operators, so called *diffusion operators*, which includes all operators (also the infinite dimensional ones) which appear in these notes, this is sufficient. More precisely, if there exists a symmetrizing measure μ for such a diffusion operator, then there exists a "(co–)tangent bundle" (in a generalized sense) and a gradient ∇ so that L is a corresponding μ-Laplacian, $L = \mathrm{div}_\mu \nabla$ with $\mathrm{div} = -\nabla^{*,\mu}$. This result in its most general form (valid even in non–symmetric cases) is due to Andreas Eberle and we shall briefly explain this result in more detail in the next subsection. Subsequently, we shall see concrete infinite dimensional examples and applications.

3.2 Representation of symmetric diffusion operators

The material in this subsection is taken from [Eb 98, Ch. 1, Appendix B and Ch. 3 Appendix D, 7)].

Let $(E, \mathcal{B}, m)$ be a σ–finite measure space and fix $p \in [1, \infty)$.

Definition 3.1. A densely defined linear operator (L, D) on $L^p(E; m)$ is called *(abstract) diffusion operator* if

(i) $\varphi(u_1, \ldots, u_k) \in D$ for all $k \in \mathbb{N}$; $u_1, \ldots, u_k \in D$, and all $\varphi \in C^\infty(\mathbb{R}^k)$ such that $\varphi(0) = 0$, and, in this case

$$L\varphi(u_1, \ldots, u_k) = \sum_{i=1}^k \partial_i \phi(u_1, \ldots, u_k) \, Lu_i + \sum_{i,j=1}^k \partial_i \partial_j \phi(u_1, \ldots, u_k) \, \Gamma(u_i, u_j) \,,$$

where

$$\Gamma(u, v) := \frac{1}{2} \left(L(uv) - uLv - vLu \right) \,; \quad u, v \in D \,.$$

(ii) $\Gamma(u, u) \geq 0$ for all $u \in D$.

The bilinear operator $\Gamma : D \times D \to L(E; m)$ is called the *square field operator* of L, (where $L(E; m)$ denotes all m–equivalence classes of functions on E).

For the rest of this section we assume $L^2(E; m)$ to be separable and fix a diffusion operator (L, D) on $L^2(E; m)$. Note that by (i), D is automatically an algebra. We additionally assume that that $D \subset L^\infty(E; m)$, that D is dense in $L^2(E; m)$, and for simplicity that $1 \in D$ (in particular, $m(E) < \infty$). Note that (i) (applied to $k = 1$, $u_1 = 1$, $\varphi \equiv 1$) implies that $L1 = 0$.

Definition 3.2. Let $T'E = (T'_z E)_{z \in E}$ be a measurable field of Hilbert spaces over E (cf. e.g. [Eb 98, Ch. 3, Appendix D, 6)]). A map $d : D \to L^2(E \to T'E; m)$ (:= space of m–square integrable sections in $T'E$) is called L^2–*differential* (w.r.t. the co–tangent bundle $T'E$) if:

(i) The span of $\{u \, dv | u, v \in D\}$ is dense in $L^2(E \to T'E; m)$.
(ii) d is linear.
(iii) $d(u \cdot v) = u \, dv + v \, du$ for all $u, v \in D$.

Theorem 3.3. *(A. Eberle) Assume that m is a symmetrizing measure for L (i.e., $\int Lu \, v \, dm = \int u \, Lv \, dm$ for all $u, v \in D$). Then there exists a measurable field of Hilbert spaces $T'E = (T'_z E)_{z \in E}$, an L^2–differential $d : D \to L^2(E \to T'E; m)$ such that*

$$Lu = -d^* du \,, \quad u \in D \,, i.e.,$$

$$- \int Lu \, v \, dm = \int_E (d_z u, d_z v)_{T'_z E} \, m \, (dz) \quad \text{for all } u, v \in D \,,$$

where $d^* : \operatorname{dom}(d^*) \subset L^2(E \to T'E; m) \to L^2(E; m)$ is the adjoint of d.

Proof. [Eb 98, Ch. 3, Theorem 3.11]. □

In [Eb 98] it is also proved that the above representation of L is unique "up to isomorphisms" and that there is a non–symmetric variant of Theorem 2.3 above.

In applications however, it is necessary to find natural and suitable representatives of $T'E$ and d. This will be done in quite simple "flat" cases in the next section. There, the choice of the tangent bundle and the gradient will be more or less obvious, so the only issue will be to find the symmetrizing measure μ. An important class of "non–flat" infinite dimensional examples for which the natural tangent bundle and gradient were only identified recently, will be discussed in Section 6.

3.3 Ornstein–Uhlenbeck type operators.

Let $(H, \langle \ , \ \rangle_H)$ be a Hilbert space and $(A, D(A))$ a self–adjoint operator on H such that for some $\varepsilon > 0$, $\langle Ah, h \rangle_H \geq \varepsilon \langle h, h \rangle_H$ for all $h \in H$. For $u \in \mathcal{F}C_b^\infty(D(A))$ (cf. (2.11) with $H = E$, $K := D(A) \subset H \equiv H'$)

$$Lu(x) := \mathrm{Trace}_H \, u''(x) - \langle x, A(\nabla_H u(x)) \rangle_H \, , \quad x \in H \, , \qquad (3.5)$$

which makes sense, since by (2.12), $\nabla_H u(x) \in D(A)$. Here u'' denotes the second Fréchet derivative of u. Following the same idea as in Subsection 3.1 the corresponding symmetrizing measure should be given by the following heuristic formula

$$\mu(dx) = \text{``}\exp\left[-\tfrac{1}{2}\langle Ax, x \rangle\right] \, d^\infty x\text{''} \qquad (3.6)$$

(where "$d^\infty x$" denotes "infinite dimensional Lebesgue measure"), i.e., μ should be a mean–zero Gaussian measure with covariance operator A^{-1}. Of course, μ will only exist on H if A^{-1} is trace class. But it always exists on a larger Hilbert (or Banach) space E as we shall see below. So, to find a symmetrizing measure in general we have to consider L on a larger state space E. In fact, E can be chosen in such a way that, as we shall see later, the solution of the corresponding Kolmogorov equation is given explicitly by a Mehler formula. More precisely, we have the following result:

Theorem 3.4. *There exists a separable real Hilbert space E and a C^0–semigroup $\left(e^{-tA_E}\right)_{t \geq 0}$ of linear operators on E with generator $(-A_E)$ such that*

   (i) $H \subset D(A_E)$ with dense Hilbert-Schmidt embedding (where $D(A_E)$ is equipped with the graph norm given by A_E).

   (ii) e^{-tA} is the restriction of e^{-tA_E} to H for all $t \geq 0$.

This result was first obtained in [BRS 97, Theorem 1.6] (even for the non–symmetric case, i.e., arbitrary generators A of C^0–semigroups on H). The simplest proof is given in [FR 97, Prop. 2.3]. Let A'_E be the adjoint of the generator A_E. Then A'_E is a linear operator on E' and if $K := D(A'_E)$ (i.e., K denotes its domain), then

$$K \subset E' \subset D(A) \subset H \subset E ,\qquad(3.7)$$

and K is dense in E'. So, looking at test functions in the smaller space $\mathcal{F}C_b^\infty(K)$, we see that they have unique extensions to E, and L with domain $\mathcal{F}C_b^\infty(K)$ becomes an operator on the state space E, i.e., for $u \in \mathcal{F}C_b^\infty(K)$

$$Lu(x) := \Delta_H u(x) - {}_E\langle x, A'_E(\nabla_H u(x))\rangle_{E'} ,\ x \in E ,\qquad(3.8)$$

where for $x \in E$

$$\begin{aligned}\Delta_H u(x) &:= \mathrm{Trace}_H\ u''(x) ,\\&= \textstyle\sum_{i,j=1}^m \partial_i \partial_j f(\ell_1(x),\dots,\ell_m(x))\ \langle \ell_i, \ell_j\rangle_H ,\end{aligned}\qquad(3.9)$$

if $u = f(\ell_1,\dots,\ell_m)$, $\ell_1,\dots,\ell_m \in K$, $f \in C_b^\infty(\mathbb{R}^{>})$, i.e., Δ_H is the *Gross-Laplacian*. Clearly (3.8) is exactly (3.5), if $x \in H$. u'' in (3.9) is the second Fréchet derivative of u on E and $\nabla_H u$ in (3.8) is defined as in (2.13).

Let now μ be the Gaussain probability measure on $(E, \mathcal{B}(E))$ with Fourier transform given by

$$\int_E e^{i\,{}_{E'}\langle \ell, x\rangle_E}\ \mu(dx) = e^{-\frac{1}{2}\|A^{-1/2}\ell\|_H^2} ,\ \ell \in E' .\qquad(3.10)$$

μ exists since the embedding $H \subset E$ is Hilbert–Schmidt and $A^{-1/2}$ is bounded (cf. e.g. [Ya 89, Theorem 3.1]). Then it is easy to check that for all $u, v \in \mathcal{F}C_b^\infty(K)$

$$\begin{aligned}\int_E Lu(x)\,v(x)\,\mu(dx) &= \int_E u(x)\,Lv(x)\,\mu(dx)\\&= \int_E \langle \nabla_H u(x), \nabla_H v(x)\rangle_H\,\mu(dx)\\&=: \mathcal{E}_\mu(u, v) .\end{aligned}\qquad(3.11)$$

In particular, $(\mathcal{E}_\mu, \mathcal{F}C_b^\infty(K))$ is closable and its closure $(\mathcal{E}_\mu, D(\mathcal{E}_\mu))$ is a symmetric, positive definite, densely defined bilinear form, hence in particular a coercive form on $L^2(E; \mu)$. $(\mathcal{E}_\mu, D(\mathcal{E}_\mu))$ is even a Dirichlet form (cf. [MR 92, Ch. II, Prop. 3.8]). So, the results of Subsect. 2.4.2 (with $A \equiv Id_H$, $b = \hat{b} \equiv 0$, $c \equiv 0$, $\alpha = 0$) apply to give a solution to the corresponding Kolmogorov equation (2.4.2) via the sub–Markovian C^0–semigroup $(T_t)_{t\ge0}$ on $L^2(E; \mu)$ associated with $(\mathcal{E}_\mu, D(\mathcal{E}_\mu))$.

We note that if in this case for vector fields $V : E \to H$ of type

$$V(x) := \sum_{i=1}^n u_i\,\ell_i\qquad(3.12)$$

$\ell_i \in K$, $u_i \in \mathcal{F}C_b^\infty(K)$, $n \in \mathbb{N}$, we define

$$\operatorname{div}_\mu V := \sum_{i=1}^n \left(\langle \nabla_H u_i, \ell_i \rangle_H - u_i \,{}_{E'}\langle A'_E \ell_i, \cdot \rangle_E \right) ,$$

then

$$L = \operatorname{div}_\mu \nabla_H .$$

By our special choice of E the semigroup $(T_t)_{t \geq 0}$ is given explicitly by a *Mehler formula*. In fact, it follows by [BR 95a, Sect. 5] that for any bounded $\mathcal{B}(E)$–measurable $f : E \to \mathbb{R}$, $t \geq 0$, and μ–a.e. $x \in E$

$$T_t f(x) = \int_E f \left(e^{-tA_E} x + \sqrt{1 - e^{-2tA_E}} x' \right) \mu \, (dx')$$
$$=: p_t f(x) . \tag{3.13}$$

Remark 3.5. (i) Obviously, $(p_t)_{t \geq 0}$ from (3.13) is a *Feller semigroup*, i.e., $p_t(C_b(E)) \subset C_b(E)$ for all $t \geq 0$.
(ii) It can be checked that for all $u \in \mathcal{F}C_b^\infty(K)$

$$\frac{d}{dt} \left(p_t u \, (x) \right)_{|_{t=0}} = Lu(x) \text{ for all } x \in A_E \tag{3.14}$$

(cf. the proof of [BR 95a, Thm. 5.3]). So, restricting to H again we obtain that $p_t u$ is a solution to Kolmogorov's equations on our *original* Hilbert space H. So, if one is satisfied with initial conditions $u \in \mathcal{F}C_b^\infty(K)$ (or $u \in C_b^2(E)$), and "pointwise–differentiation" (i.e., for fixed x) in (3.14), one can avoid the machinery of sectorial forms in this special case by using Theorem 3.4 and the Mehler formula (3.13).
(iii) The above shows that really in general $H \subsetneq E$ in this Gaussian case. Therefore, the framework presented in Subsection 2.4.2 was chosen in that way.
(iv) $(T_t)_{t \geq 0}$ above is just the first quantization of $\left(e^{-tA} \right)_{t \geq 0}$ and the generator of $(\mathcal{E}_\mu, D(\mathcal{E}_\mu))$ is the second quantization of A (see e.g. [AR 91, Sect. 7 I]). Therefore, (DLS) holds for (E, H, μ) even with $\lambda_2 = 0$ (cf. [Re S 75, Thm. X.61] and [Gr 93, Thm. 3.12]). Therefore, the results of Subsect. 2.4.2 also apply, if we add small perturbations given by b, $\hat{b}$, c to L in (3.5).

3.4 Operators with non–linear drift.

Let $(E, \langle \, , \, \rangle_E)$, $(H, \langle \, , \, \rangle_H)$ be separable Hilbert spaces such that

$$H \subset E \text{ with dense Hilbert–Schmidt embedding .} \tag{3.15}$$

Again we identify H' with H so

$$E' \subset H \subset E$$

(cf. (2.10)). Let $\{\ell_n | n \in \mathbb{N}\} \subset E'$ be an orthonormal basis of H so that $\{\alpha_n \ell_n | n \in \mathbb{N}\}$, resp. $\{\gamma_n \ell_n | n \in \mathbb{N}\}$ are orthonormal bases for E' resp. E for some $\alpha_n, \gamma_n \in (0, \infty)$, $n \in \mathbb{N}$. Define the projectors

$$P_n x := \sum_{m=1}^{n} {}_{E'}\langle \ell_m, x \rangle_E \, \ell_m \, , \quad x \in E \, . \tag{3.16}$$

The following result was obtained very recently (cf. [AKRT 98]).

Theorem 3.6. *Let E_1 be a Hilbert space containing E obtained by renormalizing the basis vectors ℓ_m, $m \in \mathbb{N}$, and let $\beta : E \to E_1$ be $\mathcal{B}(E)/\mathcal{B}(E_1)$ measurable such that the following conditions are satisfied:*
(i) There exists $N \in \mathbb{N}$ such that for all $m \in \mathbb{N}$ there exists $c_m \in (0, \infty)$ such that

$$\left| {}_{E_1'}\langle \ell_m, \beta(x) \rangle_{E_1} \right| \leq c_m \left(1 + \|x\|_E \right)^N \text{ for all } x \in E \, .$$

(ii) There exist continuous $\tilde{\beta}_n = (\beta_{n1}, \ldots, \beta_{nn}) \colon \mathbb{R}^n \to \mathbb{R}^n$, $n \in \mathbb{N}$, such that for all $n \in \mathbb{N}$

(a) $\tilde{\beta}_n$ is the vector logarithmic derivative of some probability measure μ_n on $\mathbb{R}^n$, i.e.,

$$\int_{\mathbb{R}^n} \partial_m u \, d\mu_n = -\int_{\mathbb{R}^n} u \, \beta_{nm} \, d\mu_n \text{ for all } 1 \leq m \leq n \text{ and all } u \in C_0^\infty(\mathbb{R}^n) \, ;$$

(b) the maps $\beta_n : E \to E'$ defined by

$$\beta_n(x) := \sum_{m=1}^{n} \beta_{nm}(P_n x) \, \ell_m \, , \quad x \in E \, ,$$

satisfy the following coercivity condition (uniformly in n): there exist $M \in [\frac{N}{2}, \infty)$, $\delta > 0$, and $C(M, \delta) > 0$ such that for all $n \in \mathbb{N}$ and all $x \in E \setminus \{0\}$

$$-\langle \beta_n(x), x \rangle_E \geq 2M + \|P_n \, E' \subset P_n \, H\|_{H.S.}^2 + \delta - C(M, \delta) \, \|P_n x\|_E^{-2M} \, ;$$

(c) for all $m \in \mathbb{N}$, $\varepsilon > 0$ there exists $n(\varepsilon, m) \in \mathbb{N}$ such that for all $n > n(\varepsilon, m)$ and all $x \in E$

$$\left| {}_{E_1'}\langle \ell_m, \beta(x) - \beta_n(x) \rangle_{E_1} \right| \leq \varepsilon \left(1 + \|x\|_E \right)^N \, .$$

Then there exists a probability measure μ on $(E, \mathcal{B}(E))$ such that μ is a symmetrizing measure for the operator

$$L_0 u(x) := \Delta_H u(x) + {}_{E_1}\langle \beta(x), \nabla_H u(x) \rangle_{E_1'} \tag{3.17}$$

for all $u \in \mathcal{F}C_b^\infty(K)$, where $K (\subset E_1')$ is the linear span of $\{\ell_n | n \in \mathbb{N}\}$.

For the proof and further details, in particular, the relation and application to Gibbs measures in statistical mechanics we refer to [AKRT 98]. So as in the previous subsection all results in Subsection 2.4.2 with $b = \hat{b} \equiv 0$, $c \equiv 0$ apply to L_0.

We now give a concrete example to which Theorem 3.6 applies.

Example 3.7. Consider the lattice $\mathbb{Z}^d$ and the space $\ell^2(\mathbb{Z}^d)$ of square summable sequences indexed by $\mathbb{Z}^d$, so

$$\ell^2(\mathbb{Z}^d) \subset \mathbb{R}^{\mathbb{Z}^d} .$$

Define, in addition, for $p \in \mathbb{Z}$ the weighted sequence spaces

$$\ell_p^2(\mathbb{Z}^d) := \{x \in \mathbb{R}^{\mathbb{Z}^d} \mid \|x\|_{\ell_p^2}^2 := \sum_{k \in \mathbb{Z}^d} (1 + |k|)^{2p} x_k^2 < \infty\} . \tag{3.18}$$

Let $\{\ell_k \mid k \in \mathbb{Z}^d\}$ be the standard orthonormal basis in $\ell^2(\mathbb{Z}^d)$ and $A := (a_{kj})_{k,j \in \mathbb{Z}^d}$ giving a linear operator on $\ell^2(\mathbb{Z}^d)$ such that

(i) there exists $\delta > 0$ such that $\langle Ax, x \rangle_{\ell^2} \geq \delta \|x\|_{\ell^2}^2$ for all $x \in \ell^2(\mathbb{Z}^d)$;
(ii) $a_{kj} = a(k - j)$ for all $k, j \in \mathbb{Z}^d$ ("translation invariance");
(iii) there exists $r > 0$ such that

$$a(k - j) = 0 \text{ if } |k - j|_{\mathbb{R}^d} > r \text{ ("finite range")}.$$

Let P be a polynomial on $\mathbb{R}$ of even degree,

$$P(t) = c_N t^N + \ldots + c_1 t + c_0 , \ c_N > 0 , \ N \in \mathbb{N} , \ N \text{ even.} \tag{3.19}$$

Define $\beta := (\beta_k)_{k \in \mathbb{Z}^d} : \ell_{-p_1}^2(\mathbb{Z}^d) \to \ell_{-p_2}^2(\mathbb{Z}^d)$, where $p_2 > Np_1 + d/2, p_1 > d/2$, and

$$\beta_k(x) := -\sum_{j \in \mathbb{Z}^d} a(k - j)x_j - P'(x_k) , \ k \in \mathbb{Z}^d , \ x \in \ell_{-p_1}^2(\mathbb{Z}^d) . \tag{3.20}$$

Then Theorem 3.6 applies with $E := \ell_{-p_1}^2(\mathbb{Z}^d)$, $E_1 := \ell_{-p_2}^2(\mathbb{Z}^d)$, $H := \ell^2(\mathbb{Z}^d)$ to give a symmetrizing probability measure μ on E for the operator

$$L_0 u(x) = \Delta_H u(x) - {}_{\ell_{-p_2}^2}\langle Ax + (P'(x_k))_{k \in \mathbb{Z}^d} , \ \nabla_H u(x) \rangle_{\ell_{p_2}^2} ,$$

$x \in \ell_{-p_1}^2(\mathbb{Z}^d)$, $u \in \mathcal{F}C_b^\infty(K)$ with $K :=$ lin. span $\{\ell_k \mid k \in \mathbb{Z}^d\}$.

Remark 3.8. The measure μ in Example 3.7 is well–known. It is a Gibbs measure for the so–called $P(\Phi)_d$–lattice model (cf. [AKR 97, Subsections 5.1, 5.2]).

Furthermore, we have (DLS) in the case of Example 3.7 for small c_j, $1 \leq j \leq N$:

Theorem 3.9. *Let $c_0, \ldots, c_N$ in (2.18) be sufficiently small. Then (DLS) holds with $\lambda_2 = 0$ for (E, H, μ) as in Example 3.7. In particular, all results from Subsect. 2.4.2 apply.*

Proof. [AKR 95, Theorem 6]. □

The corresponding semigroup $(T_t)_{t \geq 0}$ constructed as in Subsect. 2.4.2 in the special case $A \equiv Id_H$, $\hat{b} \equiv 0$, then satisfies the Kolmogorov equation (2.4.2) with

$$Lu = \Delta_H u - {}_{\ell^2_{-p_2}}\langle Ax + (P'(x_k))_{k \in \mathbb{Z}^d} + b(x) \,,\, \nabla_H u(x)\rangle_{\ell^2_{p_2}}$$
$$-c(x)u(x) \,. \tag{3.21}$$

In this case for vector fields $V : E \to H$ as in (3.12) we define

$$\mathrm{div}_\mu V(x) := \sum_{i=1}^{n} \langle \nabla_H u_i(x), \ell_i\rangle_H - u_i(x) \, {}_{\ell^2_{-p_2}}\langle Ax + (P'(x_k))_{k \in \mathbb{Z}^d} \,,\, \ell_i\rangle_{\ell^2_{p_2}} \,.$$
$$\tag{3.22}$$

Then

$$L_0 = \mathrm{div}_\mu \nabla_H$$

and

$$L = \mathrm{div}_\mu \nabla_H - {}_{\ell^2_{-p_2}}\langle b, \nabla_H \cdot\rangle_{\ell^2_{p_2}} - c \,.$$

4. Non-sectorial cases: perturbations by divergence free vector fields

So, far we were only able to solve cases of Kolmogorov's equation on L^p (more precisely L^2) in symmetric situations or "small perturbations" (i.e., in the presence of a sector condition) thereof. In this section we shall look at a class of infinite dimensional operators which neither have a symmetrizing measure, nore are sectorial, but are perturbations of such by a divergence free vector field. This class is of special importance for applications (cf. Subsect. 4.4 and Sect. 5. below).

4.1 Diffusion operators on $L^p(E; m)$

All the material in this subsection is taken from [Eb 98, Ch. 1, Appendix A and B].

Let $(E, \mathcal{B}, m)$ be a σ-finite measure space and $p \in [1, \infty)$. Let (L, D) be a linear operator on $L^p(E; m)$ with dense domain D.

Lemma 4.1. *(i) (L, D) is dissipative if and only if*

$$\int \text{sgn}(u)\, |u|^{p-1}\, Lu\, dm \le 0 \text{ for all } u \in D ,$$

where $\text{sgn}(x) = 1_{(0,\infty)} - 1_{(-\infty,0)}$.
(ii) Let $\lambda \ge 0$. Suppose that for every increasing $\psi \in C(\mathbb{R})$ such that $\psi(0) = 0$, and $\psi(s) \le |s|^{p-1}$ for all $s \in \mathbb{R}$, we have

$$\int \psi \circ u\, Lu\, dm \le \lambda \int |u|^p\, dm \text{ for all } u \in D .$$

Then $(L - \lambda, D)$ is dissipative on $L^p(E; m)$.

Proof. (i) is obvious.
(ii) [Eb 98, Ch. 1, Lemma 1.7]. Choose smooth increasing functions ψ_n : $\mathbb{R} \to \mathbb{R}$, $n \in \mathbb{N}$, such that $\psi_n(s) = 0$ if $|s| \le n^{-1}$, $\psi_n(s) = |s|^{p-1} \cdot \text{sgn}(s)$ if $|s| \ge 2n^{-1}$, and $|\psi_n(s)| \le |s|^{p-1}$ for all s. Then for any $u \in D$, $\psi_n \circ u$ is in $L^q(E; m)$, $\frac{1}{p} + \frac{1}{q} = 1$, and

$$\int \text{sgn}(u)\, |u|^{p-1}(L - \lambda)\, u\, dm = \lim_{n\to\infty} \int \psi_n \circ u\, Lu\, dm - \lambda \int |u|^p\, dm \le 0$$

by dominated convergence. $\qquad\square$

The following is important to check dissipativity.

Proposition 4.2. *Suppose (L, D) is a diffusion operator (in the sense of Definition 3.1) and that for some $\alpha \in [0, \infty)$ m is a sub–invariant measure for $(L - \alpha, D)$, i.e.,*

$$u,\ Lu \in L^1(E; m) \text{ and } \int Lu\, dm \le \alpha \int u\, dm \text{ for all } u \in D \text{ with } u \ge 0 .$$
$$(4.1)$$

Then $\left(L - \frac{\alpha}{p}, D\right)$ is dissipative on $L^p(E; m)$. In particular, the closure $\overline{L}$ of (L, D) generates a C^0-semigroup on $L^p(E; m)$ if $(\lambda - L)(D)$ is dense in $L^p(E; m)$ for some $\lambda > \frac{\alpha}{p}$.

Proof. [Eb 98, Ch. 1, Lemma 1.8] Let $\psi : \mathbb{R} \to \mathbb{R}$ be a smooth increasing function such that $|\psi(s)| \le |s|^{p-1}$ for all s, and let $\Psi := \int_0^\cdot \psi(s)\, ds$. Then $0 \le \Psi(t) \le |t|^p/p$ for all $t \in \mathbb{R}$, and

$$L(\Psi \circ f) = \psi \circ f\, Lf + \psi' \circ f\, \Gamma(f, f) \ge \psi \circ f\, Lf \quad m\text{–a.e.}$$

for all $f \in D$. Hence, by the sub–invariance,

$$\int \psi \circ f\, Lf\, dm \le \int L(\Psi \circ f)\, dm$$
$$\le \alpha \int \Psi \circ f\, dm \le \frac{\alpha}{p} \int |f|^p\, dm$$

for all $f \in D$. This implies the dissipativity by Lemma 4.1 (ii). The remaining part of the assertion now follows by Theorem 2.3 above. $\qquad\square$

Condition (4.1) also implies that the C^0–semigroup in the last part of Proposition 4.2 is sub–Markovian since we have:

Theorem 4.3. *Let (L, D) be a diffusion operator. Suppose (4.1) holds and that $(\lambda - L)(D)$ is dense in $L^p(E; m)$ for some $\lambda > \alpha/p$. Then the semigroup $(T_t)_{t \geq 0}$ generated by the closure $\overline{L}$ of (L, D) is sub–Markovian.*

Proof. [Eb 98, Ch. 1, Lemma 1.9].   □

Below we shall mainly be working in the case $p = 1$. The corresponding results then have automatically consequences for $L^p(E; m)$ by the following:

Lemma 4.4. *Let $(T_t^{(p)})_{t \geq 0}$ be a sub–Markovian C^0–semigroup on $L^p(E; m)$ and let $r \in [p, \infty)$. Then the restrictions of $T_t^{(p)}$, $t \geq 0$, to $L^r(E; m) \cap L^p(E; m)$ are bounded w.r.t. $\| \ \|_{L^r(m)}$. The unique continuous extensions $T_t^{(r)}$, $t \geq 0$, of these operators to $L^r(E; m)$ form a sub–Markovian C^0–semigroup on $L^r(E; m)$. Furthermore, if $(L^{(r)}, D(L^{(r)}))$ denotes the generator of $(T_t^{(r)})_{t \geq 0}$ on $L^r(E; m)$, then*

$$D_1 := \{u \in D(L^{(p)}) \mid u, L^{(p)}u \in L^r(E; m)\} \subset D(L^{(r)})$$

and

$$L^{(p)}u = L^{(r)}u \text{ for all } u \in D_1 .$$

If $T_t^{(p)}$, $t \geq 0$, are contractions on $L^p(E; m)$, then so are $T_t^{(r)}$, $t \geq 0$, on $L^r(E; m)$.

Proof. [Eb 98, Ch. 1, Lemma 1.11].   □

Lemma 4.5. *Let $(T_t)_{t \geq 0}$ be a C^0–semigroup of symmetric operators on $L^2(E; m)$. Then each T_t, $t \geq 0$, is a contraction and the assertions of Lemma 4.4 hold for all $r \in [1, \infty)$.*

Proof. [Dav 89, Thm. 1.4.1].   □

4.2 Solution of Kolmogorov equations on $L^1(E; m)$.

The results of this subsection are due to W. Stannat (cf. [St 97]). We start with the finite dimensional case.

4.2.1 $E := \mathbb{R}^d$.

Let $E := \mathbb{R}^d$ and $\varphi \in H^{1,2}_{\mathrm{loc}}(\mathbb{R}^d; dx)$, i.e., φ is locally in the Sobolev space (w.r.t. Lebesgue measure dx) of order 1 in $L^2(\mathbb{R}^d; dx)$. Assume $\varphi \neq 0$ dx–a.e. and define

$$\mu := \varphi^2 \, dx \, ,$$

$$\beta^\mu := 2\frac{\nabla\varphi}{\varphi} \left(\in L^2_{\mathrm{loc}}(\mathbb{R}^d; \mu) \right) \, . \tag{4.2}$$

β^μ is hence the vector logarithmic derivative of μ. Define

$$\mathcal{E}_\mu(u, v) := \int_{\mathbb{R}^d} \langle \nabla u, \nabla v \rangle_{\mathbb{R}^d} \, d\mu \, ; \; u, v \in C_0^\infty(\mathbb{R}^d) \, . \tag{4.3}$$

Then $(\mathcal{E}_\mu, C_0^\infty(\mathbb{R}^d))$ is closable on $L^2(\mathbb{R}^d; \mu)$ and its closure $(\mathcal{E}_\mu, D(\mathcal{E}_\mu))$ is a Dirichlet form (cf. Subsect. 2.4.1 and [MR 92, Ch. I, Subsect. 2 a)]). Let $\hat{L}_0$ denote the corresponding generator (cf. Proposition 2.10). Then $\hat{L}_0$ is nothing but the Friedrichs' extension of

$$L_0 = \Delta + \langle \beta^\mu, \nabla \cdot \rangle_{\mathbb{R}^d}$$

$$= \mathrm{div}_\mu \nabla \, , \qquad D(L_0) = C_0^\infty(\mathbb{R}^d) \tag{4.4}$$

(cf. (3.5)). Let $(T_t^0)_{t \geq 0}$ be the sub–Markovian C^0–semigroup of contractions generated by $\hat{L}_0$ on $L^2(\mathbb{R}^d; \mu)$.

Remark 4.6. In this and the next subsection the roles of $(\mathcal{E}_\mu, D(\mathcal{E}_\mu))$ (resp. $\hat{L}_0, L_0$) could be replaced by a more general coercive form (resp. corresponding operators) as in Subsecton 2.4.1. Only for simplicity L_0 was chosen as in (4.4).

Now fix $b \in L^2_{\mathrm{loc}}(\mathbb{R}^d \to \mathbb{R}^d; \mu)$ satisfying the following "divergence condition"

(DC) $\qquad \int \langle b, \nabla u \rangle_{\mathbb{R}^d} \, d\mu = 0$ for all $u \in C_0^\infty(\mathbb{R}^d)$,

or shortly written

$$\mathrm{div}_\mu \, b = 0 \, . \tag{4.5}$$

Now consider the operator

$$L_b u := L_0 u - \langle b, \nabla u \rangle_{\mathbb{R}^d}$$

$$= \Delta u + \langle \beta^\mu - b, \nabla u \rangle_{\mathbb{R}^d} \, , \; u \in C_0^\infty(\mathbb{R}^d) \, . \tag{4.6}$$

Remark 4.7. (DC) is equivalent to

$$\int L_b u \, d\mu = 0 \text{ for all } u \in C_0^\infty(\mathbb{R}^d) , \qquad (4.7)$$

i.e., μ is an invariant measure for L_b in the sense of Sect. 5. below. This holds since obviously, $\int L_0 u \, d\mu = 0$ for all $u \in C_0^\infty(\mathbb{R}^d)$. Hence by Proposition 4.2, $(L_b, C_0^\infty(\mathbb{R}^d))$ is, in particular dissipative, hence closable (cf. Remark 2.2) on $L^1(E; \mu)$.

Concerning the existence of solutions to Kolmogorov's equation for at least one extension $\hat{L}_b$ of $(L_b, C_0^\infty(\mathbb{R}^d))$, i.e. to

$$\frac{du}{dt} = \hat{L}_b u , \; u(0, \cdot) = f \qquad (4.8)$$

which are positivity preserving w.r.t. the initial condition f, we have the following result by Wilhelm Stannat. Below for $W \subset L^1(\mathbb{R}^d; \mu)$ we set

$$W_b := \{ f \in W \, | \, f \text{ bounded} \} ,$$

$$W_0 := \{ f \in W \, | \, \text{supp } |f| \cdot \mu \text{ compact} \} , \quad W_{0,b} := W_0 \cap W_b .$$

Theorem 4.8. *Suppose condition (DC) holds. Then there exists a closed extension $\hat{L}_b$ of $(L_b, C_0^\infty(\mathbb{R}^d))$ on $L^1(\mathbb{R}^d; \mu)$ which generates a sub–Markovian C^0–semigroup $(T_t^b)_{t \geq 0}$ of contractions on $L^1(\mathbb{R}^d; \mu)$. $\hat{L}_b$ has the following properties:*
(i) $D(\hat{L}_b)_b \subset D(\mathcal{E}_\mu)$ and

$$\mathcal{E}_\mu(u, v) + \int \langle b, \nabla u \rangle v \, d\mu = - \int \hat{L}_b u \; v \, d\mu \qquad (4.9)$$

$$\text{for all } u \in D(\hat{L}_b)_b, \; v \in D(\mathcal{E}_\mu)_{0,b} .$$

In particular, $\hat{L}_b$ is uniquely determined by the (non-coercive) bilinear form on the left hand side of (4.9) (since $(D(\hat{L}_b)_b$ is dense in $D(\hat{L}_b)$ w.r.t. graph norm).

(ii) $$\mathcal{E}_\mu(u, u) \leq - \int \hat{L}_b u \; u \, d\mu \text{ for } u \in D(\hat{L}_b)_b . \qquad (4.10)$$

Proof. [St 97, Prop. 1.1].   □

As a consequence of Theorem 4.8 we get the desired solution to (4.8). However, it is not at all clear whether $(\hat{L}_b, D(\hat{L}_b))$ is unique. There could be many such extensions of the operator L_b explicitly given on $C_0^\infty(\mathbb{R}^d)$ in (4.6). The same drawback we also have in the case $b \equiv 0$, where by (4.9) $\hat{L}_0$ is just the Friedrichs' extension of $(L_0, C_0^\infty(\mathbb{R}^d))$ considered on $L^1(\mathbb{R}^d; \mu)$ (cf. Lemma 4.5). Also there, we could have many extensions that are generators. This problem (related to Theorem 2.6) will be addressed in Subsect. 4.3 below.

4.2.2 $E :=$ topological vector space.

Let E, H, μ and K be as in Subsect. 2.4.2, hence in particular $\mu(E) < \infty$. Define

$$\mathcal{E}_\mu(u, v) := \int_E \langle \nabla_H u(x), \nabla_H v(x) \rangle_H \, \mu(dx) \; ; \; u, v \in \mathcal{F}C_b^\infty(K) . \qquad (4.11)$$

Suppose that

(C) $(\mathcal{E}_\mu, \mathcal{F}C_b^\infty(K))$ is closable on $L^2(E; \mu)$.

Then the closure is a Dirichlet form (cf. Subsect. 2.4.2 and [MR 92, Ch. I, Proposition 3.8]). Let $\hat{L}_0$ be the associated generator and $(T_t^0)_{t \geq 0}$ the corresponding sub–Markovian C^0–semigroup of contractions on $L^2(E; \mu)$. Note that since obviously, $1 \in D(\hat{L}_0)$ and $\hat{L}_0 1 = 0$, it follows that $T_t^0 1 = 1$ for all $t \geq 0$. So, $(T_t^0)_{t \geq 0}$ is even Markovian.

Now let $b \in \vec{L}^2(E \to H; \mu)$ again satisfying the "divergence condition"

$$\text{div}_\mu \, b = 0 , \qquad (4.12)$$

more precisely,

(DC) $\int_E \langle b, \nabla_H u \rangle_H \, d\mu = 0$ for all $u \in \mathcal{F}C_b^\infty(K)$.

We assume for simplicity

$$\mathcal{F}C_b^\infty(K) \subset D(\hat{L}_0) \qquad (4.13)$$

(see [St 97, Sect. 4] how to avoid this condition) and consider the operator

$$L_b u := \hat{L}_0 u - \langle b, \nabla_H u \rangle_H , \; u \in \mathcal{F}C_b^\infty(K) . \qquad (4.14)$$

Remark 4.9. We note that also in this case (DC) is equivalent to

$$\int L_b u \, d\mu = 0 \text{ for all } u \in \mathcal{F}C_b^\infty(K) . \qquad (4.15)$$

Consequently, $(L_b, \mathcal{F}C_b^\infty(K))$ is dissipative, hence closable on $L^1(E; \mu)$ (cf. Proposition 4.2, Remark 2.2 respectively).

Theorem 4.10. *Suppose conditions (C), (DC), and (4.13) hold. Then there exists a closed extension $\hat{L}_b$ of $(L_b, \mathcal{F}C_b^\infty(K))$ on $L^1(E; \mu)$ generating a Markovian C^0–semigroup $(T_t^b)_{t \geq 0}$ of contractions on $L^1(E; \mu)$. $\hat{L}_b$ has the following properties.*
(i) It is the closure of L_b, but considered with the larger domain $D(\hat{L}_0)_b$.
(ii) $D(\hat{L}_b)_b \subset D(\mathcal{E}_\mu)$ and

$$\mathcal{E}_\mu(u, v) + \int_E \langle b, \nabla_H u \rangle_H \, v \, d\mu = - \int_E \hat{L}_b u \, v \, d\mu$$
$$\text{for all } u \in D(\hat{L}_b)_b, \; v \in D(\mathcal{E}_\mu)_b . \qquad (4.16)$$

In particular, $\hat{L}_b$ is uniquely determined by the (non-coercive) bilinear form on the left hand side of (4.16).

(iii) $$\mathcal{E}_\mu(u,u) \leq -\int \hat{L}_b u \, u \, d\mu \text{ for all } u \in D(\hat{L}_b)_b \, .$$

$$(4.17)$$

Proof. [St 97, Prop. 4.1]. □

As in Subsect. 4.2.1 we obtain the desired solution of the corresponding Kolmogorov equation with the same drawback concerning uniqueness.

Remark 4.11. We emphasize that conditions (C) and (4.13) hold for μ defined as in the cases discussed in Subsect. 3.3 and Example 3.7.

4.3 Uniqueness problem.

Let $(E, \mathcal{B}, m)$ be a σ–finite measure space and $p \in [1, \infty)$. Let (L, D) be a linear operator on $L^p(E; m)$ with dense domain D. In the preceding subsection we presented results (for $p = 1$) ensuring that (L, D) has a closed extension $\hat{L}$ which generates a C^0–semigroup on $L^p(E; m)$. In this subsection we are concerned with the question whether $\hat{L}$ is the only extension generating a C^0–semigroup on $L^p(E; m)$. By Theorem 2.6 this is equivalent to the question whether the closure $\overline{L}$ of (L, D) on $L^p(E; m)$ generates a C^0–semigroup on $L^p(E; m)$.

Most of the results below are again taken from [St 97].

We recall that a C^0–semigroup $(T_t)_{t \geq 0}$ on $L^1(E; m)$ is called *conservative*, if for its dual semigroup $(T_t^*)_{t \geq 0}$ on $L^\infty(E; m)$ we have that $T_t^* 1 = 1$ for one (hence all) $t \geq 0$.

4.3.1 $E := \mathbb{R}^d$.

We consider the situation in Subsection 4.2.1 adopting all notations introduced there.

Remark 4.12. Obviously, $(T_t^0)_{t \geq 0}$ is conservative if $\mu(\mathbb{R}^d) < \infty$.

Theorem 4.13. *The closure $\overline{L}_0$ on $L^1(\mathbb{R}^d; \mu)$ of $(L_0, C_0^\infty(\mathbb{R}^d))$ generates a C^0–semigroup if and only if $(T_t^0)_{t \geq 0}$ (i.e., the C^0–semigroup generated by the Friedrichs' extension of $(L_0, C_0^\infty(\mathbb{R}^d))$ on $L^1(E, \mu))$ is conservative. This, in particular holds if $\mu(\mathbb{R}^d) < \infty$.*

Proof. [St 97, Thm. 2.1] (generalizing [Dav 85] to the case of non–smooth coefficients). □

As in Subsection 4.2.1 consider now

$$L_b = \Delta + \langle \beta^\mu - b, \nabla \cdot \rangle_{\mathbb{R}^d} \text{ on } C_0^\infty(\mathbb{R}^d)$$

with $b \in L_{loc}^2(\mathbb{R}^d \to \mathbb{R}^d; \mu)$ satisfying (DC). Define $D(\mathcal{E}_\mu)_{loc}$ to be the set of all $u : \mathbb{R}^d \to \mathbb{R}$ such that $\chi \cdot u \in D(\mathcal{E}_\mu)$ for all $\chi \in C_0^\infty(\mathbb{R}^d)$. Then the conservativity of $(T_t^b)_{t \geq 0}$ can be characterized as follows:

Proposition 4.14. *The following are equivalent*

(i) $(T_t^b)_{t \geq 0}$ *is conservative.*
(ii) There exist $\chi_n \in D(\mathcal{E}_\mu)_{loc}$, $n \in \mathbb{N}$, *and* $\alpha \in (0, \infty)$ *such that* $(\chi_n - 1)^- \in D(\mathcal{E}_\mu)_{0,b}$, $\lim_{n \to \infty} \chi_n = 0$ μ-*a.e. and*

$$\mathcal{E}_{\mu,\alpha}(\chi_n, v) \geq \int \langle b, \nabla \chi_n \rangle_{\mathbb{R}^d} v \, d\mu \text{ for all } v \in D(\mathcal{E}_\mu)_{0,b}, \; v \geq 0 . \quad (4.18)$$

(iii) The closure on $L^1(\mathbb{R}^d; \mu)$ *of* L_b, *but considered on the* **larger** *domain* $D(L_0)_b$, *generates a* C^0-*semigroup on* $L^1(\mathbb{R}^d; \mu)$

Proof. [St 97, Proposition 1.9]. □

The following lemma provides simple conditions which ensure that the equivalent properties (i) – (iii) in Proposition 4.14 hold:

Lemma 4.15. *Proposition 4.14 (i) – (iii) hold if one of the following conditions are satisfied:*

(i) $b \in L^1(\mathbb{R}^d; \mu)$.
(ii) $\beta^\mu - b \in L^2(\mathbb{R}^d; \mu)$ *and* $\mu(\mathbb{R}^d) < \infty$.
(iii) There exists a Lyapunov–type function for L_b, *i.e., a* C^2-*function* $V : \mathbb{R}^d \to \mathbb{R}$ *such that*

$$\lim_{|x|_{\mathbb{R}^d} \to \infty} V(x) = \infty \text{ and } \lim_{|x|_{\mathbb{R}^d} \to \infty} L_b V(x) = -\infty . \quad (4.19)$$

This is particularly the case if there exists $M \in [0, \infty)$ *such that*

$$\langle (\beta^\mu - b)(x), x \rangle_{\mathbb{R}^d} \leq M \left(\ell n(|x|_{\mathbb{R}^d}^2 + 1) + 1 \right) \text{ for all } x \in \mathbb{R}^d . \quad (4.20)$$

Proof. For (i) see [St 97, Prop. 1.10 (a)].
(ii): By Theorem 5.8 (iii) below, $\beta^\mu - b \in L^2(\mathbb{R}^d; \mu)$ implies $\beta^\mu, b \in L^2(\mathbb{R}^d; \mu)$. In particular, $b \in L^1(\mathbb{R}^d; \mu)$, so (i) implies the assertion.
(iii): See [St 97, Proofs of Prop. 1.10 (b) and Prop. 2.8]. □

Now we have the following generalization of Theorem 4.13 to the non–symmetric case.

Theorem 4.16. *The following are equivalent:*
(i) The closure of $(L_b, C_0^\infty(\mathbb{R}^d))$ *on* $L^1(\mathbb{R}^d; \mu)$ *generates a* C^0-*semigroup on* $L^1(\mathbb{R}^d; \mu)$.
(ii) One of the equivalent conditions in Proposition 4.14 holds.

Proof. By [St 97, Lemma 2.3] the proof is word by word the same as that of [St 97, Thm. 2.1].   □

Remark 4.17. The equivalent properties (i) and (ii) in Theorem 4.16 may not hold. Indeed, in [St 97, Example 1.12] it is proved that they do not hold in the following situation where μ is even Gaussian : $\varphi(x) := e^{-x^2/2}$, $b(x) := 6e^{x^2}$, $x \in \mathbb{R}^d$.

So far, we have discussed uniqueness results only in $L^1(\mathbb{R}^d; \mu)$, which hold under the weakest assumptions. Uniqueness in $L^p(\mathbb{R}^d; \mu)$ for $p \geq 1$ (in particular, for $p = 2$, i.e., if $b \equiv 0$, the question of *essential self-adjointness*) has also been studied by many authors. We refer to [Eb 98] for numerous new results and also for a survey, as well as [L 98] for a more recent special result which is not discussed in [Eb 98]. To give an idea of at least one type of results proved so far, we recall

Theorem 4.18. *Suppose* $\mu(\mathbb{R}^d) < \infty$ *and that* $|\beta^\mu - b| \in L^p_{loc}(\mathbb{R}^d; dx)$ *for some* $p > d \geq 2$ *such that*

$$|b| \in L^q(\mathbb{R}^d; \mu) \text{ for some } q \in [1, \infty] .$$

Let $r := 2 - \frac{2}{q+1}$ (*where* $\frac{1}{\infty} := 0$). *Then the closure of* $(L_b, C_0^\infty(\mathbb{R}^d))$ *on* $L^r(\mathbb{R}^d; \mu)$ *generates a* C^0*-semigroup on* $L^r(\mathbb{R}^d; \mu)$.

Proof. [ABR 97, Theorem 1.4 (iii)].   □

4.3.2 $E :=$ topological vector space.

We consider the situation of Subsection 4.2.2 adopting all notations introduced there. In particular, we assume (4.13) to hold (for simplicity).

Remark 4.19. Obviously, $(T_t^0)_{t \geq 0}$ is always conservative.

Unfortunately, there is *no* analogue of Theorem 4.13 in this infinite dimensional case.

Theorem 4.20. *The closure of* $(\hat{L}_0, \mathcal{F}C_b^\infty(K))$ *does not always generate a* C^0*-semigroup on* $L^1(E; \mu)$.

Proof. See the counterexamples in [Eb 98, Subsect. 5 b)].   □

As in Subsection 4.2.2 consider now

$$L_b := \hat{L}_0 - \langle b, \nabla_H \cdot \rangle_H \text{ on } \mathcal{F}C_b^\infty(K)$$

with $b \in L^2(E \to H; \mu)$ satisfying (DC).

In this non–symmetric situation we have:

Theorem 4.21. *Suppose the closure of* $(L_0, \mathcal{F}C_b^\infty(K))$ *on* $L^1(E; \mu)$ *generates a* C^0*-semigroup on* $L^1(E; \mu)$, *then so does the closure of* $(L_b, \mathcal{F}C_b^\infty(K))$ *on* $L^1(E; \mu)$.

Proof. [St 97, Proposition 4.3]. □

Theorem 4.21 is the infinite dimensional generalization of Theorem 4.16 in part. Indeed, under our present assumptions it follows that if $E = H = \mathbb{R}^d$, and if $\mu = \varphi^2 dx$ for some $\varphi \in H^{1,2}_{loc}(\mathbb{R}^d; dx)$ (which in fact follows from (4.13)), then Theorem 4.16 (ii) holds by Lemma 4.15 (i), since in the present subsection b is assumed to be globally μ–square integrable and $\mu(E) < \infty$.

Remark 4.22. Also in the present infinite dimensional situation the unique-ness problem has been investigated for $p \in [1, \infty)$ rather than just the case $p = 1$. We again refer to [Eb 98, Ch. 5], and also [LR 97]. The latter paper proves L^p–uniqueness results in infinite dimensions for all $p \in [1, \infty)$ which are not covered by [Eb 98, Ch. 5]. They particularly apply to the opera-tor generating the dynamics in the stochastic quantization of Euclidean field theory in finite volume (in contrast to Eberle's results also in case $p \geq 2$!).

Remark 4.23. We emphasize that the closure of $(L_0, \mathcal{F}C^\infty_b(K))$ even on $L^2(E; \mu)$ generates a C^0–semigroup on $L^2(E; \mu)$ for μ as in the cases dis-cussed in Subsect. 3.3 and Example 3.7. For the latter this follows by [AKR 97, Theorem 5.13]. For the Gaussian situation in Subsect. 3.3 this follows by the explicit form of $(T^0_t)_{t \geq 0}$ given by the Mehler formula (3.13) and Propo-sition 2.7 applied to

$$B_0 := \text{lin. span } \{\cos\left(\,_{E'}\langle \ell, \cdot \rangle_E\right)\,,\, \sin\left(\,_{E'}\langle \ell, \cdot \rangle_E\right) | \ell \in K\}$$

(cf. [BRS 96, Prop. 6.1]). So, Theorem 4.21 applies to all these cases.

4.4 Concluding remarks.

If we want to apply the results in Subsections 4.2, 4.3 above to an operator of type

$$L = \Delta + \langle B, \nabla \cdot \rangle_{\mathbb{R}^d} \text{ on } C^\infty_0(\mathbb{R}^d)$$

for some vector field $B : \mathbb{R}^d \to \mathbb{R}^d$ (or to its infinite dimensional analogue) in order to solve the corresponding Kolmogorov equation, we have to find a measure μ so that the decomposition

$$B = \beta^\mu - b$$

is such that all our assumptions are satisfied. By Remarks 4.7, resp. 4.9, we, therefore, have to look for invariant measures μ for L. If we can prove regularity for μ, i.e., that $\mu = \rho\, dx$ with ρ regular enough, then we can take

$$b := \frac{\nabla \rho}{\rho} - B\,.$$

This program will be pursued in the next section.

5. Invariant measures: regularity, existence and uniqueness

Let $(E, \mathcal{B})$ be a measurable space and let $L(E, \mathcal{B})$ denote the linear space of all $\mathcal{B}$–measurable real–valued functions on E. Let $D \subset L(E, \mathcal{B})$ be a linear subspace and $L : D \to L(E, \mathcal{B})$ a linear operator.

Definition 5.1. A probability measure μ on $(E, \mathcal{B})$ is called an *invariant measure for* (L, D) (or shortly for L if D is fixed) if

$$Lu \in L^1(E; \mu) \text{ and } \int Lu \, d\mu = 0 \text{ for all } u \in D . \qquad (5.1)$$

We abbreviate (5.1) by $L^*\mu = 0$.

Remark 5.2. Suppose μ is an invariant measure for (L, D) and that (L, D) is a linear operator on $L^1(E; \mu)$ such that it has a closed extension $\hat{L}$ which generates a C^0–semigroup $(T_t)_{t \geq 0}$ on $L^1(E; \mu)$. Then obviously μ is $(T_t)_{t \geq 0}$–invariant, i.e.,

$$\int T_t f \, d\mu = \int f \, d\mu \text{ for all } f \in L^1(E; \mu) , \qquad (5.2)$$

if and only if $(T_t)_{t \geq 0}$ is conservative (cf. Subsect. 4.3). So, by Remark 4.17, (5.1) in general does not imply (5.2). Clearly, however,

$$(5.2) \quad \Longleftrightarrow \quad \int \hat{L}u \, d\mu = 0 \text{ for all } u \in D(\hat{L}) . \qquad (5.3)$$

So, by Theorem 4.16 for $(L, D) := (L_b, C_0^\infty(\mathbb{R}^d))$, as considered there, (5.2) holds if and only if $\hat{L} = $ closure of (L, D) on $L^1(E; \mu)$.

5.1 Sectorial case

We consider the situation of Subsect. 2.4.2 with $\hat{b} \equiv 0$, $c \equiv 0$. We assume that (C) and (SFP) hold. Let L_0 denote the generator of $(Q, D(Q))$ (defined as in Subsect. 2.4.2 with $c \equiv 0$). As in Subsect. 4.2 for simplicity we assume that

$$\mathcal{F}C_b^\infty(K) \subset D(L_0) \qquad (5.4)$$

and set

$$L_b u := L_0 u - \langle b, \nabla_H u \rangle_H , \quad u \in \mathcal{F}C_b^\infty(K) . \qquad (5.5)$$

So, L_b is nothing but the restriction of $L := L^{(\alpha)} + \alpha$ to $\mathcal{F}C_b^\infty(K)$ where $L^{(\alpha)}$ is the generator of the sectorial form $(\mathcal{E}_\alpha, D(\mathcal{E}_\alpha))$ defined in Subsect.

2.4.2 with $A = Id_H$ $\hat{b} \equiv 0$, $c \equiv 0$. Let $(T_t^{(\alpha)})_{t\geq0}$ be the corresponding C^0–semigroup and set $T_t := e^{\alpha t}T_t^{(\alpha)}$, $t \geq 0$.

Since $L_b u$ is only defined as a class of functions being μ–a.e. equal, invariant measures in the sense of Definition 5.1 are only well–defined if they are absolutely continuous w.r.t. μ. In the next subsection we shall present results about existence and uniqueness of such measures.

5.1.1 Results.

Theorem 5.3. *Suppose (C), (E), (SFP) and (5.4) hold. Assume furthermore:*

(i) (DLS) holds (cf. Subsect. 2.4.2).

(ii) There exist $l_n \in E'$, $n \in \mathbb{N}$, forming an orthonormal basis of H such for all $N \in \mathbb{N}$

$$\mu_N := \mu \circ (\ell_1, \ldots, \ell_N)^{-1} = \rho_N \, dx$$

for some $\mathcal{B}(\mathbb{R}^N)$–measurable $\rho_N : \mathbb{R}^N \to [0, \infty)$ such that

$$R(\rho_N) := \left\{ y \in \mathbb{R}^N \mid \int_{\{x \mid |x-y|_{\mathbb{P}^N} < \varepsilon\}} \frac{1}{\rho_N(x)} \, dx < \infty \text{ for some } \varepsilon > 0 \right\} \quad (5.6)$$

has full μ_N–measure.

Then there exists $\rho \in D(\mathcal{E})(= D(\mathcal{E}_\mu))$, $\rho \geq 0$, such that $\nu := \rho \cdot \mu$ is an invariant measure for L_b.

Proof. [BRZ 97, Theorem 3.6 and Remark 2.9] □

5.1.2 Application to the uniqueness problem.

We still consider the situation of the previous subsection

Lemma 5.4. *In the situation of Theorem 5.3 we have that $\left|\frac{\nabla_H \rho}{\rho}\right|_H \in L^2(E; \nu)$.*

Proof. Let $k \in \mathbb{N}$. Then $ln(\rho + \frac{1}{k}) \in D(\mathcal{E})(= D(\mathcal{E}_\mu))$ and hence by [BRZ 97, Lemma 2.10] for all $k \in \mathbb{N}$

$$\int_E \langle \frac{\nabla_H \rho}{\rho + \frac{1}{k}}, \frac{\nabla_H \rho}{\rho + \frac{1}{k}} \rangle_H \left(\rho + \frac{1}{k}\right) \, d\mu$$

$$= \mathcal{E}_\mu \left(ln\left(\rho + \frac{1}{k}\right), \rho\right)$$

$$= -\lim_{n\to\infty} \int_E \langle b, \nabla_H \left(ln\left(\rho + \frac{1}{k}\right)\right)\rangle_H \inf(\rho, n) \, d\mu$$

$$\leq \int_E |b|_H |\nabla_H \rho|_H \, d\mu$$

$$\leq \left(\int_E |b|_H^2 \, d\mu\right)^{1/2} \mathcal{E}_\mu(\rho, \rho)^{1/2} < \infty .$$

Now the assertion follows by Fatou's Lemma. □

Let us assume that μ has the following property:

(U) If for all $\rho \in D(\mathcal{E}_\mu)$, with $|\nabla_H \rho / \rho|_H \in L^2(E; \rho \cdot \mu)$ and $\rho \geq 0$, we define

$$L_\rho u := L_0 u + \langle \frac{\nabla_H \rho}{\rho} , \nabla_H u \rangle_H \; ; \; u \in \mathcal{F}C_b^\infty(K) ,$$

then $(L_\rho, \mathcal{F}C_b^\infty(K))$ is well–defined on $L^1(E; \rho \cdot \mu)$ and its closure generates a C^0–semigroup on $L^1(E; \rho \cdot \mu)$.

Remark 5.5. (i)There is a lot of examples of measures μ known to satisfy condition (U). For instance it follows by [Eb 98, Corollary 5.4] that (U) holds for many Gaussian measures as in Subsect. 3.3. At present we do not know, however, whether (U) holds for Example 3.7, though we expect it is the case.
(ii) In condition (U) we set $\frac{\nabla_H \rho}{\rho} := 0$ on $\{\rho = 0\}$, and to be precise we have to consider L_ρ on the $(\rho \cdot \mu)$–equivalence classes $\mathcal{F}C_b^\infty(K)^\sim$ of $\mathcal{F}C_b^\infty(K)$. Since possibly $\mu(\{\rho = 0\}) > 0$, there might be several different representatives in $\mathcal{F}C_b^\infty(K)$ for the same class in $\mathcal{F}C_b^\infty(K)^\sim$. Therefore, we have to assume that L_ρ is well–defined representativewise.

Theorem 5.6. *Assume μ satisfies (U) and consider the situation of Theorem 5.3. Let $\nu := \rho \cdot \mu$ be as defined there. Suppose $|b|_H$, $L_0 u \in L^4(E; \mu)$ for all $u \in \mathcal{F}C_b^\infty$. Then the closure of $(L_b, \mathcal{F}C_b^\infty(K))$ on $L^1(E; \nu)$ generates a C^0–semigroup $(T_t^\nu)_{t \geq 0}$ on $L^1(E; \nu)$. Furthermore, $T_t^\nu f = T_t f$ ν–a.e. for all $t \geq 0$, $f \in L^1(E; \mu)$.*

Proof. Applying Theorem 4.21 with $\mu := \nu$, $L_0 := L_\rho$, $b := \frac{\nabla_H \rho}{\rho} - b$ which is possible by Lemma 5.4, we obtain the first part of the assertion. The second part is obvious.   □

Remark 5.7. (i) Theorem 5.6 assures us that even if we might loose uniqueness on $L^1(E; \mu)$ by passing from $(L_0, \mathcal{F}C_b^\infty(K))$ to $(L_b, \mathcal{F}C_b^\infty(K))$, we keep at least uniqueness on $L^1(E; \nu)$ where $\nu = \rho \cdot \mu$ is an invariant measure for L_b.
(ii) Again we note that Theorem 5.6 applies to all cases in Subsect. 3.3.
(iii) If (E, H, μ) is an abstract Wiener space and if $e^{|b|_H^2} \in \bigcup_{p>2} L^p(E; \mu)$, it has recently been proved in [Sh. 98, Sect. 3], that the closure of $(L_b, \mathcal{F}C_b^\infty(K))$ on $L^2(E; \mu)$ generates a C^0–semigroup on $L^2(E; \mu)$ (hence on $L^1(E, \mu)$). So, in this case one can analyze L_b on L^1 also w.r.t. the initial Gaussian measure μ, instead of the invariant measure.

5.2 Non–sectorial cases

In this section for simplicity we shall assume that the second order part of our operator is the Laplacian, resp. the Gross–Laplacian in the infinite dimensional case. For the general case we refer to the corresponding underlying literature [BR 95], [BKR 96], [BDPR 96], [ABR 97], [BKR 97], [BR 98].

5.2.1 Regularity and applications to Kolmogorov equations.

a) Let us start with the finite dimensional case: $E := \mathbb{R}^d$, $\mathcal{B} := \mathcal{B}(\mathbb{R}^d)$. Let $B : \mathbb{R}^d \to \mathbb{R}^d$ be Borel–measurable and define

$$L_B u := \Delta u + \langle B, \nabla u \rangle_{\mathbb{R}^d}, \ u \in C_0^\infty(\mathbb{R}^d) . \tag{5.7}$$

Note that if μ is a probability measure on $(\mathbb{R}^d, \mathcal{B}(\mathbb{R}^d))$ such that $L_B^* \mu = 0$ in the sense of Definition 5.1, then, in particular, $L_B u \in L^1(\mathbb{R}^d; \mu)$ for all $u \in C_0^\infty(\mathbb{R}^d)$, hence necessarily

$$B \in L^1_{\text{loc}}(\mathbb{R}^d; \mu) . \tag{5.8}$$

We recall the following special case of the regularity result [BKR 97, Theorem 1].

Theorem 5.8. *Suppose μ is a probability measure on $(\mathbb{R}^d, \mathcal{B}(\mathbb{R}^d))$ such that*

$$L_B^* \mu = 0 . \tag{5.9}$$

Let for $p > d$, $|B|_{\mathbb{R}^d} \in L^p_{\text{loc}}(\mathbb{R}^d; dx)$ or $|B|_{\mathbb{R}^d} \in L^p_{\text{loc}}(\mathbb{R}^d; \mu)$. Then $\mu \ll dx$ and $\rho := \frac{d\mu}{dx}$ has the following properties:

(i) $\rho \in H^{1,p}_{\text{loc}}(\mathbb{R}^d; dx)$ (i.e., is locally in the Sobolev space (w.r.t. dx) of order 1 in $L^p(\mathbb{R}^d; dx)$). In particular, ρ is Hölder continuous of order $1 - d/p$.
(ii) $\inf_K \rho > 0$ for all compact $K \subset \mathbb{R}^d$.
(iii) If $|B|_{\mathbb{R}^d} \in L^2(\mathbb{R}^d; \mu)$, then $\frac{\nabla \rho}{\rho} \in L^2(\mathbb{R}^d; \mu)$.

Proof. (i): [BKR 97, Theorem 1].

(ii): This is a consequence of the Harnack inequality for L_B proved in [Tr 73].
(iii): [BR 95, Theorem 3.1].   □

Consider the situation of Theorem 5.8 and decompose L_B as follows

$$L_B u := \Delta u + \langle \beta^\mu, \nabla u \rangle - \langle b, \nabla u \rangle \ ; \ u \in C_0^\infty(\mathbb{R}^d) \tag{5.10}$$

where $\beta^\mu := \frac{\nabla \rho}{\rho}$, $b := \beta^\mu - B$. Then by Remark 4.7 and Theorem 4.13 we know that Theorem 4.16 applies. In particular, in case $|B|_{\mathbb{R}^d} \in L^2(\mathbb{R}^d; \mu)$, we hence conclude by Proposition 4.14 (ii) that the closure of $(L_B, C_0^\infty(\mathbb{R}^d))$ on $L^1(\mathbb{R}^d; \mu)$ generates a C^0–semigroup on $L^1(\mathbb{R}^d; \mu)$.

b) Consider the situation in Subsect. 3.3 with $A := Id_H$, hence (E, H, μ) is an abstract Wiener space and $K = E'$. Define as in (3.8)

$$L_0 u(x) := \Delta_H u(x) - {}_E\langle x, \nabla_H u(x)\rangle_{E'} \ , \ u \in \mathcal{F}C_b^\infty(K) \ , \ x \in E \ . \quad (5.11)$$

Let $B : E \to H$ be $\mathcal{B}(E)/\mathcal{B}(H)$ measurable and define

$$L_B u := L_0 u + \langle B, \nabla_H u\rangle_H \ , \ u \in \mathcal{F}C_b^\infty(K) \ . \quad (5.12)$$

Then we have the following regularity result for invariant measures (see also [ABR 97, Theorem 4.3] for a generalization).

Theorem 5.9. *Suppose ν is a probability measure on $(E, \mathcal{B}(E))$ such that $L_B^* \nu = 0$. Assume in addition that ${}_{E'}\langle \ell, \cdot\rangle_E \in L^2(E; \nu)$ for all $\ell \in E'$ and that $|B|_H \in L^2(E; \nu)$. Then $\nu \ll \mu$ and $\rho := \frac{d\nu}{d\mu} \in D(\mathcal{E}_\mu)$.*

Proof. [BR 95, Theorem 3.5].   □

Consider the situation of Theorem 5.9 and decompose L_B as follows

$$L_B u := L_0 u + \langle \tfrac{\nabla_H \rho}{\rho}, \nabla_H u\rangle_H - \langle \tfrac{\nabla_H \rho}{\rho} - B, \nabla_H u\rangle_H \ , \ u \in \mathcal{F}C_b^\infty(K) \ . \quad (5.13)$$

since by Remark 5.5 condition (U) is satisfied we conclude by Remark 4.9 and Theorem 4.21 that the closure of $(L_B, \mathcal{F}C_b^\infty(K))$ on $L^1(E; \nu)$ generates a C^0–semigroup on $L^1(E; \mu)$.

5.2.2 Existence and uniqueness $E := \mathbb{R}^d$.

Consider the situation of part a) of Subsect. 5.2.1. We recall the following special case of the main result in [BR 98].

Theorem 5.10. *Suppose $|B| \in L_{loc}^p(\mathbb{R}^d; dx)$ for $p > d$ and assume that there exists a Lyapunov-type function of L_B, i.e., there exists $V \in C^2(\mathbb{R}^d)$ such that*

$$\lim_{|x|\to\infty} V(x) = \infty \ , \ \lim_{|x|\to\infty} L_B V(x) = -\infty \ .$$

Then there exists a unique probability measure μ on $(\mathbb{R}^d, \mathcal{B}(\mathbb{R}^d))$ such that $L_B^* \mu = 0$.

Proof. [BR 98, Theorem 1.6 and Remark 1.10 (ii)].   □

There is a close connection between uniqueness of solutions to $L_B^* \mu = 0$ and whether the closure of $(L_B, C_0^\infty(\mathbb{R}^d))$ on $L^1(\mathbb{R}^d; \mu)$ generates a C^0–semigroup on $L^1(\mathbb{R}^d; \mu)$. More precisely, we have the following.

Theorem 5.11. *Suppose $|B| \in L_{loc}^p(\mathbb{R}^d; dx)$ for $p > d$. Let K be a convex set of probability measures μ on $(\mathbb{R}^d; \mathcal{B}(\mathbb{R}^d))$ such that for every $\mu \in K$*

(i) $L_B^* \mu = 0$;

(ii) the closure of $(L_B, C_0^\infty(\mathbb{R}^d))$ on $L^1(\mathbb{R}^d; \mu)$ generates a C^0–semigroup on $L^1(\mathbb{R}^d; \mu)$.

Then $\#\mathcal{K} \leq 1$. In particular,

$$\# \left\{ \mu \mid L_B^* \mu = 0 \text{ and } B - \frac{\nabla\rho}{\rho} \in L^1(\mathbb{R}^d; \mu) \text{ with } \rho := \frac{d\mu}{dx} \right\} \leq 1 .$$

Proof. [ABR 97, Theorem 1.2] and the discussion following it. □

Finally, we shall look at the special case where B is a gradient. More precisely, assume

$$\begin{aligned} &B \text{ is a (fixed) Borel–measurable } dx\text{–version of} &(5.14)\\ &2\frac{\nabla\varphi}{\varphi} \text{ for some } \varphi \in H_{\text{loc}}^{1,2}(\mathbb{R}^d; dx) , \ \int \varphi^2 \, dx = 1 . \end{aligned}$$

Here as usual we set $\frac{\nabla\varphi}{\varphi} := 0$ on $\{\varphi = 0\}$. Then obviously $\mu_0 := \varphi^2 \, dx$ belongs to the following set of probability measures on $(\mathbb{R}^d, \mathcal{B}(\mathbb{R}^d))$

$$\begin{aligned} \mathcal{M}_1 := \{\mu \mid \ &L_B^* \mu = 0 \text{ such that } |B|_{\mathbb{R}^d} \in L_{\text{loc}}^2(\mathbb{R}^d; \mu)\\ &\text{and } |B - \tfrac{\nabla\rho}{\rho}| \in L^2(\mathbb{R}^d; \mu)\} . \end{aligned} \qquad (5.15)$$

Clearly, μ_0 is symmetrizing for $(L_B, C_0^\infty(\mathbb{R}^d))$, and, in fact, by the following result so are all other measures in $\mathcal{M}_1$. In some cases we even have uniqueness.

Theorem 5.12. Let B be as in (5.14) and $\mu_0 := \varphi^2 \, dx$. Then:

(i) Every $\mu \in \mathcal{M}_1$ is a symmetrizing measure for $(L_B, C_0^\infty(\mathbb{R}^d))$.

(ii) If $B \in L_{\text{loc}}^1(U; dx)$ for some connected open set $U \subset \mathbb{R}^d$ such that $dx(\mathbb{R}^d \setminus U) = 0$, then

$$\mathcal{M}_1 = \{\mu_0\} .$$

Remark 5.13. The condition in Theorem 5.12 (ii) cannot be dropped. Indeed, let $d = 1$ and $\varphi(x) := x \ (2\pi)^{-1/4} e^{-x^2/4}$, $x \in \mathbb{R}$. Then $B(x) = -x + \frac{2}{x}$, $x \in \mathbb{R}$, by (5.14), and $B \notin L_{\text{loc}}^1(\mathbb{R}; dx)$. Let $\varphi_1 := (\int_0^\infty \varphi \, dx)^{-1} 1_{[0,\infty)} \varphi$ and $\varphi_2 := (\int_{-\infty}^0 \varphi \, dx)^{-1} 1_{(-\infty,0]} \varphi$. Then $\alpha \varphi_1 \, dx + (1 - \alpha)\varphi_2 \, dx \in \mathcal{M}_1$ for all $\alpha \in [0,1]$ (cf. [BR 95, Example 6.1]).

5.2.3 Existence and uniqueness for $E :=$ topological vector space.

Consider E, H, K as in Subsect. 2.4.2 and let $B : E \to E$ be $\mathcal{B}(E)/\mathcal{B}(E)$–measurable. Define

$$L_B u := \Delta_H u + {}_E\langle B, \nabla_H u\rangle_{E'}\ , \ u \in \mathcal{F}C_b^\infty(K)\ . \tag{5.16}$$

Assume that also E is a Hilbert space with inner product $\langle\ ,\ \rangle_E$.

In this situation we have the following existence result:

Proposition 5.14. *Assume that*

$$\lim_{|x|_E \to \infty} \langle B(x), x\rangle_E = -\infty\ , \tag{5.17}$$

that $x \mapsto {}_{E'}\langle \ell, B(x)\rangle_E$ is weakly continuous on E for all $\ell \in E'$, and that there exist C_1, C_2, $\alpha \in (0, \infty)$ such that

$$|B(x)|_E \leq C_1 + C_2\ |x|_E^\alpha \ for\ all\ x \in E\ . \tag{5.18}$$

Then there exists a probability measure μ on $(E, \mathcal{B}(E))$ such that $L_B^* \mu = 0$.

Proof. [BR 95, Theorem 5.2]. □

The above result is not optimal. There is work in progress towards generalizations. The question, which conditions imply uniqueness for invariant measures of $(L_B, \mathcal{F}C_b^\infty(K))$, are in this general case largely open.

Concerning the question under what conditions invariant measures are symmetrizing in this infinite dimensional case, we have the following result.

Proposition 5.15. *Let (E, H, μ) be an abstract Wiener space and let $(L_B, \mathcal{F}C_b^\infty(K))$ be as defined in (5.12) with $K := E'$. Suppose $B = -Id_E + b$ such that*

$$b \ is \ a \ \mu\text{-version of}\ 2\frac{\nabla_H \varphi_0}{\varphi_0}\ for\ some\ \varphi_0 \in D(\mathcal{E}_\mu)\ with$$
$$\int \varphi_0^2\ d\mu = 1\ (where\ \nabla_H \varphi_0/\varphi_0 := 0\ on\ \{\varphi_0 = 0\})\ . \tag{5.19}$$

(i) Assume ν is a probability measure on $(E, \mathcal{B}(E))$ such that ${}_{E'}\langle \ell, \cdot\rangle_E \in L^2(E; \nu)$ for all $\ell \in E'$, $|b|_H \in L^2(E; \nu)$ and $L_B^* \nu = 0$. Then $\nu = \varphi^2 \cdot \mu$ for some $\varphi \in D(\mathcal{E}_\mu)$ and

$$\frac{\nabla_H \varphi}{\varphi} = \frac{\nabla_H \varphi_0}{\varphi_0}\ \mu\text{-}a.e.$$

(Obviously, $\nu_0 := \varphi_0^2 \cdot \mu$ is such a measure). In particular, ν is symmetrizing for $(L_B, \mathcal{F}C_b^\infty(K))$.

(ii) If $|\frac{\nabla_H \varphi_0}{\varphi_0}|_H \in L^2(E; \mu)$, then $\nu_0 := \varphi_0 \cdot \mu$ is the only probability measure satisfying the assumptions in (i).

Proof. [ABR 97, Theorem 4.5]. □

6. Corresponding diffusions and relation to Martingale problems

In this section we show that the positivity preserving C^0–semigroups constructed above, in most cases are really transition semigroups of Markov processes. So, to call the equation

$$\frac{d}{dt} T_t u = L T_t u \qquad (6.1)$$

a "Kolmogorov equation" is indeed justified. Though we do not use these processes at all to construct solutions to our Kolmogorov equations, this section is enclosed to complete the picture (however, in a slightly sketchy style). In addition, the process can be used subsequently to get more specific information about $(T_t)_{t\geq 0}$ and L.

6.1 Existence of associated diffusions

6.1.1 Sectorial case.

Let E be a topological space and $\mathcal{B}(E)$ its Borel σ–algebra. Let m be a σ–finite positive measure on $(E, \mathcal{B}(E))$. Let $(\mathcal{E}, D(\mathcal{E}))$ be a sectorial form on $L^2(E; m)$ with associated generator $(L, D(L))$ and C^0–semigroup $(T_t)_{t\geq 0}$.

Definition 6.1. (i) An increasing sequence $(E_k)_{k\in\mathbb{N}}$ of closed subsets of E is called an $\mathcal{E}$–nest if

$$\{u \in D(\mathcal{E}) \mid u = 0 \ m\text{–a.e. on } E \setminus E_k \text{ for some } k \in \mathbb{N}\}$$

is dense in $D(\mathcal{E})$ w.r.t. $\tilde{\mathcal{E}}_1^{1/2}$.

(ii) $N \subset E$ is called $\mathcal{E}$–exceptional, if $N \subset E \setminus \bigcup_{k=1}^{\infty} E_k$ for some $\mathcal{E}$–nest $(E_k)_{k\in\mathbb{N}}$. A property of points $x \in E$ is said to hold $\mathcal{E}$–quasi-everwhere (abbreviated $\mathcal{E}$–q.e.) if it holds outside an $\mathcal{E}$–exceptional set.

(iii) A function $f : A \to \mathbb{R}$, $A \subset E$, is called $\mathcal{E}$–quasi-continuous if $f_{|E_k}$ is continuous for all $k \in \mathbb{N}$ for some $\mathcal{E}$–nest $(E_k)_{k\in\mathbb{N}}$ such that $\bigcup_{k=1}^{\infty} E_k \subset A$.

Definition 6.2. $(\mathcal{E}, D(\mathcal{E}))$ is said to be *quasi-regular* if the following conditions hold:

(i) There exists an $\mathcal{E}$–nest $(E_k)_{k\in\mathbb{N}}$ consisting of compact sets.

(ii) There exists an $\tilde{\mathcal{E}}_1^{1/2}$–dense subset of $D(\mathcal{E})$ whose elements have $\mathcal{E}$–quasi-continuous m–versions.

(iii) There exist $u_n \in D(\mathcal{E})$, $n \in \mathbb{N}$, having $\mathcal{E}$–quasi-continuous m–versions $\tilde{u}_n$, $n \in \mathbb{N}$, and an $\mathcal{E}$–exceptional set $N \subset E$ such that $\{\tilde{u}_n \mid n \in \mathbb{N}\}$ separates the points of $E \setminus N$.

Definition 6.3. Let $(\mathcal{E}, D(\mathcal{E}))$ be a Semi–Dirichlet form. A *diffusion process* $\mathbb{M} = (\Omega, \mathcal{F}, (\mathcal{F}_t), (X_t)_{t \geq 0}, (P_x)_{x \in E_\Delta})$ on E with lifetime ζ (i.e., a strong Markov process with continuous sample paths on $[0, \zeta)$) is called *associated to* $(\mathcal{E}, D(\mathcal{E}))$ if for every $f \in L^2(E; m) \cap L^\infty(E; m)$ and any m–version $\hat{f}$ of f and all $\alpha > 0$

$$x \mapsto E_x \left[\int_0^\infty e^{-\alpha t} \hat{f}(X_t) \, dt \right] \tag{6.2}$$

is an $\mathcal{E}$–quasi–continuous m–version of $G_\alpha f := \int_0^\infty e^{-\alpha t} T_t f \, dt$.

Remark 6.4. $\mathbb{M}$ is associated to $(\mathcal{E}, D(\mathcal{E}))$ if and only if for every $f \in L^2(E; m) \cap L^\infty(E; m)$ and any m–version $\hat{f}$ or f and all $t \geq 0$

$$x \mapsto E_x \left[\hat{f}(X_t) \right] \,, \ x \in E \,, \tag{6.3}$$

is an $\mathcal{E}$–quasi–continuous m–version of $T_t f$.

Proof. [MR 92, Ch. IV, Prop. 2.8]. □

Theorem 6.5. *Let* $(\mathcal{E}_\alpha, D(\mathcal{E}_\alpha))$ *be as in Subsections 2.4.1 or 2.4.2 with* $\alpha > 0$ *as defined there. Assume that* E *in Subsect. 2.4.2 is a separable Banach space. Then:*
(i) $(\mathcal{E}_\alpha, D(\mathcal{E}_\alpha))$ *has the local property, i.e.,*

$$\mathcal{E}_\alpha(u, v) = 0 \text{ whenever } u, v \in D(\mathcal{E}) \text{ such that}$$
$$\text{supp } (|u| \cdot m) \cap \text{ supp } (|v| \cdot m) = \emptyset \,,$$

(ii) $(\mathcal{E}_\alpha, D(\mathcal{E}_\alpha))$ *is quasi–regular.*
(iii) Suppose $(\mathcal{E}_\alpha, D(\mathcal{E}_\alpha))$ *is a Semi–Dirichlet form (which is e.g. the case if* $\hat{b} \equiv 0$*). Then there exists a diffusion process* $\mathbb{M}$ *on* E *associated with* $(\mathcal{E}_\alpha, D(\mathcal{E}_\alpha))$*.*

Proof. (i): [MR 92, Ch. V, Examples 1.12 (i)].
(ii): [MR 92, Ch. IV, Sect. 4].
(iii): By the main result in [MR 92, Ch. IV] more precisely its "Semi–Dirichlet form –version" in [MOR 95] we know that quasi–regularity is equivalent to the existence of a so–called m–special standard process $\mathbb{M}$ on E associated with $(\mathcal{E}_\alpha, D(\mathcal{E}_\alpha))$. Because of (i) it follows that $\mathbb{M}$ is indeed a diffusion (cf. [MR 92, Ch. V, Sect. 1]). A proof that $(\mathcal{E}_\alpha, D(\mathcal{E}_\alpha))$ is a Semi–Dirichlet form if $\hat{b} \equiv 0$, can be found in [MR 95, Remark 2.6]. □

Remark 6.6. Even if Theorem 6.5 only gives a process $\mathbb{M}$ with transition semigroup determined by $(T_t^{(\alpha)})_{t \geq 0}$ (cf. Subsect. 2.4), "unkilling" $\mathbb{M}$ by using the multiplicative functional $e^{\alpha t}$, $t \geq 0$, we obtain a process with transition semigroup determined by $(T_t)_{t \geq 0}$.

6.1.2 Non–sectorial cases.

Now we consider the situations of Subsections 4.2.1 and 4.2.2. Let $(T_t^b)_{t\geq 0}$ be the sub–Markovian C^0–semigroups of contractions on $L^1(\mathbb{R}^d;\mu)$ resp. $L^1(E;\mu)$ appearing in Theorems 4.8 and 4.10 respectively.

In [St 96] a theory has been developed, called the *"Theory of Generalized Dirichlet forms"*, which includes the theory of sectorial forms, time dependent versions of it, as well as the cases studied in Subsections 4.2.1 and 4.2.2 above. Also the notion of quasi–regularity extends to this more general framework, yielding (as in the sectorial case in [MR 92]) a complete analytic characterization of all generalized Semi–Dirichlet forms associated with (minimally regular) strong Markov processes (in the sense of Definition 6.3). So, there is a complete analogue of Theorem 6.5 for the semigroups $(T_t^b)_{t>0}$ above (cf. [St 97, Thm. 3.5, Prop. 3.6 resp. Thm. 4.6, Prop. 4.7]). We only mention here that in the situation of Subsect. 4.2.2 we even have that $\zeta = \infty$. The only draw–back is that Remark 6.4 above does not hold in these cases. The function

$$x \mapsto E_x[\hat{f}(X_t)] \, , \ x \in E \, ,$$

is still an m–version of $T_t f$, but in general no longer $\mathcal{E}$–quasi–continuous.

6.2 Solution of the martingale problem

The following is a special case of a general result in [T 97, 98].

Theorem 6.7. *(i) The diffusion process* M *in Theorem 6.5 solves the martingale problem for* (L, D) *with* $D := C_0^\infty(\mathbb{R}^d)$ *resp.* $D := \mathcal{F}C_b^\infty(K)$ *in the situations of Subsections 2.4.1 resp. 2.4.2, i.e., for all* $u \in D$
(a) $\int_0^t Lu(X_s) \, ds$, $t \geq 0$, *is* P_x*–a.s. independent of the* μ*–version for* Lu *for* μ*–a.e.* $x \in E$.
(b) $u(X_t) - u(X_0) - \int_0^t Lu(X_s) \, ds$, $t \geq 0$, *is an* $(\mathcal{F}_t)$*–martingale under* P_x *for* μ*–a.e.* $x \in E$.
(ii) An analogous statement holds for the cases discussed in Subsect. 6.1.2 with L_b *replacing* L.

6.3 Uniqueness

We recall that if $\mathbb{M} = (\Omega, \mathcal{F}, (\mathcal{F}_t)_{t\geq 0}, (X_t)_{t\geq 0}, (P_x)_{x\in E_\Delta})$ is a diffusion process on E, a probability measure is called *subinvariant* for $\mathbb{M}$ if for all $f : E \to \mathbb{R}_+$, $\mathcal{B}(E)$–measurable

$$\int E_x[f(X_t)] \, \mu(dx) \leq \int f(x) \, \mu(dx) \text{ for all } t \geq 0 \, . \tag{6.4}$$

The entire Sect. 4. of these notes was devoted to obtain criteria to ensure that the closure of a diffusion operator (L, D) on $L^1(E; \mu)$ generates a C^0–semigroup on $L^1(E; \mu)$. One application of this is to obtain uniqueness for Markov processes solving the corresponding martingale problem:

Theorem 6.8. *(i) Consider the situation of Theorem 6.7(i). Assume that the closure of (L, D) on $L^1(E; \mu)$ generates a C^0–semigroup on $L^1(E; \mu)$. Let $\mathbb{M}' = (\Omega', \mathcal{F}', (\mathcal{F}'_t)_{t \geq 0}, (X'_t)_{t \geq 0}, (P'_x)_{x \in E_\Delta})$ be a diffusion (or even only a right) process on E solving the martingale problem for (L, D) such that μ is subinvariant for $\mathbb{M}'$ (M from Theorem 6.5 is such a process). Then $x \mapsto E'_x[\hat{f}(X'_t)]$ is a μ–version of $T_t f$ for all $f \in L^1(\mathbb{R}^d; \mu) \cap L^\infty(\mathbb{R}^d; \mu)$ and all μ–versions $\hat{f}$ of f, i.e., $\mathbb{M}' = \mathbb{M}$ up to μ–equivalence.*
(ii) An analogous statement holds for the cases discussed in Subsect 6.1.2 with L_b replacing L.

Proof. [AR 95] and [St 97, Prop. 2.6]. ☐

7. Appendix

7.1 Kolmogorov equations in $L^2(E; \mu)$ for infinite dimensional manifolds E: a case study from continuum statistical mechanics

The purpose of this section is to give an important *"non–flat"* example where the measurable field of Hilbert spaces, to represent an m–symmetric diffusion operator L as in Theorem 3.3, can be constructed explicitly in a natural way giving rise to numerous applications. Let us first describe our framework.

7.1.1 Framework and relevant operators L.

Let X be a connected, oriented C^∞ Riemannian manifold such that $m(X) = \infty$ where m is the volume element. Let $\langle \, , \, \rangle_{TX}$ denote the Riemannian metric and ∇^X, Δ^X the corresponding gradient resp. Laplacian. There is a natural and simple infinite dimensional structure associated with X, the so–called *configuration space* Γ_X over X defined as follows:

$$\Gamma_X := \{\gamma \subset X \mid \gamma \cap K \text{ is a finite set for each compact } K \subset X\} . \quad (7.1)$$

$\gamma \in \Gamma_X$ is identified with the $\mathbb{Z}_+ \cup \{+\infty\}$–valued Radon measure

$$\gamma := \sum_{x \in \gamma} \varepsilon_x , \quad (7.2)$$

where ε_x denotes Dirac measure at x. Γ_X can therefore be endowed with the vague topology. Let $\mathcal{B}(\Gamma_X)$ denote the corresponding Borel σ–algebra.

Functions $f \in C_0^\infty(X)$ on X are *lifted to Γ_X* as follows:

$$\langle f, \gamma \rangle := \sum_{x \in \gamma} f(x) = \int_X f(x) \, \gamma(dx) . \tag{7.3}$$

This gives rise to the following algebra of *test functions* on Γ_X:

$$\mathcal{F}C_b^\infty := \{ g(\langle f_1, \cdot \rangle, \ldots, \langle f_N, \cdot \rangle) \mid N \in \mathbb{N}, \; f_1, \ldots, f_N \in C_0^\infty(X), \; g \in C_b^\infty(\mathbb{R}^N) \} .$$

Typical differential operators of interest on Γ_X are (cf. [AKR 98 a, b], [R 98] and in particular the references there in):

$$
\begin{aligned}
& LF(\gamma) \\
&= \sum_{i,j=1}^N \partial_i \partial_j g(\langle f_1, \gamma \rangle, \ldots, \langle f_N, \gamma \rangle) \int_X \langle \nabla^X f_i(x), \nabla^X f_j(x) \rangle_{T_x X} \; \gamma(dx) \\
&\quad + \sum_{i=1}^N \partial_i g(\langle f_1, \gamma \rangle, \ldots, \langle f_N, \gamma \rangle) \int_X \Delta^X f_i(x) \; \gamma(dx) \\
&\quad - \sum_{i=1}^N \partial_i g(\langle f_1, \gamma \rangle, \ldots, \langle f_N, \gamma \rangle) \sum_{\{x,y\} \subset \gamma} \langle \nabla_x^X \phi(x,y), \nabla^X f_i(x) \rangle_{T_x X} \tag{7.4}
\end{aligned}
$$

where $F = g(\langle f_1, \cdot \rangle, \ldots, \langle f_N, \cdot \rangle) \in \mathcal{F}C_b^\infty$ and $\phi : X \times X \to \mathbb{R}$ is a given function satisfying suitable conditions. Below for simplicity we shall only handle the case $\phi \equiv 0$ and refer for the case with non–trivial *interaction potential* ϕ to [AKR 98 b] and [R 98].

One immediately checks that L in (7.4) is a diffusion operator in the sense of Definition 3.1. There are various ways to find a corresponding natural "differential geometric" structure on Γ_X giving a concrete model for the representation in Theorem 3.3. We shall obtain such a structure by lifting the geometry on X to Γ_X.

7.1.2 A Riemannian–type structure on configuration space.

a) Lifting of flows

Let $v \in V_0(X)$ (:= all smooth vector fields on X with compact support). Let ϕ_t^v, $t \in \mathbb{R}$, be the corresponding flow on X, i.e., the unique solution to

$$\frac{d}{dt} \phi_t^v(x) = v(\phi_t^v(x)) , \quad \phi_0^v(x) = x \in X . \tag{7.5}$$

We define a corresponding *flow on Γ_X* by

$$\phi_t^v(\gamma) := \{ \phi_t^v(x) \mid x \in \gamma \} = \sum_{x \in \gamma} \varepsilon_{\phi_t^v(x)} . \tag{7.6}$$

We immediately get from this the following:

b) Lifting of directional derivatives

On X for $v \in V_0(X)$, $f \in C_0^\infty(X)$ we have

$$\nabla_v^X f(x) := \frac{d}{dt} \, f(\phi_t^v(x))_{|t=0} = \langle \nabla^X f(x), v(x) \rangle_{T_x X} \, . \tag{7.7}$$

The corresponding *directional derivative on* Γ_X is hence for $F = g(\langle f_1, \cdot \rangle, \ldots,$ $\langle f_N, \cdot \rangle) \in \mathcal{F}C_b^\infty$ and $\gamma \in \Gamma_X$

$$\begin{aligned}
\nabla_v^\Gamma F(\gamma) &:= \frac{d}{dt} \, F\left(\phi_t^v(\gamma)\right)_{|t=0} \\
&= \sum_{i=1}^N \partial_i g\left(\langle f_1, \gamma \rangle, \ldots, \langle f_N, \gamma \rangle\right) \int_X \langle \nabla_v^X f(x), v(x) \rangle_{T_x X} \, \gamma(dx) \, .
\end{aligned} \tag{7.8}$$

Comparing (7.7) and (7.8) we obtain:

c) Lifting of gradients and tangent bundle

Define for $F = g(\langle f_1, \cdot \rangle, \ldots, \langle f_N, \cdot \rangle) \in \mathcal{F}C_b^\infty$, $\gamma \in \Gamma_X$,

$$\nabla^\Gamma F(\gamma) := \sum_{i=1}^N \partial_i g(\langle f_1, \gamma \rangle, \ldots, \langle f_N, \gamma \rangle) \nabla^X f_i \quad (\in V_0(X)) \, . \tag{7.9}$$

Then by (7.7), (7.8) for $v \in V_0(X)$

$$\begin{aligned}
\nabla_v^\Gamma F(\gamma) &= \int \langle \nabla^\Gamma F(\gamma)(x), v(x) \rangle_{T_x X} \, \gamma(dx) \\
&= \langle \nabla^\Gamma F(\gamma), v \rangle_{L^2(X \to TX; \gamma)} \, ,
\end{aligned} \tag{7.10}$$

where $L^2(X \to TX; \gamma)$ denotes the space of (γ–classes of) γ–square integrable sections in the tangent bundle $TX = (T_x X)_{x \in X}$. Hence the appropriate "tangent bundle" on Γ_X is given by

$$T_\gamma \Gamma_X := L^2(X \to TX; \gamma) \, , \quad \gamma \in \Gamma_X \, , \tag{7.11}$$

and the metric $\langle \, , \, \rangle_{T_\gamma \Gamma_X}$ is just given by the usual inner product in this vector–valued L^2–space.

7.1.3 L as a μ–Laplacian.

Let $\mathcal{V}\mathcal{F}C_b^\infty$ denote the set of all vector fields V on Γ_X (i.e., all sections in $T\Gamma_X := (T_\gamma \Gamma_X)_{\gamma \in \Gamma_X}$) of the form

$$\gamma \mapsto V(\gamma) = \sum_{i=1}^N F_i(\gamma) \, v_i \quad (\in T_\gamma \Gamma_X), \; F_i \in \mathcal{F}C_b^\infty \, , \tag{7.12}$$

$$v_i \in V_0(X), \; N \in \mathbb{N} \, .$$

For such V define

$$\gamma \mapsto \mathrm{div}^\Gamma V(\gamma) := \sum_{i=1}^N \left(\nabla_{v_i}^\Gamma F_i(\gamma) + F_i(\gamma) \langle \, \mathrm{div}^X v_i, \gamma \rangle \right) \, . \tag{7.13}$$

Note that this is the exact lifting of the divergence on X, since by (7.3) we have necessarily

$$\mathrm{div}^\Gamma v(\gamma) := \langle\, \mathrm{div}^X v, \gamma \,\rangle \,, \ \gamma \in \Gamma_X \,,$$

hence requiring the usual product rule for div^Γ we arrive at (7.13) for vector fields V as in (7.12).

Obviously, for the operator $(L, \mathcal{F}C_b^\infty)$ defined in (7.4) with $\phi \equiv 0$ we have

$$L = \mathrm{div}^\Gamma \nabla^\Gamma =: \Delta^\Gamma \text{ on } \mathcal{F}C_b^\infty \,. \tag{7.14}$$

But so far we have no measure μ so that L is really a μ–Laplacian, i.e., $-\,\mathrm{div}^\Gamma$ is the adjoint of $(\nabla^\Gamma, \mathcal{F}C_b^\infty)$ on $L^2(\Gamma_X; \mu)$, or shortly so that

$$-\left(\nabla^\Gamma\right)^{*,\mu} = \mathrm{div}^\Gamma \text{ on } \mathcal{V}\mathcal{F}C_b^\infty \,. \tag{7.15}$$

In fact, one can characterize all such measures.

Recall there exists a unique measure π_m on $(\Gamma_X, \mathcal{B}(\Gamma_X))$ with Laplace transform

$$\int_{\Gamma_X} e^{<f,\gamma>} \pi_m(d\gamma) = e^{\int_X (e^f - 1)\, dm} \text{ for all } f \in C_0^\infty(X) \tag{7.16}$$

called *(pure) Poisson measure* with intensity (or mean) measure m (cf. e..g. [R 98, Subsect. 2.3]).

Theorem 7.1. *Let μ be a probability measure on $(\Gamma_X, \mathcal{B}(\Gamma_X))$ such that*

$$\int_{\Gamma_X} |\langle f, \gamma\rangle|\, \mu(d\gamma) < \infty \text{ for all } f \in C_0^\infty(X) \tag{7.17}$$

(i.e., the mean measure of μ is Radon). Then the following are equivalent:
(i) $(\nabla^\Gamma)^{*,\mu} = -\,\mathrm{div}^\Gamma$.
(ii) μ is a mixed Poisson measure, i.e., there exists a probability measure λ on $([0,\infty), \mathcal{B}([0,\infty)))$ such that

$$\mu = \int_{[0,\infty)} \pi_{z\cdot m}\, \lambda(dz) \,. \tag{7.18}$$

In particular, in this case for all $F, G \in \mathcal{F}C_b^\infty$

$$-\int_{\Gamma_X} LF(\gamma)\, G(\gamma)\, \mu(d\gamma) = \int_{\Gamma_X} \langle \nabla^\Gamma F(\gamma), \nabla^\Gamma G(\gamma)\rangle_{T_\gamma \Gamma}\, \mu(d\gamma) \,, \tag{7.19}$$

i.e., $L = \mathrm{div}^\Gamma \nabla^\Gamma$ is a μ–Laplacian.

Proof. [R 98, Theorem 2.2] and [AKR 98 b, Theorem 4.1]. $\square$

Remark 7.2. (i) Note that since we a–priori assume (7.17), λ in (7.18) must satisfy

$$\int_{[0,\infty)} z\,\lambda(dz) < \infty\ . \tag{7.20}$$

(ii) Tracing through the proof of [AKR 98 b, Theorem 4.1, (i) $\Rightarrow$ (ii)] one sees that the set of all measures, which have Radon mean and *symmetrize* L, is indeed exactly the set of all mixed Poisson measures with λ satisfying (7.20).

There is a well–known theorem in differential geometry characterizing the volume element m on a Riemannian manifold X as the up to a constant unique Radon measure σ on X so that div^X is the adjoint of ∇^Γ on $L^2(X;\sigma)$. In this sense Theorem 7.1 can be interpreted as follows: it characterizes the mixed Poisson measures with λ satisfying (7.20) as exactly the *volume elements* on our "manifold" Γ_X (with tangent bundle $(T_\gamma \Gamma_X)_{\gamma \in \Gamma}$).

Theorem 7.1 has its analogue for non–trivial interaction potentials ϕ (cf. [AKR 98 b]) with so–called *Ruelle measures* as symmetrizing measures. It also has a number of applications. One concerns ergodicity and will be presented in the next subsection.

7.2 Ergodicity

The results of this subsection are of a very general nature and hold for arbitrary diffusion operators. We shall explain them here only in the ("non–flat"!) situation of the previous subsection. We refer to [Eb 98, Ch. 3, Theorem 3.8] for the general case and to [AKR 97 a, Sect. 5], [AKR 98, Sect. 4], [AKR 98 b, Sect. 6] for cases of various types of Gibbs measures on the lattice or in continuum including *infinite volume* Euclidean quantum fields and also including the situation of Example 3.7 above.

We introduce the following notation

$$\mu_\lambda := \int_{[0,\infty)} \pi_{z\cdot m}\,\lambda(dz)\ , \tag{7.21}$$

where $\pi_{z\cdot m}$ is as defined in (7.16) with $z\cdot m$ replacing m for $z \in [0,\infty)$ and λ is a probability measure on $([0,\infty), \mathcal{B}([0,\infty)))$ satisfying (7.20). Let $\mathcal{M}$ denote the set of all such μ_λ. By Theorem 7.1, $\mathcal{M}$ is exactly the set of all volume elements on Γ_X.

7.2.1 Sobolev spaces on Γ_X.

Fix $\mu_\lambda \in \mathcal{M}$ and consider

$$\mathcal{E}^\Gamma_{\mu_\lambda}(F,G) := \int_{\Gamma_X} \langle \nabla^\Gamma F(\gamma), \nabla^\Gamma G(\gamma)\rangle_{T_\gamma \Gamma_X}\,\mu_\lambda(d\gamma)\,,\quad F,\,G \in \mathcal{F}C_b^\infty\ . \tag{7.22}$$

By Theorem 7.1 (ii) $\Rightarrow$ (i), $(\mathcal{E}^\Gamma_{\mu_\lambda}, \mathcal{F}C^\infty_b)$ is closable on $L^2(\Gamma_X; \mu_\lambda)$. Let $(\mathcal{E}^\Gamma_{\mu_\lambda}, D(\mathcal{E}^\Gamma_{\mu_\lambda}))$ denote its closure. Then we set

$$H^{1,2}_0(\Gamma_X; \mu_\lambda) := D\left(\mathcal{E}^\Gamma_{\mu_\lambda}\right), \tag{7.23}$$

i.e., we consider $D(\mathcal{E}^\Gamma_{\mu_\lambda})$ as the $(1,2)$–Sobolev space on Γ_X, equipped with the inner product

$$\langle F, G \rangle_{H^{1,2}_0(\Gamma_X; \mu_\lambda)} := \int_{\Gamma_X} \langle \nabla^\Gamma F, \nabla^\Gamma G \rangle_{T\Gamma_X} \, d\mu_\lambda + \int FG \, d\mu_\lambda$$
$$\left(= \mathcal{E}^\Gamma_{\mu_\lambda, 1}(F, G)\right); \quad F, G \in H^{1,2}_0(\Gamma_X; \mu_\lambda) \tag{7.24}$$

and corresponding norm $\| \cdot \|_{H^{1,2}_0(\Gamma_X; \mu_\lambda)}$. We are now going to construct an extension of $H^{1,2}_0(\Gamma_X; \mu_\lambda)$ which is the exact analogue on Γ_X of a weak Sobolev space on a finite dimensional manifold. This is more or less a special case of a construction by A. Eberle in [Eb 98, Ch. 3 b)].

We first note that, as is easy to check, the μ_λ–equivalence classes determined by $\mathcal{V}\mathcal{F}C^\infty_b$ (cf. (6.13)) are dense in $L^2(\Gamma_X \to T\Gamma_X; \mu_\lambda)$, i.e., the Hilbert space of (μ–classes of) μ–square integrable sections in $T\Gamma_X$. Let $((\mathrm{div}^\Gamma)^{*,\mu_\lambda}, D((\mathrm{div}^\Gamma)^{*,\mu_\lambda}))$ be the adjoint of $(\mathrm{div}^\Gamma, \mathcal{V}\mathcal{F}C^\infty_b)$ as an operator from $L^2(\Gamma_X \to T\Gamma_X; \mu_\lambda)$ to $L^2(\Gamma_X; \mu_\lambda)$.

Remark 7.3. As an adjoint the operator $((\mathrm{div}^\Gamma)^{*,\mu_\lambda}, D((\mathrm{div}^\Gamma)^{*,\mu_\lambda}))$ is automatically closed, and by definition it is an operator from $L^2(\Gamma_X; \mu_\lambda)$ to $L^2(\Gamma_X \to T\Gamma_X; \mu_\lambda)$. Furthermore, again by definition, $G \in L^2(\Gamma_X; \mu_\lambda)$ belongs to $D((\mathrm{div}^\Gamma)^{*,\mu_\lambda})$ if and only if there exists $V_G \in L^2(\Gamma_X \to T\Gamma_X; \mu_\lambda)$ such that

$$\int G \, \mathrm{div}^\Gamma V \, d\mu_\lambda = -\int \langle V_G, V \rangle_{T\Gamma_X} \, d\mu \quad \text{for all } V \in \mathcal{V}\mathcal{F}C^\infty_b . \tag{7.25}$$

In this case $(\mathrm{div}^\Gamma)^{*,\mu_\lambda} G = V_G$.

Because of Remark 7.3 we set

$$W^{1,2}(\Gamma_X; \mu_\lambda) := D((\mathrm{div}^\Gamma)^{*,\mu_\lambda}), \quad d^{\mu_\lambda} := (\mathrm{div}^\Gamma)^{*,\mu_\lambda} \tag{7.26}$$

and think of $W^{1,2}(\Gamma_X; \mu_\lambda)$ as a *weak* $(1,2)$–*Sobolev space on Γ_X* with inner product

$$\langle F, G \rangle_{W^{1,2}(\Gamma_X; \mu_\lambda)} := \int_{\Gamma_X} \langle d^{\mu_\lambda} F, d^{\mu_\lambda} G \rangle_{T\Gamma_X} \, d\mu_\lambda + \int_{\Gamma_X} FG \, d\mu_\lambda , \tag{7.27}$$
$$F, G \in W^{1,2}(\Gamma_X; \mu_\lambda)$$

and corresponding norm $\| \cdot \|_{W^{1,2}(\Gamma_X; \mu_\lambda)}$. Remark 7.3 immediately implies that

$$\mathcal{F}C^\infty_b \subset W^{1,2}(\Gamma_X; \mu_\lambda) \text{ and } d^{\mu_\lambda} = \nabla^\Gamma \text{ on } \mathcal{F}C^\infty_b , \tag{7.28}$$

hence

$$H_0^{1,2}(\Gamma_X;\mu_\lambda) \subset W^{1,2}(\Gamma_X;\mu_\lambda) \text{ and } d^\mu = \nabla^\Gamma \text{ on } H_0^{1,2}(\Gamma_X;\mu_\lambda) . \quad (7.29)$$

The above construction can be carried out for a large class of probability measures μ on $(\Gamma_X, \mathcal{B}(\Gamma_X))$ (cf. [R 98, Subsect. 4.1]) and in general we *do not* have equality in (7.29) with such μ replacing μ_λ. It is a remarkable fact that we do have equality for mixed Poisson measures μ_λ in many cases.

Proposition 7.4. *Suppose X is complete, and $\lambda = \varepsilon_z$ for some $z \in [0,\infty)$ or Brownian motion on X is conservative. Then*

$$H_0^{1,2}(\Gamma_X;\mu_\lambda) = W^{1,2}(\Gamma_X;\mu_\lambda) . \quad (7.30)$$

Proof. [R 98, Proposition 4.6]. □

7.2.2 Ergodicity and extremality.

Obviously, the set $\mathcal{M}$ introduced at the beginning of Subsect. 7.2 is convex. Let $\mathcal{M}_{ex}$ denote the set of its extreme points.

Remark 7.5. It is not hard to check that

$$\mathcal{M}_{ex} = \{\pi_{z \cdot m} \mid z \in [0,\infty)\} . \quad (7.31)$$

Now we can formulate the main result of Subsect. 7.2.

Theorem 7.6. *Suppose the conditions in Proposition 7.4 hold and let $\mu \in \mathcal{M}$. Then the following are equivalent:*

(i) $\mu \in \mathcal{M}_{ex}$ *(i.e., by Remark 7.5, $\mu = \pi_{z \cdot m}$ for some $z \in [0,\infty)$).*

(ii) $\{\nu \in \mathcal{M} \mid \nu = \rho \cdot \mu$ *for some $\rho : \Gamma_X \to [0,\infty)$, bounded, $\mathcal{B}(\Gamma_X)$-measurable$\} = \{\mu\}$.*

(iii) $(\mathcal{E}_\mu^\Gamma; D(\mathcal{E}_\mu^\Gamma))$ *is irreducible (i.e., $F \in D(\mathcal{E}_\mu^\Gamma)$, $\mathcal{E}_\mu^\Gamma(F,F) = 0$ implies $F = const.$).*

(iv) *If $(L_\mu, D(L_\mu))$ denotes the generator of the Dirichlet form $(\mathcal{E}_\mu^\Gamma, D(\mathcal{E}_\mu^\Gamma))$ (i.e., $(L_\mu, D(L_\mu))$ is the Friedrichs' extension of $(\Delta^\Gamma, \mathcal{F}C_b^\infty)$, cf. (7.14)), then: $F \in D(L_\mu)$, $L_\mu F = 0$ implies $F = const.$ ("uniqueness of ground state").*

(v) $(T_t^\mu)_{t\geq 0} := (e^{tL_\mu})_{t\geq 0}$ *is irreducible, (i.e., if $F \in L^2(\Gamma_X;\mu)$ such that $T_t^\mu(FG) = FT_t^\mu G$ for all $G \in L^\infty(\Gamma_X;\mu)$, $t \geq 0$, then $F = const.$).*

(vi) $F \in L^2(\Gamma_X;\mu)$, $T_t^\mu F = F$, $t \geq 0$, *implies $F = const.$*

(vii) $(T_t^\mu)_{t\geq 0}$ *is L^2-ergodic (i.e., for all $F \in L^2(\Gamma_X;\mu)$*

$$\int \left(T_t^\mu F - \int F \, d\mu \right)^2 d\mu \to 0 \text{ as } t \to \infty) .$$

Remark 7.7. It can be shown that $(\mathcal{E}_\mu, D(\mathcal{E}_\mu))$ as in Theorem 7.6 has a diffusion process associated to it in the sense of Definition 6.3 and that (i) - (vii) above is equivalent to the time–ergodicity of that diffusion if it is started with distribution μ (cf. [R 98, Prop. 3.17]).

Proof. (Proof of Theorem 7.6 cf. [R 98 , Theorem 9.4]).

The equivalence of (iii) – (vii) is standard and is proved e.g. entirely analogously to [AKR 97a, Proposition 2.3]. So, we only prove (i) $\Leftrightarrow$ (ii) $\Leftrightarrow$ (iii).

(i) $\Rightarrow$ (ii): Assume (i) holds. Let $\rho : \Gamma_X \to [0, \infty)$ be bounded and $\mathcal{B}(\Gamma_X)$–measurable such that $\nu := \rho \cdot \mu \in \mathcal{M}$, and let $M := \sup_{\gamma \in \Gamma_X} \rho(\gamma)$. Define

$$\mu_1 := \frac{M - \rho}{M - 1} \mu .$$

Then $\mu_1 \in \mathcal{M}$ and $\mu = \frac{M-1}{M} \mu_1 + \frac{1}{M} \nu$. By assumption (i) it follows that $\rho = 1$, and (ii) is proved.

(ii) $\Rightarrow$ (i): Assume (ii) holds. Let $\mu_1, \mu_2 \in \mathcal{M}$ and $t \in (0, 1)$ such that $\mu = t\mu_1 + (1 - t)\mu_2$. Then they are both absolutely continuous w.r.t. μ with bounded densities. By assumption (ii) it follows that $\mu_1 = \mu = \mu_2$. Consequently $\mu \in \mathcal{M}$.

(ii) $\Rightarrow$ (iii): Assume (ii) holds. Let $G \in H_0^{1,2}(\Gamma_X; \mu)(= D(\mathcal{E}_\mu))$ such that $\mathcal{E}_\mu(G, G) = 0$. Let $G_n := \sup(\inf(G, n), -n)$. Then

$$\lim_{n \to \infty} \|G - G_n\|_{H_0^{1,2}(\Gamma_X; \mu)} = 0$$

(cf. e.g. [MR 92, Ch. I, Prop. 4.17]) and

$$\mathcal{E}_\mu(G_n, G_n) \le \mathcal{E}_\mu(G, G) = 0$$

(cf. e.g. [MR 92, Ch. I, Theorem 4.12]). Since $G =$ const. if all $G_n =$ const., we may assume that G is bounded and, since $1 \in D(\mathcal{E}_\mu)$ and $\mathcal{E}_\mu(1, 1) = 0$, also that $G \ge 0$ (otherwise we add a large constant) and that $\int G \, d\mu = 1$. Define $\nu := G \cdot \mu$. Then, since $\nabla^\Gamma G = 0$ and $\mu \in \mathcal{M}$, we have

$$
\begin{aligned}
\int_{\Gamma_X} \langle \nabla^\Gamma F, V \rangle_{T\Gamma_X} \, d\nu &= \int_{\Gamma_X} \langle \nabla^\Gamma (FG), V \rangle_{T\Gamma_X} \, d\mu \\
&= -\int F \, G \, \mathrm{div}^\Gamma V \, d\mu \\
&= -\int F \, \mathrm{div}^\Gamma V \, d\nu
\end{aligned}
$$

for all $F \in \mathcal{F}C_b^\infty$, $V \in \mathcal{V}\mathcal{F}C_b^\infty$. Hence $\nu \in \mathcal{M}$ and thus by assumption (ii), $G = 1$.

(iii) $\Rightarrow$ (ii): Assume (iii) holds. Let $\rho : \Gamma_X \to [0, \infty)$, $\mathcal{B}(\Gamma_X)$–measurable and bounded so that $\nu := \rho \cdot \mu \in \mathcal{M}$. Then there exist $F_n \in \mathcal{F}C_b^\infty$, $n \in \mathbb{N}$, such that

$$\lim_{n \to \infty} \|\rho - F_n\|_{H_0^{1,2}(\Gamma_X; \mu)} = 0 .$$

Hence for all $V \in \mathcal{V}\mathcal{F}C_b^\infty$, since $\nu \in \mathcal{M}$,

$$\int \mathrm{div}^\Gamma V \, \rho \, d\mu = 0 .$$

It follows by Remark 7.3, (7.25) and (7.26) that

$$\rho \in W^{1,2}(\Gamma_X; \mu) \text{ and } d^\mu \rho = 0 .$$

Proposition 7.4 and (7.29) now imply that $\rho \in H_0^{1,2}(\Gamma_X; \mu)$ and that $\nabla^\Gamma \rho = d^\mu \rho = 0$, i.e., $\mathcal{E}_\mu(\rho, \rho) = 0$. By assumption (iii) we conclude that $\rho \equiv 1$. $\qquad\square$

Remark 7.8. If we are dealing with measures μ where $H_0^{1,2}(\Gamma_X;\mu) \subsetneqq W^{1,2}(\Gamma_X;\mu)$, there is an analogue of Theorem 6.6 where $H_0^{1,2}(\Gamma_X;\mu)$ is replaced by $W^{1,2}(\Gamma_X;\mu)$. We refer to [R 98, Sect. 9] (or [AKR 98 b, Sect. 6]) for details and examples.

References

[ABR 97] S. Albeverio, V.I. Bogachev, M. Röckner: *On uniqueness of invariant measures for finite and infinite dimensional diffusions.* SFB–343–Preprint 1997. To appear in: Commun. Pure and Appl Math., 46 Seiten.

[AKR 96] S. Albeverio, Y.G. Kondratiev, M. Röckner: *Dirichlet operators via stochachstic analysis.* J. Funct. Anal. 128, 102–138 (1995).

[AKR 97a] S. Albeverio, Y.G. Kondratiev, M. Röckner: *Ergodicity of L^2-semigroups and extremality of Gibbs states.* J. Funct. Anal. 144, 394–423 (1997).

[AKR 98] S. Albeverio, Y.G. Kondratiev, M. Röckner: *Ergodicity for the stochastic dynamics of quasi–invariant measures with applications to Gibbs states.* J. Funct. Anal. 149, 415–469 (1997).

[AKR 98a] S. Albeverio, Y.G. Kondratiev, M. Röckner: *Analysis and Geometry on configuration spaces.* J. Funct. Anal. 154, 444–500 (1998).

[AKR 98b] S. Albeverio, Y.G. Kondratiev, M. Röckner: *Analysis and Geometry on configuration spaces. The Gibbsian case.* J. Funct. Anal. 157, 242–291 (1998).

[AKRT 98] S. Albeverio, Y.G. Kondratiev, M. Röckner, T. Tsikalenko: *Existence and exponential moment bounds for symmetrizing measures and applications to Gibbs states.* Preprint (1998). Publication in preparation.

[AR 90] S. Albeverio, M. Röckner: *Dirichlet forms on topological vector spaces – closability and a Cameron–Martin formula.* J. Funct. Anal. 88, 395–436 (1990).

[AR 91] S. Albeverio, M. Röckner: *Stochastic differential equations in infinite dimensions: solutions via Dirichlet forms.* Probab. Th. Rel. Fields 89, 347–386 (1991).

[AR 95] S. Albeverio, M. Röckner: *Dirichlet form methods for uniqueness of martingale problems and applications.* In: Stochastic Analysis. Proceedings of Symposia in Pure Mathematics Vol. 57, 513–528. Editors: M.C. Cranston, M.A. Pinsky. Am. Math. Soc.: Providence, Rhode Island 1995.

[Ar 86] W. Arendt: *The abstract Cauchy problem, special semigroups and perturbation* In: One–parameter semigroups of positive operators, Edited by R. Nagel. Berlin: Springer 1986.

[BDPR 96] V.I. Bogachev, G. Da Prato, M. Röckner: *Regularity of invariant measures for a class of perturbed Ornstein–Uhlenbeck operators.* NoDEA 3, 261–268 (1996).

[BKR 96] V.I. Bogachev, N. Krylov, M. Röckner: *Regularity of invariant measures: the case of non–constant diffusion part.* J. Funct. Anal. 138, 223–242 (1996).

[BKR 97] V.I. Bogachev, N. Krylov, M. Röckner: *Elliptic regularity and essential self–adjointness of Dirichlet operators on $\mathbb{R}^d$.* Ann. Scuola Norm. Sup. Pisa. Cl. Sci., Serie IV, Vol. XXIV. Fasc. 3, 451–461 (1997).

[BR 95] V.I. Bogachev, M. Röckner: *Regularity of invariant measures on finite and infinite dimensional spaces and applications.* J. Funct. Anal. 133, 168–223 (1995).

[BR 95a] V.I. Bogachev, M. Röckner: *Mehler formula and capacities for infinite dimensional Ornstein–Uhlenbeck processes with general linear drift*. Osaka J. Math. 32, 237–274 (1995).

[BR 98] V.I. Bogachev, M. Röckner: *A generalization of Hasminski's theorem on existence of invariant measures for locally integrable drifts*. SFB–343–Preprint 1998. To appear in: Theory Prob. Appl., 18 Seiten.

[BRS 96] V.I. Bogachev, M. Röckner, B. Schmuland: *Generalized Mehler semi-groups and applications*. Probab. Th. Rel. Fields 105, 193–225 (1996).

[BRZ 97] V.I. Bogachev, M. Röckner, T.S. Zhang: *Existence and uniqueness of invariant measures: an approach via sectorial forms*. SFB–343–Preprint 1997. To appear in: Appl. Math. Optim. 28 Seiten.

[Dav 85] E.B. Davies: L^1-*Properties of second order elliptic operators*. Bull London Math. Soc. 17, 417–436 (1985).

[Dav 89] E.B. Davies: *Heat kernels and spectral theory*. Cambridge University Press 1989.

[Eb 97] A. Eberle: *Uniqueness and non–uniqueness of singular diffusion operators*. Doctor–Thesis, Bielefeld 1997, SFB–343–Preprint (1998), 291 pages, publication in preparation.

[EthKur 86] S.N. Ethier, T.G. Kurtz: *Markov processes. Characterization and convergence*. New York: Wiley 1986.

[FR 97] M. Fuhrman, M. Röckner: *Generalized Mehler semigroups: The non–Gaussian case*. FSP–Universität Bielefeld–Preprint 1997, 37 Seiten. To appear in: Potential Analysis.

[Gr 93] L. Gross: *Logarithmic Sobolev inequalities and contractive properties of semigroups*. Lect. Notes Math. 1563, 54–82. Berlin: Springer 1993.

[Ko 37] A.N. Kolmogorov: *Zur Umkehrbarkeit der statistischen Naturgesetze*. Math. Ann. 113, 766–772 (1937).

[L 98] V. Liskevich: *On the uniqueness problem for Dirichlet operators*. Preprint 1998.

[LR 97] V. Liskevich, M. Röckner: *Strong uniqueness for a class of infinite dimensional Dirichlet operators and applications to stochastic quantization*. SFB–343–Preprint 1997. To appear in: Ann. Scuola Norm. di Pisa, 25 Seiten.

[MOR 95] L. Overbeck, Z.M. Ma, M. Röckner: *Markov processes associated with Semi–Dirichlet forms*. Osaka J. Math. 32, 97–119 (1995).

[MR 92] Z.M. Ma, M. Röckner: *An introduction to the theory of (non–symmetric) Dirichlet forms*. Berlin: Springer 1992.

[MR 95] Z.M. Ma, M. Röckner: *Markov processes associated with positivity preserving forms*. Can. J. Math. 47, 817–840 (1995).

[Pa 85] A. Pazy: *Semigroups of linear operators and applications to partial differential equations*. Berlin: Springer 1985.

[ReS 75] M. Reed, B. Simon: *Methods of modern mathematical physics II. Fourier Analysis*. New York – San Francisco – London: Academic Press 1975.

[R 98] M. Röckner: *Stochastic analysis on configuration spaces: basic ideas and recent results*. In: New directions in Dirichlet forms, 157–231. Editors: J. Jost et al. Studies in Advanced Mathematics, International Press, 1998.

[Sh 98] I. Shigekawa: *A non–symmetric diffusion process on the Wiener space*. Preprint 1998.

[St 96] W. Stannat: *The theory of generalized Dirichlet forms and its applications in analysis and stochastics*. Doctor–Thesis, Bielefeld 1996, SFB–343–Preprint (1996), 100 pages. To appear in: Memoirs of the AMS.

[St 97] W. Stannat: *(Nonsymmetric) Dirichlet operators on L^1: existence, uniqueness and associated Markov processes*. SFB–343–Preprint (1997), 38 pages. Publication in preparation.

[Tr 73] N.S. Trudinger: *Linear elliptic operators with measurable coefficients.* Ann. Scuola Normale Sup. Pisa 27, 265–308 (1973).

[T 97] G. Trutnau: *Stochastic calculus of generalized Dirichlet forms and applications to stochastic differential equations in infinite dimensions.* SFB–343–Preprint (1998), 29 pages. Publication in preparation.

[T 98] G. Trutnau: *Doctor–Thesis,* Bielefeld, in preparation.

[Ya 89] J.A. Yan: *Generalizations of Gross' and Minlos' theorems.* In: Azema, J., Meyer, P.A., Yor, M. (eds.) Séminaire de Probabilités. XXII (Lect. Notes Math., vol. 1372, pp. 395–404) Berlin – Heidelberg –New York: Springer 1989.

Parabolic equations on Hilbert spaces

J. Zabczyk

1. Preface

The notes are an expanded version of 8 lectures given to a School on Kolmogorov Equations organized by Prof. G. Da Prato in Cetraro, Italy, August 20– September 2, 1998. The lectures were intended as an introduction to second order parabolic equations on a separable Hilbert space H.

Let L be a second order elliptic operator on R^d:

$$L\varphi(x) = \frac{1}{2} \sum_{i,j=1}^{d} q_{ij}(x) \frac{\partial^2 \varphi}{\partial x_i \partial x_j}(x) + \sum_{i=1}^{d} f_i(x) \frac{\partial \varphi}{\partial x_i}(x), \quad x \in R^d, \qquad (1.1)$$

where, for each $x \in R^d$, $Q(x) = (q_{ij}(x))$ is a nonnegative definite matrix and $F(x) = (f_1(x), \ldots, f_d(x))$ a vector from R^d. Under appropriate regularity conditions imposed on Q, F and a function φ, there exists, see [Fr] and [Kr], a unique solution $u(t,x)$, $t \geq 0$, $x \in R^d$ of the following parabolic equation:

$$\frac{\partial u}{\partial t}(t,x) = Lu(t,x), \quad t > 0, \qquad (1.2)$$
$$u(0,x) = \varphi(x), \quad x \in R^d .$$

The equation 1.2 is one of the most studied equations in mathematics. The theory of its infinite dimensional version, with R^d replaced by a separable Banach space H, was initiated by L. Gross[Gr] and Yu. Daleckij [Dal]. The papers [Gr], [Dal] used, in an essential way, connections between parabolic PDEs and probability theory. Let us recall that the solution u of 1.2 is of the form:

$$u(t,x) = \int_{R^d} p_t(x,y)\varphi(y)\, dy, \quad t > 0, \ x \in R^d. \qquad (1.3)$$

where the positive function $p_t(x,y)$, $t > 0$, $x, y \in R^d$, satisfies the relations,

$$\int_{R^d} p_t(x,y)\, dy = 1, \quad x \in R^d, \qquad (1.4)$$

$$p_{t+s}(x,y) = \int_{R^d} p_t(x,z) p_s(z,y)\, dz, \quad t, s > 0, \ x, y \in R^d. \qquad (1.5)$$

The research supported by the KBN grant No. 2 PO3A 037-16 "Równania paraboliczne w przestrzeniach Hilberta"

Thus the function p, called also the *fundamental solution* of 1.2, is a density of a probability measure, for any $t > 0$ and $x \in R^d$. It was proved by Kolmogorov [Ko] that the fundamental solutions are in fact transition densities of a Markov process and this is why the equation 1.2 is often called the *Kolmogorov equation*. The relation 1.5 is known as the Chapman–Kolmogorov equation. Although the formula 1.3 has a probabilistic meaning it does not allow to *construct* a solution to 1.2. A constructive way, based on the theory of stochastic processes, was proposed by K.Itô [It]. Assume that the diffusion matrix $Q(x)$ can be factorized:

$$Q(x) = G(x)G^*(x), \quad x \in R^d,$$

where $G(x)$, $x \in R^d$ is a $d \times k$ matrix valued function. *Stochastic characteristics* of the equation 1.2 are solutions $X^x(t)$, $t \geq 0$ of the following Itô stochastic equation:

$$dX = F(X)dt + G(X)dW(t), \quad X(0) = x, \tag{1.6}$$

where W is an R^k valued Wiener process on a probability space $(\Omega, \mathcal{F}, \mathbb{P})$. Under rather general conditions imposed on the coefficients the equation 1.6 has a unique solution, which can be constructed by successive approximations. If in addition, the function φ is bounded and smooth then solution to 1.1 is given by the stochastic formula:

$$u(t, x) = \mathbb{E}(\varphi(X^x(t))), \quad t \geq 0, x \in R^d \tag{1.7}$$

in which $\mathbb{E}$ stands for the integral with respect to the measure $\mathbb{P}$. For each $t > 0$ and $x \in R^d$, $p_t(x, \cdot)$ is the density of the distribution of the random variable $X^x(t)$. The Chapman–Kolmogorov equation is a consequence of the Markovian character of the stochastic solutions to 1.6. The regular dependence of the function u on t and x follows from a regular dependence of the process X on t and on the initial data x. This way several results on the regularity and existence of solutions to 1.2 can be obtained by probabilistic methods. Stochastic constructions can be generalized to the infinite dimensional situation and several results in L. Gross[Gr] and in Yu. Daleckij [Dal], as well as in a number of recent papers, were obtained this way. The same approach will be followed in the present notes. Although the infinite dimensional theory is, in many respect, similar to the classical one, there exist also many new infinite dimensional phenomena, which make the subject specially interesting, see e.g. §§3.5 and §5. A different approach to PDEs on Hilbert spaces was proposed by [Le].

The lectures constitute only an *introduction* to the theory and analytical techniques like the continuation method and Schauder estimates in infinite dimensions, see [DaPr2], are not discussed here. We have not covered also important results by A. Piech [Pie] and H.H. Kuo [Ku], as well as variational inequalities on Hilbert spaces see [Za3]. Familiarity with elementary concepts of analysis and functional analysis and in particular with differential calculus on Hilbert and on Banach spaces is taken for granted. Special results on linear operators and on stochastic processes are gathered in the Preliminaries. For more details we often refer to [DaPrZa2] and [DaPrZa4].

The lectures are divided into several chapters. Chapter 2 contains preliminary results on linear operators, probability measures, Wiener process and stochastic integration. Chapter 3 deals with the infinite dimensional heat equation. Majority of the results presented here are due to L. Gross. Chapter 4 is devoted to the semigroup treatment of the parabolic equations with an emphasis on the heat semigroup. The heat equation perturbed by first order terms is the subject of Chapter 5. The existence of solutions is the main subject here. Chapter 6 is on parabolic equations with the state dependent coefficients. It starts from a construction of stochastic characteristics and examines the regularity of the generalized solutions given by 1.7. The uniqueness question is addressed in Chapter 7. Parabolic equations with Dirichlet boundary conditions are discussed in Chapter 8. Applications to nonlinear parabolic equations on Hilbert spaces are treated in Chapter 9. In the same chapter a financial model leading to an infinite dimensional parabolic equation is described. The Appendix is devoted to theorems on implicit functions. They allow to show existence and regular dependence on initial data of solutions to stochastic equations in a simple, functional analytic, way.

Acknowledgements All topics discussed in the notes were an object of numerous discussions with Prof. G. Da Prato and I thank him for his help. Thanks go also to A. Chojnowska-Michalik, B. Goldys, E. Priola and A. Talarczyk, who read some parts of the manuscript and made valuable comments.

2. Preliminaries

Results on linear operators, measures and on stochastic processes, which will be used throughout the lecture notes, are gathered here.

2.1 Linear operators

The space of all linear bounded operators from a Banach space H into a Banach space U, equipped with the operator norm, will be denoted by $L(H, U)$.

2.1.1 Operators of l^p classes. A linear bounded operator $R : H \to H$, defined on a separable Hilbert space H, is said to be *trace class* if it can be represented in a form:

$$Ry = \sum_{k=1}^{+\infty} \langle y, a_k \rangle b_k, \quad y \in H$$

for some sequences (a_k), (b_k) such that

$$\sum_{k=1}^{+\infty} \|a_k\| \, \|b_k\| < +\infty$$

and the trace norm $\|R\|_1$ is the infimum of $\sum_{k=1}^{+\infty} \|a_k\| \, \|b_k\|$ over all possible representations. If an operator R is trace class then its trace, $\text{Tr}\, R$, is defined by the formula:

$$\text{Tr}R = \sum_{j=1}^{+\infty} < Re_j, e_j >,$$

where (e_j) is an orthonormal and complete basis on H. The definition is independent on the choice of the basis and

$$|\text{Tr}R| \le \|R\|_1 .$$

A linear operator $S : H \to H$ is Hilbert–Schmidt if for arbitrary orthonormal and complete basis (e_k)

$$\sum_{k,j=1}^{+\infty} |\langle Se_k, e_j \rangle|^2 < +\infty \tag{2.1}$$

If 2.1 holds for some, then also for all, complete orthonormal basis, and the Hilbert–Schmidt norm $\|S\|_{HS}$ is given by,

$$\begin{aligned}
\|S\|_{HS}^2 &= \sum_{k,j=1}^{+\infty} |\langle Se_k, e_j \rangle|^2 = \sum_{k,j=1}^{+\infty} |\langle e_k, S^* e_j \rangle|^2 \\
&= \sum_{k=1}^{+\infty} \|Se_k\|^2 = \sum_{j=1}^{+\infty} \|S^* e_j\|^2 = \|S^*\|_{HS}^2.
\end{aligned}$$

Proposition 2.1. *Assume that S and T are Hilbert–Schmidt operators on a separable Hilbert space H. Then the operator $R = ST$ is trace class and its trace class norm $\|R\|_1$ satisfies the inequality*

$$\|R\|_1 \le \|S\|_{HS}\|T\|_{HS}$$

Proof. Let (e_k) be a fixed, complete and orthonormal basis, then

$$\begin{aligned}
Ty &= \sum_{k=1}^{+\infty} \langle Ty, e_k \rangle e_k = \sum_{k=1}^{+\infty} \langle y, T^* e_k \rangle e_k \\
STy &= \sum_{R=1}^{+\infty} \langle y, T^* e_k \rangle Se_k.
\end{aligned}$$

Consequently

$$\begin{aligned}
\|ST\|_1 &\le \sum_{k=1}^{+\infty} \|T^* e_k\| \, \|Se_k\| \le \left(\sum_{k=1}^{+\infty} \|T^* e_k\|^2 \right)^{1/2} \left(\sum_{k=1}^{+\infty} \|Se_k\|^2 \right)^{1/2} \\
&\le \|T\|_{HS}\|S\|_{HS}. \quad \square
\end{aligned}$$

More generally let S be a compact operator on a separable Hilbert space H. Denote by (λ_k) the sequence of all positive eigenvalues of the operator $(S^*S)^{1/2}$, repeated according to their multiplicity. Denote by $l^p(H)$, $p > 0$, the set of all operator S such that

$$\|S\|_p = \left(\sum_{k=1}^{+\infty} \lambda_k^p \right)^{1/p} < +\infty \tag{2.2}$$

Operators belonging to $l^1(H)$ and $l^2(H)$ are precisely the trace class and the Hilbert–Schmidt operators. The trace class and the Hilbert-Schmidt norms coincide with $l^1(H)$ and $l^2(H)$ norms. The space $l^2(H)$ is denoted also by $L_{HS}(H)$. The following result holds, see [DS]:

Proposition 2.2. *If $S \in l^p(H)$, $T \in l^q(H)$ and*

$$\frac{1}{r} = \frac{1}{p} + \frac{1}{q}, \quad p, q, r > 0, \tag{2.3}$$

then $ST \in l^r(H)$ and

$$\|ST\|_r \le 2^{1/r} \|S\|_p \|T\|_q. \tag{2.4}$$

2.1.2 Range inclusions and pseudoinverses. The following result is due to R. Douglas [Do] and is a consequence of the closed graph theorem, see [Za2].

Theorem 2.3. *Assume that H_1, H_2, H are Hilbert-spaces and $S_1 \in L(H_1, H)$, $S_2 \in L(H_2, H)$. Then*

$$\text{Range}\, S_1 \subset \text{Range}\, S_2 \tag{2.5}$$

if and only if there exists a constant $c > 0$ such that

$$\|S_1^* h\|_{H_1} \le c \|S_2^* h\|_{H_2}, \quad \text{for all } h \in H. \tag{2.6}$$

If U, H are Hilbert spaces and $S \in L(U, H)$, then for arbitrary $y \in \text{Range}S$, there exists a unique element $\hat{x} \in U$ such that,

$$S(\hat{x}) = y \text{ and if } S(x) = y \text{ then } \|\hat{x}\| \le \|x\|.$$

The element $\hat{x}$ is denoted by $S^{-1}y$ and the transformation S^{-1}, from $\text{Range}S$ into U is linear, called the *pseudoinverse* of S. The graph of S^{-1} is a linear, closed subset of $H \times H$.

2.1.3 Semigroups of linear operators. Any family P_t, $t \geq 0$ of bounded linear operators on a Banach space E such that $P_0 = I$ and

$$P_{t+s} = P_t(P_s), \quad t, s > 0. \tag{2.7}$$

is called a *semigroup* of operators. The semigroup (P_t) is said to be *strongly continuous* or a C_0-semigroup if for arbitrary $\varphi \in E$, the E-valued function

$$t \to P_t\varphi, \quad t \geq 0 \tag{2.8}$$

is continuous at $t = 0$, see [Yo]. It turns out that then the functions given by 2.8 are continuous at all $t \geq 0$ and there exist constants $M > 0$, $\omega \in R^1$, such that,

$$\|P_t\| \leq Me^{\omega t}, \quad t \geq 0. \tag{2.9}$$

The fundamental characteristic of any strongly continuous semigroup (P_t) is its *infinitesimal generator A*. It is a linear, usually unbounded operator, defined on the *domain $D(A)$* consisting of all $\varphi \in E$ such that there exists the limit,

$$\frac{P_t\varphi - \varphi}{t} \quad \text{as } t \downarrow 0.$$

For $\varphi \in D(A)$ one sets,

$$A\varphi = \lim_{t \downarrow 0} \frac{1}{t}(P_t\varphi - \varphi).$$

The set $D(A)$ is a linear and dense subspace of E moreover the graph of the operator $A : \{(\varphi, A\varphi); \ \varphi \in D(A)\}$ is a closed subspace of $E \times E$. If 2.9 holds and $\lambda > \omega$ then the operators

$$A_\lambda = \lambda A(\lambda I - A)^{-1},$$

are bounded and called *Yosida approximations* of A. Moreover if one denotes:

$$P_t^\lambda = e^{tA_\lambda} = \sum_{n=0}^{+\infty} \frac{(tA_\lambda)^n}{n!},$$

then for arbitrary $\varphi \in E$,

$$P_t^\lambda \varphi \to P_t\varphi, \quad \text{as } \lambda \to +\infty,$$

uniformly in t from bounded subsets of R_+^1.

There exists a one-to-one correspondence between C_0-semigroups and their generators however only very seldom one can find an explicit description of the domain $D(A)$ and an explicit formula for A. A linear subspace $K \subset D(A)$ is said to be a *core* of the generator A if K is dense in the space $D(A)$ equipped with the graph norm:

$$\|\varphi\|_{D(A)}^2 = \|\varphi\|^2 + \|A\varphi\|^2.$$

Note that the generator A is completely determined by its values on a core and for many applications the knowledge of the values of A on a core is sufficient. If a linear subspace $K \subset D(A)$ is dense in E and invariant with respect to (P_t) then K is a core for (P_t), see [Da].

2.2 Measures and random variables

We start from recalling basic concepts of the probability theory.

2.2.1 Probability spaces. A *measurable space* $(E, \mathcal{E})$ consists of a set E and of a σ-field $\mathcal{E}$. If μ is a nonnegative measure on a measurable space $(E, \mathcal{E})$ such that $\mu(E) = 1$ then μ is called a *probability measure*, shortly probability, and the triplet $(E, \mathcal{E}, \mu)$ is called a *probability space*, see [Bi]. If $(\Omega, \mathcal{F})$ and $(E, \mathcal{E})$ are two measurable spaces then any measurable transformation X : $\Omega \to E$ is called a *random variable*. Assume that $(\Omega, \mathcal{F}, \mathbb{P})$ is a probability space and X a random variable with values in E. The image μ of the measure $\mathbb{P}$ by the transformation X,

$$\mu(\Gamma) = \mathbb{P}(\omega : X(\omega) \in \Gamma), \quad \Gamma \in \mathcal{E},$$

is called the *law* or the *distribution* of X and denoted by $\mathcal{L}(X)$. If a probability measure μ on $(E, \mathcal{E})$ is given then $(E, \mathcal{E}, \mu)$ is a probability space and the mapping $X(x) = x$, $x \in E$ is called the *canonical* random variable. Its law is identical with the measure μ. Random variables $X_1, \ldots, X_n$ with values in $E_1, \ldots, E_n$ are said to be *independent* if $\mathcal{L}(X_1, \ldots, X_n)$ on $(E_1 \times \ldots \times E_n, \mathcal{E}_1 \otimes \ldots \otimes \mathcal{E}_n)$ is identical with the product $\mathcal{L}(X_1) \times \ldots \times \mathcal{L}(X_n)$ of the laws:

$$\mathbb{P}(X_1 \in \Gamma_1, \ldots, X_n \in \Gamma_n) = \prod_{k=1}^{n} \mathbb{P}(X_k \in \Gamma_k), \quad \Gamma_k \in \mathcal{E}_k, \ k = 1, \ldots, n.$$

An arbitrary family of random variables is said to be independent if any finite sequence of its elements is independent.

Assume that H is a separable Hilbert space equipped with the Borel σ-field $\mathcal{B}(H)$. Probability measures on H will be always regarded as defined on $\mathcal{B}(H)$. If μ is a probability measure on H then its *characteristic function* $\widehat{\mu}$ is a complex valued function on H of the form:

$$\widehat{\mu}(\lambda) = \int_H e^{i\langle \lambda, x \rangle} \mu(dx), \quad \lambda \in H.$$

If characteristic functions of two probability measures μ, ν on H, are identical : $\widehat{\mu}(\lambda) = \widehat{\nu}(\lambda)$, $\lambda \in H$, then also the measures are identical, $\mu = \nu$. Therefore there exists one to one correspondence between probability measures and their characteristic functions.

2.2.2 Gaussian measures. *Gaussian* probability measure on R^1 is either concentrated at one point, say $m \in R^1$ or has a density $\frac{1}{\sqrt{2\pi q}} e^{-\frac{1}{2} \frac{(x-m)^2}{q}}$, $x \in R^1$ where $q > 0$. More generally a measure μ on a Hilbert space H is Gaussian, [Ku], if all linear mappings $y \to \langle \lambda, y \rangle$ defined on H, considered as random variables on $(H, \mathcal{B}(H), \mu)$ with values in $(R^1, \mathcal{B}(R^1))$, have Gaussian laws. Random variables are Gaussian if their laws are Gaussian. Gaussian measures are completely described by the following theorem.

Theorem 2.4. *A measure μ is Gaussian if and only if*

$$\widehat{\mu}(\lambda) = e^{i\langle \lambda, m \rangle - \frac{1}{2}\langle Q\lambda, \lambda \rangle}, \quad \lambda \in H,$$

where $m \in H$ and Q is a self-adjoint, nonnegative, operator with finite trace.

The measure μ from the theorem will be denoted by $N(m, Q)$. If $m = 0$ the measure μ is called *symmetric* or *centered*. Instead of $N(0, Q)$ we will often write N_Q.

Assume that X is a Gaussian random variable $\mathcal{L}(X) = N(m, Q)$ and (e_k) is the complete system of eigenvectors of Q corresponding to the eigenvalues (γ_k):

$$Qe_k = \gamma_k e_k, k = 1, \dots$$

then,

$$\mathbb{E}\langle X, \lambda \rangle = \langle m, \lambda \rangle, \quad \lambda \in H,$$
$$\mathbb{E}\langle X - m, a \rangle \langle X - m, b \rangle = \langle Qa, b \rangle, \quad a, b \in H,$$
$$\mathbb{E}|X - m|^2 = \sum_{k=1}^{+\infty} \gamma_k.$$

Vector m is called the *mean* vector and operator Q the *covariance* of $N(m, Q)$ and of the random variable X.

If $\mathcal{L}(X) = N(m, Q)$ then

$$X = \sum_k \sqrt{\gamma_k} \xi_k e_k + m$$

where $\xi_1, \xi_2, \dots$ are independent random variables with real values, $\mathcal{L}(\xi_n) = N(0, 1)$. In fact one can define :

$$\xi_n = (\gamma_n)^{-\frac{1}{2}}\langle X, e_n \rangle \ n = 1, \dots .$$

The following proposition explains why the operator Q should be of trace class or equivalently why $\sum_k \gamma_k < +\infty$.

Proposition 2.5. *Let (e_k) be an orthonormal sequence in a Hilbert space H, (ξ_k) a sequence of independent Gaussian random variables with $\mathcal{L}(\xi_k) = N(0, 1)$, $k = 1, \dots$ and let γ_k, $k = 1, \dots$ be nonnegative numbers. The series*

$$\sum_k \sqrt{\gamma_k} \xi_k e_k \tag{2.10}$$

converges in H, $\mathbb{P}$-a.s. if and only if $\sum_{k=1}^{+\infty} \gamma_k < +\infty$.

Proof. The series 2.10 converges if and only if

$$\zeta = \sum_k \gamma_k \xi_k^2 < +\infty.$$

Since

$$\mathbb{E}(e^{-\zeta}) = \prod_{k=1}^{+\infty} \mathbb{E}(e^{-\gamma_k \xi_k^2})$$

and

$$\mathbb{E}(e^{-\gamma\xi}) = \frac{1}{\sqrt{1+2\gamma}}, \text{ if } \mathcal{L}(\xi) = N(0,1),$$

one has that

$$\mathbb{E}(e^{-\zeta}) = \frac{1}{\left(\prod_k (1+2\gamma_k)\right)^{1/2}}$$

However

$$\prod_k (1+2\gamma_k) = +\infty \text{ if and only if } \sum_{k=1}^{+\infty} \gamma_k = +\infty.$$

Thus if $\sum \gamma_k = +\infty$ then $\mathbb{E}(e^{-\zeta}) = 0$ and $\mathbb{P}(\zeta = +\infty) = 1$. If $\sum_k \gamma_k < +\infty$ then $\mathbb{E}(\zeta) = \sum_{k=1}^{+\infty} \gamma_k < +\infty$ so $\mathbb{P}(\zeta < +\infty) = 1$. □

As in the finite dimensional spaces, Gaussian measures have exponential moments also in general Hilbert spaces.

Theorem 2.6. *Assume that $\mu = N(0, Q)$. Then for arbitrary $s \in (0, \frac{1}{2 \operatorname{Tr} Q})$,*

$$\int_H e^{s\|x\|^2} \mu(dx) \leq \frac{1}{\sqrt{1 - 2s \operatorname{Tr} Q}} \tag{2.11}$$

Proof. Let (e_k) be a complete orthonormal basis which diagonalizes Q and (λ_k) the corresponding sequence of the eigenvalues of Q. For $s \in (0, \frac{1}{2 \operatorname{Tr} Q})$ and $n \in \mathbb{N}$ define

$$I_n(s) = \int_H e^{s \sum_{k=1}^n \langle x, e_k \rangle^2} \mu(dx). \tag{2.12}$$

Then

$$\begin{aligned}
I_n(s) &= \prod_{k=1}^n \int_H e^{s\langle x, e_k \rangle^2} \mu(dx) \tag{2.13} \\
&= \prod_{k=1}^n \frac{1}{\sqrt{2\pi\lambda_k}} \int_{-\infty}^{+\infty} e^{sy^2 - \frac{y^2}{2\lambda_k}} dy \\
&= \prod_{k=1}^n \frac{1}{\sqrt{1 - 2s\lambda_k}} = e^{-\frac{1}{2} \sum_{k=1}^n \log(1 - 2s\lambda_k)}.
\end{aligned}$$

However if $|y| < 1$ then

$$- \ln(1 - y) = \sum_{m=1}^{+\infty} \frac{y^m}{m}. \tag{2.14}$$

Therefore

$$I_n(s) = e^{\frac{1}{2} \sum_{m=1}^{+\infty} \sum_{k=1}^{n} \frac{(2s\lambda_k)^m}{m}}$$

and consequently, from 2.12 and 2.13,

$$\begin{aligned}
\int_H e^{s\|x\|^2} \mu(dx) &= e^{\frac{1}{2} \sum_{m=1}^{+\infty} \sum_{k=1}^{+\infty} \frac{(2s\lambda_k)^m}{m}} \\
&= e^{\frac{1}{2} \sum_{m=1}^{+\infty} \frac{(2s)^m}{m} \operatorname{Tr}(Q^m)}, \quad s \in \left(0, \frac{1}{2\operatorname{Tr}Q}\right).
\end{aligned}$$

Since $\operatorname{Tr}(Q^m) \leq (\operatorname{Tr}Q)^m$,

$$\begin{aligned}
\int_H e^{s\|x\|^2} \mu(dx) &\leq e^{\frac{1}{2} \sum_{m=1}^{+\infty} \frac{(2s\operatorname{Tr}Q)^m}{m}} \\
&\leq \frac{1}{\sqrt{1 - 2s\operatorname{Tr}Q}}. \quad \square
\end{aligned}$$

As a corollary we have the following result.

Theorem 2.7. *If ξ is a Gaussian random variable with values in a separable Hilbert space H and $\mathbb{E}\xi = 0$ then, for all $s \in (0, \frac{1}{2\mathbb{E}\|\xi\|^2})$,*

$$\mathbb{E}(e^{s\|\xi\|^2}) \leq \frac{1}{\sqrt{1 - 2s\mathbb{E}\|\xi\|^2}}$$

Proof. It is enough to remark that if $\mathcal{L}(\xi) = N(0, Q)$ then $\mathbb{E}\|\xi\|^2 = \operatorname{Tr}Q$. $\square$

We finally recall the so called *Cameron- Martin* formula and *Feldman- Hajek* theorem, see [DaPrZa2] . Denote by $Q^{1/2}$ and $Q^{-1/2}$ the positive square root of Q and its pseudo inverse. If $Qx = 0$ only for $x = 0$ then the pseudo inverse of $Q^{1/2}$ is identical with the usual inverse defined on the domain equal to the image of $Q^{1/2}$. For arbitrary $v \in H$, the function $y \to \langle v, y \rangle$ is a centered Gaussian random variable on the probability space $(H, \mathbb{B}(H), N_Q)$. Let (e_j, γ_j) be the eigensequence corresponding to Q. We set for all $y \in H$,

$$\langle g, Q^{-1/2}y \rangle = \sum_{j=1}^{+\infty} \langle g, e_j \rangle \frac{\langle y, e_j \rangle}{\sqrt{\gamma_j}}. \tag{2.15}$$

The sum in 2.15 converges in $L^2(H, \mathbb{B}(H), N_Q)$ and defines a centered, Gaussian random variable on H such that for all $g, h \in H$,

$$\int_H \langle g, Q^{-1/2}y \rangle \langle h, Q^{-1/2}y \rangle N_Q(dy) = \langle g, h \rangle. \tag{2.16}$$

If $g \in \operatorname{Im}Q^{1/2}$ then $\langle g, Q^{-1/2}y \rangle = \langle Q^{-1/2}g, y \rangle$, $y \in H$. We will need the following well known result.

Theorem 2.8. *For arbitrary $m_1, m_2 \in H$, measures $N(m_1, Q)$, $N(m_2, Q)$ are either equivalent or singular. They are equivalent if and only if $m_1 - m_2 \in \operatorname{Im} Q^{1/2}$ and then Radon–Nikodym derivative is given by the Cameron–Martin formula:*

$$\frac{dN(m_1, Q)}{dN(m_2, Q)}(y) = e^{\langle Q^{-1/2}(m_1 - m_2), Q^{-1/2}(y - m_2)\rangle - \frac{1}{2}|Q^{-1/2}(m_1 - m_2)|^2}. \qquad (2.17)$$

The next theorem is due to Feldman and Hajek.

Theorem 2.9. *For arbitrary vectors $m_1, m_2 \in H$, and arbitrary trace class selfadjoint operators Q_1, Q_2 measures $N(m_1, Q_1)$, $N(m_2, Q_2)$ are either equivalent or singular. They are equivalent if and only if $\operatorname{Im} Q_1^{1/2} = \operatorname{Im} Q_2^{1/2}$, $m_1 - m_2 \in \operatorname{Im} Q_1^{1/2}$, and the operator $(Q_1^{-1/2} Q_2^{1/2})(Q_1^{-1/2} Q_2^{1/2})^* - I$ is Hilbert–Schmidt.*

Thus Theorem 2.9 is a generalization of Theorem 2.8.

2.3 Wiener process and stochastic equations

2.3.1 Infinite dimensional Wiener processes. Let U be a separable Hilbert spaces and $(\Omega, \mathcal{F}, \mathbb{P})$ a probability space with a given increasing family of σ-fields $\mathcal{F}_t \subset \mathcal{F}$, $t \geq 0$. A family $W(t)$, $t \geq 0$ of H-valued random variables is called a *Wiener process*, see [DaPrZa2], if and only if

1) $W(0) = 0$, 2) For almost all $\omega \in \Omega$, $W(t, w)$, $t \geq 0$ is a continuous function, 3) $W(t_1)$, $W(t_2) - W(t_1), \ldots, W(t_n) - W(t_{n-1})$ are independent random variables, $0 \leq t_1 < t_2 < \ldots < t_n$, $n \in \mathbb{N}$ and 4) $\mathcal{L}(W(t) - W(s)) = \mathcal{L}(W(t - s))$, $t \geq s$.

It turns out that if W is a Wiener process then $\mathcal{L}(W(t))$ is a Gaussian measure on U with the mean vector 0:

$$\mathcal{L}(W(t)) = N(0, tQ)$$

where Q is a nonnegative operator with finite trace. Moreover,

$$\mathbb{E}\langle W(t), a\rangle_U \langle W(s), b\rangle_U = t \wedge s \langle Qa, b\rangle_U, \qquad a, b \in U.$$

Let (e_k) be the sequence of all eigenvectors of Q corresponding to the sequence of eigenvalues (γ_k). If $\gamma_k > 0$ then

$$\beta_k(t) = (\gamma_k)^{-\frac{1}{2}} \langle W(t), e_k\rangle_U, \quad t \geq 0,$$

is a 1–dimensional, standard Wiener process ($Q = 1$) and the Wiener processes β_k, $k = 1, \ldots$ are independent. It is clear that

$$W(t) = \sum_{k=1}^{+\infty} \sqrt{\gamma_k} \beta_k(t) e_k, \quad t \geq 0. \qquad (2.18)$$

Conversely, if (e_k) is an orthonormal basis in U, (γ_k) is a summable sequence of nonnegative numbers and (β_k) are independent, standard Wiener processes then the formula 2.18 defines an $U-$ valued Wiener process. We have also the following, instructive result.

Proposition 2.10. *Define*

$$W^N(t,\omega) = \sum_{k=1}^{N} \sqrt{\gamma_k}\beta_k(t,\omega)e_k, \quad N = 1,\ldots \ t \geq 0.$$

Then for arbitrary $T > 0$ there exists a sequence $N_m \to +\infty$ such that $W^{N_m}(\cdot)$ is uniformly convergent on $[0,T]$, for almost all $\omega \in \Omega$.

Proof. Let $N > M$ then

$$\mathbb{E}\left(\sup_{t\leq T}\left|\sum_{k=M+1}^{N} \sqrt{\gamma_k}\beta_k(t)e_k\right|^2\right) = \mathbb{E}\left(\sup_{t\leq T}\sum_{k=M+1}^{N} \gamma_k\beta_k^2(t)\right)$$

$$\leq \sum_{k=M+1}^{N} \gamma_k\mathbb{E}(\sup_{t\in[0,T]}\beta_k^2(t)) \leq C\sum_{k=M+1}^{N}\gamma_k,$$

where $C = \mathbb{E}(\sup_{t\in[0,T]}\beta_1^2(t))$.

Therefore one can find an increasing sequence (N_m) such that,

$$\mathbb{P}\left(\sup_{t\leq T}|W^{N_{m+1}}(t) - W^{N_m}(t)| \geq \frac{1}{2^m}\right) \leq \frac{1}{2^m}, \quad m = 1,\ldots.$$

The result now easily follows. $\square$

Example 2.11. Here is the simplest example of the infinite dimensional Wiener process. Denote by $\mathbb{Z}$ the set of all integers and by $\mathbb{Z}^d$ the d-dimensional lattice. Let $(\beta_\gamma)_{\gamma\in\mathbb{Z}^d}$ be independent real Wiener processes and $W(t) = (\beta_\gamma(t))_{\gamma\in\mathbb{Z}^d}$. Then W is U-valued Wiener process when $U = l_\varrho^2(\mathbb{Z}^d)$ with the weight $\varrho(\gamma) > 0$, $\gamma \in \mathbb{Z}^d$, $\sum_{\gamma\in\mathbb{Z}^d}\varrho(\gamma) < +\infty$. The covariance Q is the multiplication operator

$$(Qx)(\gamma) = \varrho(\gamma)x(\gamma), \quad \gamma \in \mathbb{Z}^d, \ x \in U.$$

Example 2.12. Let $B(s,t)$, $(s,t) \in [0,1] \times [0,1]$ be a Gaussian family such that

$$\mathbb{E}B(s,t)B(u,v) = (s \wedge u)(t \wedge v), \quad (s,t),(u,v) \in [0,1] \times [0,1].$$

The random field B has a continuous version and is called *Brownian sheet*. The $U = L^2(0,1)$ valued process

$$W(t) = B(t,\cdot), \quad t \in (0,1]$$

is a Wiener process with the covariance Q:

$$Qx(\xi) = \int_0^1 (\xi \wedge \eta)x(\eta)\,d\eta.$$

2.3.2 Stochastic integration. There is a natural class of operator valued processes which can be stochastically integrated with respect to a U-valued Wiener process W. Denote by U_W the image of U by the operator $Q^{1/2}$ equipped with the following scalar product:

$$\langle a, b \rangle_{U_W} = \langle Q^{-1/2}a, Q^{-1/2}b \rangle_U \tag{2.19}$$

where $Q^{-1/2}$ denotes the pseudoinverse of $Q^{1/2}$. The space U_W is a separable Hilbert space. Denote by $L_{HS}(U_W, H)$ the Hilbert space of all Hilbert–Schmidt operators from U_W into H equipped with the Hilbert–Schmidt norm. The space $L_{HS}(U_W, H)$ is again a separable Hilbert space which, from now on, will be denoted by $\mathcal{H}$. Note that operators belonging to $L_{HS}(U_W, H)$ are not, in general, defined on the whole space U and that different Wiener processes may lead to the same space $\mathcal{H}$. An $\mathcal{H}$-valued process Ψ is called measurable on $[0, t]$ if it is measurable when treated as a transformation from the set $\Omega \times [0, t]$, equipped with the $\sigma-$ field $\mathcal{F}_t \times \mathcal{B}([0, t])$, into $(\mathcal{H}, \mathcal{B}(\mathcal{H}))$. If the process Ψ, is measurable on all intervals $[0, t], t \geq 0$, then it is called *progressively measurable*. A progressively measurable process Ψ such that

$$\mathbb{P}\left(\int_0^t \|\Psi(s)\|_{\mathcal{H}}^2 \, ds < +\infty, \quad t \geq 0 \right) = 1, \tag{2.20}$$

is called *stochastically integrable*. For stochastically integrable processes the *stochastic Itô integral*

$$\int_0^t \Psi(s) \, dW(s), \quad t \geq 0, \tag{2.21}$$

is well defined, see [DaPrZa2]. Moreover:

$$\mathbb{E}\left(\| \int_0^t \Psi(s) \, dW(s) \|_H^2 \right) \leq \mathbb{E}\left(\int_0^t \|\Psi(s)\|_{\mathcal{H}}^2 \, ds \right). \tag{2.22}$$

If the right-hand side of 2.22 is finite then 2.22 becomes an identity. In addition the following Burkholder–Davis–Gundy inequalities hold. For arbitrary $p > 0$ there exists a constant $c_p > 0$ such that

$$\mathbb{E}\left(\sup_{s \leq t} \| \int_0^s \Psi(u) \, dW(u) \|_H^p \right) \leq c_p \mathbb{E}\left(\int_0^t \|\Psi(s)\|_{\mathcal{H}}^2 \, ds \right)^{p/2}, \quad t \geq 0. \tag{2.23}$$

Assume that Ψ is an $\mathcal{H}$-valued process stochastically integrable on $[0, T]$, ψ is an H-valued process, Bochner integrable on $[0, T]$, $\mathbb{P}$-a.s. and $Z(0)$ is an $\mathcal{F}_0$-measurable H valued random variable. Then the following process

$$Z(t) = Z(0) + \int_0^t \psi(s) \, ds + \int_0^t \Psi(s) \, dW(s), \quad t \in (0, T) \tag{2.24}$$

is continuous and well defined. Let $F : [0, T] \times H \to R^1$ be a function uniformly continuous on bounded subsets of $[0, T] \times H$ together with its partial derivatives F_t, F_x, F_{xx}. The following *Itô's formula* holds.

Theorem 2.13. *Under the above assumptions on the function F and the process Z, $\mathbb{P}$-a.s., for all $t \in [0, T]$,*

$$F(t, Z(t)) = F(0, Z(0)) + \int_0^t \langle F_x(s, Z(s)), \Psi(s)\, dW(s) \rangle_H \qquad (2.25)$$

$$+ \int_0^t \{F_t(s, Z(s)) + \langle F_x(s, Z(s)), \psi(s) \rangle_H$$

$$+ \frac{1}{2} \operatorname{Tr}[(\Psi(s)Q^{1/2})^* F_{xx}(s, Z(s))(\Psi(s)Q^{1/2})]\}\, ds.$$

For more details see [DaPrZa2] .

2.3.3 Stochastic equations. In the lecture notes we will often deal with solutions $X(t, x)$ $t \geq 0$, $x \in H$, of the following stochastic equation,

$$dX = (AX + F(X))\, dt + G(X(t))\, dW(t), \qquad (2.26)$$
$$X(0) = x \in H, \qquad (2.27)$$

on H. In the equation 2.26 – 2.27, F and G are mappings from H into H and from H into $L(U, H)$ respectively and A denotes a linear operator, usually unbounded. We will require that:

The operator A generates a C_0-semigroup $S(t)$, $t \geq 0$, of linear operators on H.

A solution to 2.26–2.27 is a progressively measurable H–valued process $X(t) = X(t, x)$, $t \geq 0$, such that for each $t \geq 0$, $\mathbb{P}$—a.s.:

$$X(t) = S(t)x + \int_0^t S(t-s)F(X(s))\, ds + \int_0^t S(t-s)G(X(s))\, dW(s). \quad (2.28)$$

In particular if $U = R^m$, $H = R^n$ and $A = 0$ the equation 2.26 is the classical ItÔ equation. Note also that if $F = 0$ and $G = 0$ and H is a general Hilbert space then our assumption on A is equivalent to the requirement that the Cauchy problem

$$\frac{dX}{dt} = AX, \quad X(0) = x \qquad (2.29)$$

is well posed.

The equation 2.26 is called *linear* if $F(x) = 0$, $x \in H$ and $G(x) = B$, $x \in H$ for some $B \in L(U, H)$. We write also $X^x(t)$ instead of $X(t, x)$. For more details see [DaPrZa2] .

3. Heat Equation

The chapter is devoted to the existence of solutions to the heat equation on a separable Hilbert space. Regularity of the solutions is studied as well.

3.1 Introduction

We recall first basic formulae for the solutions of the heat equations in R^d and then we pass to the heat equation on a separable Hilbert space H.

Let $Q = (q_{ij})$ be a symmetric, nonnegative definite $d \times d$-matrix. One of the most studied partial differential equations is the heat equation:

$$\frac{\partial u}{\partial t}(t, x) = \frac{1}{2} \sum_{i,j=1}^{d} q_{ij} \frac{\partial^2 u}{\partial x_i \partial x_j}(t, x), \quad t > 0, \; x = (x_1, \ldots, x_d) \in R^d \quad (3.1)$$

$$u(0, x) = \varphi(x), \quad x \in R^d. \tag{3.2}$$

It is well known, see e.g. [Fr] , and [Kr] , that if the matrix Q is invertible and φ is a bounded and continuous function on R^d then there exists a unique, continuous and bounded function u on $[0, +\infty) \times R^d$ such that partial derivatives

$$\frac{\partial u}{\partial t}(x, t) \; , \frac{D^2 u}{\partial x_i \partial x_j}(t, x) i, j = 1, ..., d,$$

are well defined and continuous on $(0, +\infty) \times R^d$ and 3.1 and 3.2 hold. Moreover the solution is given by the following explicit formula:

$$\begin{aligned}
u(t, x) &= \frac{1}{\sqrt{(2\pi t)|\det Q|}} \int_{R^d} e^{-1/2t\langle Q^{-1}(y-x), y-x\rangle} \varphi(y) \, dy \quad (3.3) \\
&= \int_{R^d} \varphi(x + y) n_t(y) dy, \quad t > 0, \; x \in R^d
\end{aligned}$$

where :

$$n_t(y) = \frac{1}{\sqrt{(2\pi t)^d |\det Q|}} e^{-1/2t \langle Q^{-1} y, y\rangle}, \; y \in R^d. \tag{3.4}$$

One can check directly that the functions $p_t(x, y) = n_t(x - y)$, $t > 0, x, y, \in R^d$, satisfy the Chapman- Kolmogorov equation.

It follows from 3.3 that the solution u is a C^∞ function, for each $t > 0$, and depends linearly on the initial function φ. In fact denote by $C_b(R^d)$ and $UC_b(R^d)$ the Banach spaces of all bounded continuous and all bounded and uniformly continuous functions on R^d equipped with the supremum norm. Then for each $t > 0$, the following formula

$$u(t, x) = P_t\varphi(x) = \int_{R^d} \varphi(x + y) n_t(y) dy, \quad x \in R^d, \tag{3.5}$$

defines a linear bounded operator P_t transforming functions φ from $C_b(R^d)$ onto functions in $C_b(R^d)$ and from $UC_b(R^d)$ onto functions in $UC_b(R^d)$. From the Chapman- Kolmogorov equation or from the fact that the formula 3.3 defines a unique solution to 3.1–3.2, it follows that the family P_t, $t > 0$ has the semigroup property:

$$P_{t+s} = P_t(P_s), \quad t, s > 0. \tag{3.6}$$

If one sets $P_0 = I$, then 3.6 is valid for all $t, s \geq 0$.

Basic properties of the solution u can be expressed in terms of the semigroup 3.5 as follows.

Theorem 3.1. i)*Assume that the matrix Q is positive definite. Then the formula 3.5 defines a semigroup of bounded operators on $C_b(R^d)$ and on $UC_b(R^d)$. The semigroup is strongly continuous on $UC_b(R^d)$ but not on $C_b(R^d)$.*

   ii) *For arbitrary $\varphi \in C_b(R^d)$ the function $P_t\varphi(x)$, $t > 0$, $x \in R^d$ is of class C^∞ with respect to variables $t > 0$ and $x \in R^d$.*

We omit the proof as more general results will be proved in the next chapter.

We assume now that Q is only nonnegative definite, symmetric matrix. Let $e_1, \ldots, e_d$ be its eigenvectors and $\gamma_1, \ldots, \gamma_m$, $m \leq d$, its positive eigenvalues. If $m = d$ then Q is invertible and we therefore assume that $m < d$. Note that

$$\sum_{i,j} q_{ij} \frac{\partial^2 u}{\partial x_i \partial x_j}(x) = \operatorname{Tr} Q D^2 u(x), \quad x \in R^d,$$

where $D^2 u(x)$ denotes the second Fréchet derivative of u at x and the trace of a quadratic matrix is the sum of its diagonal elements. Therefore, in the coordinates determined by the eigenvectors of Q,

$$\operatorname{Tr} Q D^2 u(x) = \sum_{i=1}^{m} \gamma_i \frac{\partial^2 u}{\partial x_i^2}(x) . \tag{3.7}$$

The solution $u(t, x)$, $t \geq 0$, $x \in R^d$ of

$$\frac{\partial u}{\partial t}(t, x) = \frac{1}{2} \sum_{i=1}^{m} \gamma_i \frac{\partial^2 u}{\partial x_i^2}(t, x), \quad t > 0, \; x \in R^d , \tag{3.8}$$

$$u(0, x) = \varphi(x) .$$

is given, for $x = (x_1, \ldots, x_m, x_{m+1}, \ldots, x_d)$, by the formula:

$$u(t, x) = \frac{1}{\sqrt{(2t\pi)^m \prod_{k=1}^{m} \gamma_k}} \int_{R^m} e^{-\frac{1}{2t} \sum_{i=1}^{m} \frac{1}{\gamma_i}(x_i - y_i)^2} \tag{3.9}$$

$$\times \varphi(y_1, \ldots, y_m, x_{m+1}, \ldots, x_d) dy_1 \ldots dy_m. \tag{3.10}$$

If φ is in $C_b(R^d)$ then, in general, $u(t, x)$, $t > 0$, $x \in R^d$ is not a C^∞-function. This is clear if, for instance, φ depends only on the variables $x_{m+1}, \ldots, x_d$, say $\varphi(x_1, \ldots, x_d) = \varphi_0(x_{m+1}, \ldots, x_d)$, as then

$$u(t, x) = \varphi_0(x_{d_1+1}, \ldots, x_d), \quad x \in R^d$$

and therefore the solution u has the same regularity as the initial function φ_0. We will see that *the lack of the regularizing power* for degenerate heat equation in R^d will be shared by all heat equations on infinite dimensional Hilbert spaces.

Let finally H be an infinite dimensional Hilbert space and Q a bounded self-adjoint, nonnegative operator on H. As we already know if $H = R^d$ then equation 3.1–3.2 can be written in the following way,

$$\frac{\partial u}{\partial t}(t,x) = \frac{1}{2}\operatorname{Tr} QD^2 u(t,x), \quad t > 0, \ x \in H. \tag{3.11}$$

$$u(0,x) = \varphi(x). \tag{3.12}$$

Note that the formulae 3.3–3.4 for the solutions loose their meaning if $\dim H = +\infty$. They depend on the dimension d of the space and are written in terms of the Lebesgue measure for which an exact counterpart in infinite dimensions does not exist. One way to construct a solution to 3.11–3.12 in the infinite dimensional situation would be to consider a sequence of equations

$$\frac{\partial u_n}{\partial t}(t,x) = \frac{1}{2}\operatorname{Tr} Q_n D^2 u_n(t,x), \quad t > 0, \ x \in H \tag{3.13}$$

$$u_n(0,x) = \varphi(x) \tag{3.14}$$

with finite rank nonnegative operators Q_n converging to Q. For each natural n the equation has a unique solution u_n given by formula (3.9). If the sequence of solutions u_n were convergent then its limit could be taken as a candidate for the solution to 3.11–3.12. The following theorem shows that tó define the solution this way some *restrictions* on the operator Q have to be imposed. Let us recall that if Q is a nonnegative linear operator on a Hilbert space then its trace is given by the formula:

$$\operatorname{Tr} Q = \sum_{k=1}^{+\infty} < Qe_k, e_k >$$

for some (and then for all) complete, orthonormal basis (e_n).

Theorem 3.2. *Assume that* $\varphi \in C_b(H)$ *and* $\lim_{|y| \to +\infty} \varphi(y) = 0$. *If* $\operatorname{Tr} Q = +\infty$ *and* (Q_n) *is a sequence of finite rank, nonnegative operators converging strongly to* Q, *then*

$$\lim_{n \to +\infty} u_n(t,x) = 0, \quad \text{for all } t > 0 \text{ and } x \in H.$$

Proof. It easily follows that $\lim_n \operatorname{Tr} Q_n = +\infty$. Without any loss of generality one can assume that $x = 0$. Let us consider first a special function $\hat{\varphi}$:

$$\hat{\varphi} = e^{-|x|^2}, \quad x \in H .$$

Let $e_1^n, \ldots, e_{m_n}^n$ and $\gamma_1^n, \ldots, \gamma_{m_n}^n$ be the eigenvectors corresponding to all positive eigenvalues of Q_n. Then, by (3.9), for the corresponding solutions $\hat{u}_n$,

$$\hat{u}_n(t,0) = (\prod_{k=1}^{m_n} (1 + 2t\gamma_k^n))^{-\frac{1}{2}} .$$

One can assume that for some $\alpha > 0$

$$2t\gamma_k^n \leq \alpha, \quad k = 1, 2, \ldots, m_n, \ n = 1, 2 \ldots .$$

Then, for a constant $\beta > 0$,

$$\hat{u}_n(t,0) \leq e^{-\beta(\sum_{k=1}^{m_n} \gamma_k^n)} .$$

Since $\operatorname{Tr} Q_n = \sum_{k=1}^{m_n} \gamma_k^n \to +\infty$, the result is true for the special function. If now φ is any function satisfying the conditions of the theorem then, for any $\epsilon > 0$ one can find a $\delta > 0$ and a decomposition

$$\varphi = \varphi_0 + \varphi_1 ,$$

with $\varphi_0, \varphi_1 \in C_b(H)$ such that,

$$|\varphi_0(x)| \leq \delta \, e^{-|x|^2}, \quad |\varphi_1(x)| \leq \epsilon, \quad x \in H .$$

Consequently, for the solutions u_n ,

$$|u_n(t,0)| \leq \delta \, \hat{u}_n(t,0) + \epsilon ,$$

and $\limsup |u_n(t,0)| \leq \epsilon$. Since ϵ is an arbitrary positive number the proof of the theorem is complete.   □

Theorem 3.2 indicates that if $\operatorname{Tr} Q = +\infty$ then, for a majority of initial functions φ, the equation 3.11 *does not have a continuous solution* on $[0, +\infty) \times H$. This is why we will assume that the symmetric nonnegative operator Q is of trace class:

$$\operatorname{Tr} Q < +\infty. \tag{3.15}$$

The same condition has already appeared in the Preliminaries in connection with Gaussian measures.

We go back to the heat equation. It is easy to see that if 3.15 holds and $\varphi \in C_b(H)$ then approximate solutions u_n from Theorem 3.2 are given by the formula:

$$u_n(t,x) = \int_H \varphi(x + y) N_{(tQ_n)}(dy), \quad x \in H, \ t \geq 0. \tag{3.16}$$

They converge, as $n \to +\infty$, to the function u:

$$u(t,x) = \int_H \varphi(x + y) N_{tQ}(dy), \quad x \in H, \ t \geq 0. \tag{3.17}$$

Moreover the operators P_t:

$$P_t\varphi(x) = \int_H \varphi(x+y)N_{tQ}(dy), \quad x \in H, \ \varphi \in C_b(H) \qquad (3.18)$$

form a semigroup of operators. The function

$$u(t,x) = P_t\varphi(x), \quad t \geq 0, \ x \in H \qquad (3.19)$$

where (P_t) is given by 3.18 will be called the *generalized solution to* 3.11. This is in agreement with the finite dimensional case. In fact we have the following counterpart of Theorem 3.1.

Theorem 3.3. *The formula 3.18 defines a semigroup of bounded operators on $C_b(H)$. The semigroup (P_t) is strongly continuous on $UC_b(H)$ but not on $C_b(H)$.*

We omit the proof of the theorem as more general results will be shown in the next chapter. We will call (P_t) the *heat semigroup*.

3.2 Regular initial functions

It is instructive to realize that the heat equation 3.11 is the Kolmogorov equation corresponding to the simplest ItÔ's equation:

$$dX = dW \ , X(0) = x, \qquad (3.20)$$

on a Hilbert space H, where W is a Wiener process on H, with the covariance operator Q, see the Preliminaries. However this fact will be not used in the present chapter.

It will be shown in 3.6 that for some initial functions $\varphi \in UC_b(H)$ the generalized solution $u(t,x)$, $t \geq 0$, $x \in H$ of 3.11 is not Lipschitz in the state variable, for $t > 0$, so it can not satisfy 3.11 in the classical sense. In this subsection we show that if the function φ is more regular then 3.17 defines a classical solution to 3.11.

A function $u(t,x)$, $t \geq 0$, $x \in H$ is said to be a *strict solution* to 3.11 if its first time derivative and its second Fréchet derivative, in the space variable, exist for all $t \geq 0$, $x \in H$, are continuous on $[0, +\infty) \times H$ and satisfy 3.11–3.12. We denote by $UC_b^n(H)$ the space of all real functions on H which are n-times Fréchet differentiable with all derivatives of order $k = 0, 1, \ldots, n$ bounded and uniformly continuous. We have the following lemma.

Lemma 3.4. *If $\varphi \in UC_b^2(H)$, then for arbitrary $t \geq 0$, $u(t,\cdot) \in UC_b^2(H)$ and*

$$Du(t,x) = \int_H D\varphi(x + \sqrt{t}y)N_Q(dy) \qquad (3.21)$$

$$D^2u(t,x) = \int_H D^2\varphi(x + \sqrt{t}y)N_Q(dy), \quad t \geq 0, \ x \in H. \qquad (3.22)$$

Proof. The integral in 3.21 is of Bochner type and in 3.22 is strong Bochner, see [DaPrZa2]. Let $g, h \in H$ then

$$Du(t, x; g) = \int_H \langle D\varphi(x + \sqrt{t}y), g \rangle N_Q(dy)$$

$$D^2 u(t, x; g, h) = \int_H \langle D^2\varphi(x + \sqrt{t}y)g, h \rangle N_Q(dy),$$

by the rules of classical analysis and therefore 3.21–3.22 follow. The uniform continuity of the functions $u(t, \cdot)$, $Du(t, \cdot)$, $D^2 u(t, \cdot)$ is a consequence of the formulae 3.21–3.22.   □

The following theorem is taken from [DaPrZa2].

Theorem 3.5. *If $\varphi \in UC_b^2(H)$, then the function u given by 3.17 is a strict solution to 3.11–3.12.*

Proof. By the mean value theorem

$$u(t, x) = \varphi(x) + \sqrt{t} \int_H \langle D\varphi(x), y \rangle N_Q(dy)$$

$$+ \frac{1}{2} t \int_H \langle D^2\varphi(x + \sigma(y)\sqrt{t}y)y, y \rangle N_Q(dy),$$

where σ is a Borel function from H into $[0, 1]$. Note that

$$\int_H \langle D\varphi(x), y \rangle N_Q(dy) = 0,$$

$$\int_H \langle D^2\varphi(x)y, y \rangle N_Q(dy) = \operatorname{Tr} Q D^2\varphi(x), \quad x \in H$$

and consequently

$$\frac{u(t, x) - u(0, x)}{t} - \frac{1}{2} \operatorname{Tr} Q D^2 u(0, x)$$

$$= \frac{1}{2} \int_H \langle [D^2\varphi(x + \sigma(y)\sqrt{t}y) - D^2\varphi(x)]y, y \rangle N_Q(dy).$$

Therefore

$$\left| \frac{u(t, x) - u(0, x)}{t} - \frac{1}{2} \operatorname{Tr} Q D^2 u(0, x) \right| \leq$$

$$\leq \frac{1}{2} \left[\int_H |D^2\varphi(x + \sigma(y)\sqrt{t}y) - D^2\varphi(x)|^2 N_Q(dy) \right]^{1/2} \left[\int_H |y|^4 N_Q(dy) \right]^{1/2}$$

and

$$\frac{u(t, x) - u(0, x)}{t} \to \frac{1}{2} \operatorname{Tr} Q D^2 u(0, x), \quad \text{as } t \to 0,$$

for all $x \in H$.

This way we have shown that

$$\frac{\partial^+}{\partial t} u(0, x) = \frac{1}{2} \operatorname{Tr} Q D^2 u(0, x), \quad x \in H.$$

Now fix $s > 0$. Since $P_{s+t}\varphi = P_t(P_s\varphi)$, applying the previous argument with φ replaced by $P_s\varphi$, we obtain for all $x \in H$,

$$\frac{\partial^+}{\partial t} u(s, x) = \frac{1}{2} \operatorname{Tr} Q D^2 u(s, x), \quad s \geq 0, \ x \in H. \qquad (3.23)$$

The right hand side of 3.23 is continuous on $[0, +\infty) \times H$. In particular, for every x, the right derivative in time of u is a bounded and continuous function in $s \in [0, +\infty)$. From the elementary calculus, $u(\cdot, x)$ is continuously differentiable and the result follows. $\qquad \square$

3.3 Gross Laplacian

We will see in §3.6 that for $\varphi \in UC_b(H)$ the generalized solution u might be even not Lipschitz in x for positive $t > 0$. However if φ is twice Fréchet differentiable then,

$$\operatorname{Tr} Q D^2\varphi(x) = \sum_{i=1}^{+\infty} \gamma_i \frac{\partial^2 \varphi}{\partial x_i^2}(x), \ x \in H \ ,$$

where γ_i are eigenvalues of Q and $\frac{\partial^2 \varphi}{\partial x_i^2}(x)$ stand for second derivatives of φ in the directions of eigenvectors e_i. Therefore to give a meaning to the right hand side of the heat equation on H one does not need that u is twice Fréchet differentiable in space but only that some of its directional derivatives exist and are properly summable. This way we are lead to the concept of Gross Laplacian, see [Gr].

Let $(B, \|\cdot\|_B)$ be a separable Banach space and G a linear dense subspace of B equipped with a scalar product $\langle\cdot,\cdot\rangle_G$ and the Hilbertian norm: $\|g\|_G = \sqrt{\langle h, g\rangle_G}$, $g \in G$. It is assumed $(G, \|\cdot\|_G)$ is a separable Hilbert space and for some $c > 0$

$$\|g\|_B \leq c\|g\|_G, \quad g \in G. \qquad (3.24)$$

Identifying G with its dual G^* and taking into account that the embedding $G \subset B$ is continuous one can identify B^* with a subset of G. Thus

$$B^* \subset G^* = G \subset B. \qquad (3.25)$$

Let E be another Banach space and u a transformation from B into E. If there exists $T \in L(G, E)$ such that

$$\lim_{\|g\|_G \to 0} \frac{\|u(x + g) - u(x) - Tg\|_E}{\|g\|_G} = 0$$

then T is called the *G-derivative* of u at x and is denoted by $D_G u(x)$. Replacing the space E with $L(G, E)$ one can define the same way $D_G^2 u(x)$ as an element of $L(G, L(G, E))$. Identifying $L(G, L(G, E))$, in the usual way, with the Banach space $L^2(G, E)$ of all bilinear transformations from G into E, one gets that $D_G^2 u(x) \in L^2(G, E)$. In a similar way $D_G^n u(x)$ can be defined and if $D_G^n u(x)$ exists then $D_G^n u(x) \in L^n(G, E)$. Note that in particular if $f_1, \ldots, f_n \in G$ and

$$v(t_1, \ldots, t_n) = u(x + t_1 f_1 + \ldots + t_n f_n), \quad t_1, \ldots, t_u \in R^1,$$

then

$$\frac{\partial^n (0, \ldots, 0)}{\partial t_n \ldots \partial t_1} = D_G^n u(x; f_1, \ldots, f_n). \tag{3.26}$$

If $E = R^1$ then $D_G^2 u(x)$ is said to be of trace class if,

$$\sum_{m=1}^{+\infty} |D_G^2 u(x; g_m, g_m)| < +\infty \tag{3.27}$$

for some complete orthonormal system (g_m). Then 3.26 holds for all orthonormal bases and the *Gross Laplacian* $\Delta_G u(x)$ is defined by the formula

$$\Delta_G u(x) = \sum_{m=1}^{+\infty} D_G^2 u(x; g_m, g_m). \tag{3.28}$$

Let now B be a separable Hilbert space H and $Q : H \to H$ a self–adjoint nonnegative, trace class operator such that $Qh = 0$ only if $h = 0$. Define $G = Q^{1/2}(H)$ and,

$$\langle g_1, g_2 \rangle_G = \langle Q^{-1/2} g_1, Q^{-1/2} g_2 \rangle_H, \quad g_1, g_2 \in G.$$

Then $H^* \subset G^* = G \subset H$ and $H^* = Q(H) = Q^{1/2}(G)$ with the induced norms. Assume that $u : H \to R$ is twice Fréchet differentiable at $x \in H$ with $Du(x)$, $D^2 u(x)$ its first and second Fréchet derivatives. Then for arbitrary $g, g_1, g_2 \in G$:

$$\langle Du(x), g \rangle_H = \langle D_G u(x), g \rangle_G, \tag{3.29}$$

$$\langle D^2 u(x) g_1, g_2 \rangle_H = \langle D_G^2 u(z) g_1, g_2 \rangle_G, \tag{3.30}$$

where bilinear forms $D^2 u(x)$, $D_G^2 u(x)$ were identified with linear operators on H and G respectively. Since

$$\langle D_G u(x), g \rangle_G = \langle Q^{-1/2} D_G u(x), Q^{-1/2} g \rangle_H, \tag{3.31}$$

$$\langle D_G^2 u(x) g_1, g_2 \rangle_G = \langle Q^{-1/2} D_G^2 u(z) g_1, Q^{-1/2} g_2 \rangle_H, \tag{3.32}$$

one arrives at the following relations:

$$D_G u(x) = Q Du(x), \quad D_G^2 u(x) = Q D^2 u(x). \tag{3.33}$$

Moreover if (h_n) is an orthonormal, complete basis in H then $g_m = Q^{1/2}h_m$, $m = 1, \ldots,$ is an orthonormal and complete basis in G and

$$
\begin{aligned}
\Delta_G u(x) &= \sum_{m=1}^{+\infty} \langle D_G^2 u(x) g_m, g_m \rangle_G \qquad (3.34) \\
&= \sum_{m=1}^{+\infty} \langle Q^{-1/2}(QD^2 u(x))Q^{1/2} h_m, Q^{-1/2}Q^{1/2}h_m \rangle_H \\
&= \sum_{m=1}^{+\infty} \langle Q^{1/2} D^2 u(x) Q^{1/2} h_m, h_m \rangle_H.
\end{aligned}
$$

Therefore

$$
\Delta_G u(x) = \text{Trace} Q^{1/2} D^2 u(x) Q^{1/2}, \qquad (3.35)
$$

and for regular functions u the Gross Laplacian is identical, up to the constant 2, with the operator in the right hand side of the heat equation 3.11.

If $G = Q^{1/2}(H)$ we define the Gross derivatives $D_Q u(x) \in H$, $D_Q^2 u(x) \in L(H, H)$ by the formulae

$$
\begin{aligned}
D_Q u(x) &= Q^{-1/2} D_G u(x), \\
D_Q^2 u(x) &= Q^{-1/2} D_G^2 u(x(Q^{1/2}.
\end{aligned}
$$

Note that

$$
\text{Trace } D_G^2 u(x) = \text{Trace } D_Q^2 u(x),
$$

where the traces are calculated with respect to the spaces G and H respectively.

3.4 Heat equation with general initial functions

In the lecture notes we always assume that the space B, from the previous section, is a separable Hilbert space H. This way our presentation is less general than that of L. Gross [Gr], nevertheless it covers interesting applications and allows essential simplifications.

Theorem 3.6. *Let u be the generalized solution corresponding to $\varphi \in B_b(H)$. Then Gross derivatives $D_Q u(t, x)$, $D_Q^2 u(t, x)$ exist for all $t > 0$, $x \in H$ and for $g, h \in H$,*

$$
\langle D_Q u(t, x), g \rangle = \frac{1}{\sqrt{t}} \int_H \langle g, (tQ)^{-1/2} y \rangle \varphi(x + y) N_{tQ}(dy), \qquad (3.36)
$$

$$
\langle D_Q^2 u(t, x) h, g \rangle \qquad (3.37)
$$

$$
= \frac{1}{t} \int_H [\langle g, (tQ)^{-1/2} y \rangle \langle h, (tQ)^{-1/2} y \rangle - \langle g, h \rangle] \varphi(x + y) N_{tQ}(dy).
$$

Moreover

$$\|D_Q u(t,x)\| \leq \frac{1}{\sqrt{t}}\|\varphi\|_0 \tag{3.38}$$

$$\|D_Q^2 u(t,x)\| \leq \frac{2}{\sqrt{t}}\|\varphi\|_0 \tag{3.39}$$

Proof. We will only derive the formula 3.36–3.37. In the following calculations we will use easy to check result that if (ξ,η) is a Gaussian, two dimensional random vector, ζ is a bounded random variable on $(\Omega, F, \mathbb{P})$ and

$$\psi(\alpha,\beta) = \mathbb{E}(e^{\alpha\xi+\beta\eta}), \quad \alpha,\beta \in R^1$$

then ψ is a C^∞ and

$$\frac{\partial\psi}{\partial\alpha}(\alpha,\beta) = \mathbb{E}(\xi\zeta e^{\alpha\xi+\beta\eta}), \quad \frac{\partial\psi}{\partial\beta\partial\alpha}(\alpha,\beta) = \mathbb{E}(\xi\eta\zeta e^{\alpha\xi+\beta\eta}). \tag{3.40}$$

Fix $t > 0$, $x \in H$, vectors $g, h \in H$ and define,

$$\begin{aligned}
v(\alpha,\beta) &= u(t, x + \alpha Q^{1/2}g + \beta Q^{1/2}h) \tag{3.41}\\
&= \int_H \varphi(x + \alpha Q^{1/2}g + \beta Q^{1/2}h + y)N_{tQ}(dy)\\
&= \int_H \varphi(x + y)N(\alpha Q^{1/2}g + \beta Q^{1/2}h, tQ)(dy), \quad \alpha,\beta \in R^1.
\end{aligned}$$

It is clear that

$$\langle D_Q u(t,x), g \rangle = \frac{\partial v}{\partial\alpha}(0,0), \tag{3.42}$$

$$\langle D_Q^2 u(t,x)h, g \rangle = \frac{\partial v}{\partial\beta\partial\alpha}(0,0). \tag{3.43}$$

By the Cameron–Martin formula

$$\begin{aligned}
v(\alpha,\beta) &= \int_H \varphi(x + y)e^{\frac{1}{\sqrt{t}}\langle\alpha g+\beta h,(tQ)^{-1/2}y\rangle - \frac{1}{2t}|\alpha g+\beta h|^2}N_{tQ}(dy) \tag{3.44}\\
&= e^{-\frac{1}{2t}[\alpha^2|g|^2+\beta^2|h|^2+2\alpha\beta\langle g,h\rangle]}\\
&\quad\times \int_H \varphi(x + y)e^{\frac{1}{\sqrt{t}}[\alpha\langle g,(tQ)^{-1/2}y\rangle+\beta\langle h,(tQ)^{-1/2}y\rangle]}N_{tQ}(dy)\\
&= v_1(\alpha,\beta)v_2(\alpha,\beta), \quad \alpha,\beta \in R^1.
\end{aligned}$$

By 3.40

$$\begin{aligned}
\frac{\partial v_1}{\partial\alpha}(\alpha,\beta) &= -\frac{1}{t}[\alpha|g|^2 + \beta\langle g,h\rangle]v_1(\alpha,\beta),\\
\frac{\partial v_2}{\partial\alpha}(\alpha,\beta) &= \frac{1}{\sqrt{t}}\int_H \varphi(x+y)\langle g,(tQ)^{-1/2}y\rangle\\
&\quad\times e^{\frac{1}{\sqrt{t}}[\alpha\langle g,(tQ)^{-1/2}y\rangle+\beta\langle h,(tQ)^{-1/2}y\rangle]}N_{tQ}(dy),\\
\frac{\partial^2 v_2}{\partial\beta\partial\alpha}(\alpha,\beta) &= \frac{1}{t}\int_H \varphi(x+y)\langle g,(tQ)^{-1/2}y\rangle\langle h,(tQ)^{-1/2}y\rangle\\
&\quad\times e^{\frac{1}{\sqrt{t}}[\alpha\langle g,(tQ)^{-1/2}y\rangle+\beta\langle h,(tQ)^{-1/2}y\rangle]}N_{tQ}(dy).
\end{aligned}$$

Since

$$\frac{\partial v}{\partial \alpha}(0,0) = v_1(0,0)\frac{\partial v_2}{\partial \alpha}(0,0),$$

$$\frac{\partial^2 v}{\partial \beta \partial \alpha}(0,0) = v_1(0,0)\frac{\partial^2 v_2}{\partial \beta \partial \alpha}(0,0) + \frac{\partial^2 v_1}{\partial \beta \partial \alpha}(0,0)v_2(0,0),$$

the required formula follows. □

Theorem 3.7. *For arbitrary $\varphi \in B_b(H)$, $t > 0$ and $x \in H$, the operator $D_Q^2 u(t,x)$ is of Hilbert–Schmidt type and*

$$\|D_Q^2 u(t,x)\|_{HS} \leq \frac{\sqrt{2}}{t}\|\varphi\|_0 \qquad (3.45)$$

Proof. Let (e_n) be an orthonormal and complete basis in H. For $n, m \in \mathbb{N}$ define

$$f_{n,m}(y) = \langle e_n, (tQ)^{-1/2}y\rangle\langle e_m, (tQ)^{-1/2}y\rangle - \langle e_n, e_m\rangle, \quad y \in H.$$

Then

$$\|D_Q^2 u(t,x)\|_{HS}^2 = \frac{1}{t^2}\sum_{n,m\in\mathbb{N}}\left|\int_H f_{n,m}(y)\varphi(x+y)N_{tQ}(dy)\right|^2$$

$$= \frac{1}{t^2}\left[\sum_{n\in\mathbb{N}}\left|\int_H f_{n,n}(y)\varphi(x+y)N_{tQ}(dy)\right|^2\right.$$

$$\left. + \sum_{n\neq m}\left|\int_H f_{n,m}(y)\varphi(x+y)N_{tQ}(dy)\right|^2\right].$$

Taking into account that $\langle e_n, (tQ)^{-1/2}(\cdot)\rangle$, $n = 1,2$ are independent, Gaussian random variables on $(H, \mathcal{B}(H), N_{tQ})$ with mean 0 and variances 1 one calculates that

$$\int_H f_{n,m}(y)f_{n',m'}(y)N_{tQ}(dy) = \begin{cases} 0 & \text{if } (n,m) \neq (n',m'), , \\ 1 & \text{if } (n,m) = (n',m'),\ n \neq m, \\ 2 & \text{if } (n,m) = (n',m'),\ n = m \end{cases}$$

Therefore the system of functions $\tilde{f}_{n,m}$ $n,m \in \mathbb{N}$,

$$\tilde{f}_{n,n} = \frac{f_{n,n}}{\sqrt{2}}, n \in \mathbb{N}, \tilde{f}_{n,m} = f_{n,m}, n,m \in \mathbb{N},\ n \neq m,$$

is orthonormal in the Hilbert space $\mathcal{H} = L^2(H, \mathcal{B}(H), N_{tQ})$. Since

$$\|D_Q^2 u(t,x)\|_{HS}^2 \leq \frac{2}{t^2}\sum_{n,m}|(\tilde{f}_{n,m}, \varphi(x+\cdot))_\mathcal{H}|^2,$$

therefore

$$\|D_Q^2 u(t,x)\|_{HS}^2 \leq \frac{2}{t^2}\int_H \varphi^2(x+y)N_{tQ}(dy) \leq \frac{2}{t^2}\|\varphi\|_0^2. \ \square$$

For the theory of the heat equation a slightly stronger result, that the trace norm of $D_Q^2 u(t, x)$ is finite, would be needed. We will show in the final section of the present chapter that even if the initial functions is in $UC_b(H)$ one can not expect that this is true. However if the initial function is more regular, the trace of $D_Q^2 u(t, x)$ is finite, see [Gr].

Theorem 3.8. *If a bounded function φ has a continuous and bounded Fréchet derivative $D\varphi$, then the operator $D_Q^2 u(t, x)$ is trace class for any $t > 0$ and $x \in H$ and*

$$\operatorname{Tr} D_Q^2 u(t, x) = \frac{1}{t} \int_H \langle y, D\varphi(x + y) \rangle N_{tQ}(dy). \tag{3.46}$$

Moreover

$$|\operatorname{Tr} D_Q^2 u(t, x)| \leq \frac{1}{\sqrt{t}} (\operatorname{Tr} Q)^{1/2} \|D\varphi\|_0 \tag{3.47}$$

and the heat equation holds:

$$\frac{\partial}{\partial t} u(t, x) = \frac{1}{2} \operatorname{Tr} D_Q^2 u(t, x), \quad t > 0, \ x \in H. \tag{3.48}$$

Proof. By differentiating expression 3.36 in direction $Q^{1/2}h$ one gets

$$\langle D_Q^2 u(t, x)h, g \rangle = \frac{1}{\sqrt{t}} \int_H \langle g, (tQ)^{-1/2}y \rangle \langle D\varphi(x + y), Q^{1/2}h \rangle N_{tQ}(dy). \tag{3.49}$$

By 3.37 the operator $D_Q^2 u(t, x)$ is self-adjoint. From 3.49,

$$\begin{aligned}
D_Q^2 u(t, x)g &= Q^{1/2} \frac{1}{\sqrt{t}} \int_H D\varphi(x + y) \langle g, (tQ)^{-1/2}y \rangle N_{tQ}(dy) \\
&= Q^{1/2} G g(x), \quad g \in H.
\end{aligned}$$

Since $Q^{1/2}$ is a Hilbert–Schmidt operator to show that $D_Q^2 u(t, x)$ is trace class it is enough to prove that also G is a Hilbert–Schmidt operator. Note that

$$\|G\|_{HS}^2 = \frac{1}{t} \sum_{i,j=1}^{+\infty} \left| \int_H \langle e_j, D\varphi(x + y) \rangle \langle e_i, (tQ)^{-1/2}y \rangle N_{tQ}(dy) \right|^2,$$

where (e_k) is an orthonormal and complete basis in H. However the sequence $(\langle e_k, (tQ)^{-1/2}(\cdot) \rangle)$ is orthonormal in $(H, \mathcal{B}(H), N_{tQ})$ and therefore by the Parseval identity,

$$\|G\|_{HS}^2 \leq \frac{1}{t} \sum_{j=1}^{+\infty} \int_H \langle e_j, D\varphi(x + y) \rangle^2 N_{tQ}(dy).$$

Consequently

$$\|G\|_{HS}^2 \leq \frac{1}{t} \int_H |D\varphi(x + y)|^2 N_{tQ}(dy),$$

and

$$\|D_Q^2 u(t,x)\|_1 \;\leq\; \|Q^{1/2}\|_{HS} \cdot \|G\|_{HS}$$
$$\leq\; (\text{Trace}\,Q)^{1/2} \cdot \frac{1}{\sqrt{t}}\|D\varphi\|_0.$$

This way inequality 3.47 has been established and in particular $D_Q^2 u(t,x)$ is a trace class operator. To show that 3.46 holds we use 3.49. If (e_k, λ_k) is an orthonormal sequence determined by Q, then

$$\sum_k \langle D_Q^2 u(t,x)e_k, e_k\rangle = \frac{1}{\sqrt{t}}\int_H \sum_{k=1}^{+\infty} \langle (tQ)^{-1/2}e_k, y\rangle\langle D\varphi(x+y), Q^{1/2}e_k\rangle] N_{tQ}(dy)$$

$$= \frac{1}{\sqrt{t}}\int_H \left[\sum_{k=1}^{+\infty} \frac{1}{\sqrt{t}}\frac{1}{\sqrt{\lambda_k}}\langle e_k, y\rangle\langle D\varphi(x+y), \sqrt{\lambda_k}e_k\rangle\right] N_{tQ}(dy)$$

$$= \frac{1}{t}\int_H \langle y, D\varphi(x+y)\rangle N_{tQ}(dy)$$

and 3.46 holds. Since

$$u(t,x) \;=\; \int_H \varphi(x + \sqrt{t}y)N_Q(dy),$$

$$\frac{\partial u}{\partial t}(t,x) \;=\; \int_H \left\langle D\varphi(x+\sqrt{t}y), \frac{1}{2\sqrt{t}}y\right\rangle N_Q(dy)$$

$$=\; \frac{1}{2t}\int_H \left\langle D\varphi(x+\sqrt{t}y), \sqrt{t}y\right\rangle N_Q(dy)$$

$$=\; \frac{1}{2t}\int_H \langle D\varphi(x+y), y\rangle N_{tQ}(dy),$$

the equation 3.48 holds by 3.46.   □

Remark 3.9. It is known, see [Ph], that the set of all those points were a Lipschitz continuous functions on Hilbert is not Gateaux differentiable is of measure zero with respect to any nondegenerate Gaussian measure. Using this result one can extend the theorem to Lipschitz φ , see[Gr] for a different proof.

3.5 Generators of the heat semigroups

As we have already said the heat semigroup:

$$P_t\varphi(x) = \int_H \varphi(x+y)N_{tQ}(dy), \quad \varphi \in UC_b(H), \quad x \in H.$$

is C_0 continuous on $UC_b(H)$ In this section we construct cores for this semigroup and show that its infinitesimal generator is equal to the Gross Laplacian. For this purpose we have to introduce several function spaces.

The space $UC_Q^1(H)$ consists of all $\varphi \in UC_b(H)$ such that the first Gross derivative $D_Q\varphi(x)$ is well defined for each $x \in H$ and the mapping $x \to D_Q\varphi(x)$ belongs to $UC_b(H, H)$. In a similar way $C_Q^2(H)$ consists of all those $\varphi \in UC_Q^1(H)$ such that the second Gross derivative $D_Q^2\varphi(x)$ is well defined for all $x \in H$ and the mapping $x \to D_Q^2\varphi(x)$ belongs to $UC_b(H, L(H))$. The spaces $UC_Q^1(H)$, $UC_Q^2(H)$ are Banach with respect to the norms:

$$\|\varphi\|_{1,Q} = \|\varphi\|_0 + \|D_Q\varphi\|_0, \quad \|\varphi\|_{2,Q} = \|\varphi\|_{1,Q} + \|D_Q^2\varphi\|_0.$$

By $L_1(H)$ we denote the space of all trace class operators on H equipped with the trace norm $\|\cdot\|_1$. The following operator A_0:

$$A_0\varphi(x) = \frac{1}{2}\operatorname{Tr} D_Q^2\varphi(x), \quad x \in H \qquad (3.50)$$

defined on the set $D(A_0)$:

$$D(A_0) = \{\varphi \in UC_Q^2(H); \ D_Q^2\varphi \in UC_b(H, L_1(H))\} \qquad (3.51)$$

will play an important role. If (e_k, λ_k) is the eigensequence determined by Q, then

$$
\begin{aligned}
A_0\varphi(x) &= \frac{1}{2}\sum_{k=1}^{+\infty}\langle D_Q^2\varphi(x)e_k, e_k\rangle \qquad (3.52)\\
&= \frac{1}{2}\sum_{k=1}^{+\infty}\lambda_k D_k^2\varphi(x), \quad x \in H,
\end{aligned}
$$

where $D_k^2\varphi(x)$ denotes the second directional derivative of φ at x in the directions e_k.

The following theorem is taken from [Pr].

Theorem 3.10. *Let (P_t) be the heat semigroup on a separable Hilbert space H and A its infinitesimal generator. Then*

i) *A is an extension of A_0,*
ii) *$D(A_0) \cap UC_b^1(H)$ is an invariant subspace for (P_t), dense in $UC_b(H)$,*
iii) *sets $D(A_0)$ and $D(A_0) \cap UC_b^1(H)$ are cores for A.*

Proof. i) We will show that for $\varphi \in D(A_0)$, $t > 0$ and $x \in H$,

$$P_t\varphi(x) = \varphi(x) + \int_0^t P_s(A_0\varphi)(x)ds. \qquad (3.53)$$

Since $P_s(A_0\varphi) \to A_0\varphi$ as $s \downarrow 0$ in $UC_b(H)$, the identity implies that $\varphi \in D(A)$ and $A_0\varphi = A\varphi$. Define,

$$R_n x = \sum_{k=1}^{n} \langle x, e_k \rangle e_k,$$

$$Q_n x = \sum_{n=1}^{n} \lambda_k \langle x, e_k \rangle e_k, \qquad x \in H.$$

Let (P_t^n) be the heat semigroup corresponding to Q_n and A_n its infinitesimal generator. By the finite dimensional theory, $D(A_0) \subset D(A_n)$ and

$$P_t^n \varphi(x) = \varphi(x) + \int_0^t P_s^n (A_n \varphi)(x) ds. \tag{3.54}$$

Moreover, for $\varphi \in UC_b(H)$,

$$P_t^n \varphi(x) = \int_H \varphi(x + R_n y) N_{tQ}(dy)$$

and consequently $P_t^n \varphi(x) \to P_t \varphi(x)$ as $n \to +\infty$. Note also, that

$$\begin{aligned}
A_n \varphi(x) &= \frac{1}{2} \sum_k \langle D_{Q_n}^2 \varphi(x) e_k, e_k \rangle \\
&= \frac{1}{2} \sum_{k=1}^{n} \langle D_Q^2 \varphi(x) e_k, e_k \rangle \\
&\to \frac{1}{2} \sum_{k=1}^{+\infty} \langle D_Q^2 \varphi(x) e_k, e_k \rangle = A_0 \varphi(x), \quad \text{as } n \to +\infty.
\end{aligned}$$

In addition,

$$|A_n \varphi(x)| \le \frac{1}{2} \sum_{k=1}^{+\infty} |\langle D_Q^2 \varphi(x) e_k, e_k \rangle| \le \frac{1}{2} \| D_Q^2 \varphi(x) \|_1, \tag{3.55}$$

as for self-adjoint operators S and arbitrary orthonormal basis (f_k),

$$\sum_{k=1}^{+\infty} |\langle S f_k, f_k \rangle| \le \|S\|_1$$

with the identity taking place if (f_k) is determined by S. Thus the sequence $(A_n \varphi)$ is uniformly bounded. To establish 3.53 it is enough to prove that

$$\lim_{n \to +\infty} \int_0^t P_s^n (A_n \varphi)(x) ds = \int_0^t P_s(A_0 \varphi)(x) ds.$$

Since the sequence of functions $P_s^n (A_n \varphi)(x)$, $s \in [0, t]$ is bounded it is sufficient to verify that for $s \in [0, t]$,

$$\lim_n P_s^n (A_n \varphi)(z) = P_s(A_0 \varphi)(x).$$

However

$$
\begin{aligned}
|P_s^n(A_n\varphi)(x) \quad &- \quad P_s(A_0\varphi)(x)| \le |P_s^n(A_n\varphi)(x) - P_s(A_n\varphi)(x)| \\
&+ |P_s(A_n\varphi)(x) - P_s(A_0\varphi)(x)| \\
\le \quad &\left| \int_H [A_n\varphi(x + R_n y) - A_n\varphi(x + y)] N_{sQ}(dy) \right| \\
&+ \left| \int_H [A_n\varphi(x + y) - A_0\varphi(x + y)] N_{sQ}(dy) \right| \le I_1 + I_2.
\end{aligned}
$$

The term I_2 converges to zero by the Dominated Convergence Theorem. By 3.55, for arbitrary $x, z \in H$ and $n \in \mathbb{N}$,

$$
\begin{aligned}
|A_n\varphi(x) \quad &- \quad A_n\varphi(x + z)| \\
&\le \quad \frac{1}{2} \|D_Q^2\varphi(x) - D_Q^2\varphi(x + z)\|_1.
\end{aligned}
$$

Since $D_Q^2\varphi \in UC_b(H, L_1(H))$, also $I_1 \to 0$, by the same Dominated Convergence Theorem. This way the proof of i) is complete.
The proof of ii) starts from the lemma $\square$

Lemma 3.11. If $\varphi \in UC_b^1(H)$, $t > 0$, $\lambda > 0$ then $P_t\varphi \in D(A_0) \cap UC_b^1(H)$ and $R_\lambda\varphi \in D(A_0) \cap UC_b^1(H)$ where

$$
R_\lambda\varphi(x) = \int_0^{+\infty} e^{-\lambda s} P_s\varphi(x) d.
$$

Proof. It follows from Theorem 3.8 and from its proof that $\|D_Q^2 P_t\varphi(x)\|_1 < +\infty$, $x \in H$ and

$$
\|D_Q^2 P_t\varphi(x) - D_Q^2 P_t\varphi(z)\|_1 \le \frac{1}{\sqrt{t}} (\text{Tr } Q)^{1/2} \int_H (|D\varphi(x+y) - D\varphi(z+y)|^2) N_{tQ}(dy)
$$

So $P_t\varphi \in D(A_0)$. Since

$$
DP_t\varphi(x) = \int_H D\varphi(x + y) N_{tQ}(dy), \quad x \in H,
$$

$P_t\varphi \in UC_b^1(H)$ as well. In exactly the same way one shows that $R_\lambda\varphi \in D(A_0) \cap UC_b^1(H)$.

Lemma 3.11 implies that the subspace $D(A_0) \cap UC_b^1(H)$ is invariant. To show that it is dense in $UC_b(H)$ we use the fact that $D(A)$ is dense in $UC_b(H)$ and take $f \in D(A)$ and set $g = f - Af$. Since $UC_b^1(H)$ is dense in $UC_b(H)$, there exists a sequence (g_n) of elements from $UC_b^1(H)$ such that $g_n \to g$ in $UC_b(H)$. By lemma 3.11,

$$
f_n = R_1 g_n \in D(A_0) \cap UC_b^1(H), \quad n = 1, 2, \dots
$$

Since $f = R_1 g = \lim_n R_1 g_n = \lim_n f_n$, the density of $D(A_0) \cap UC_b^1(H)$ in $D(A)$ and therefore also in $UC_b(H)$ follows. The proof of ii) is complete.
Since $D(A_0) \cap UC_b^1(H)$ is invariant with respect to (P_t) and dense therefore $D(A_0) \cap UC_b^1(H)$ is a core for the generator A. Since $UC_b^1(H) \cap D(A_0) \subset D(A_0) \subset D(A)$ the set $D(A_0)$ is also a core. $\square$

3.6 Nonparabolicity

We present here new results showing that the theory of the heat equation on infinite dimensional spaces differs, in many respects, from that on the finite dimensional one.

If $\varphi \in C_b(H)$ then the function u given by 3.19 is continuous on $[0, +\infty) \times H$. If $H = R^d$ and Q is nondegenerate then u is even a smooth function on $(0, +\infty) \times R^d$. But if the nonnegative matrix Q is degenerate then, as we have seen in §3.1, the solution u might be non differentiable at some points $x \in R^d$ for some $t > 0$. The following theorem shows t hat if dim $H = +\infty$ then for *all* heat equations the generalized solution is not smooth on $(0, +\infty) \times H$, for some $\varphi \in UC_b(H)$. If $\psi \in UC_b(H)$ then $\partial_v \psi(x)$ will denote the directional derivative of ψ at x in the direction v.

Theorem 3.12. *Assume that* dim $H = +\infty$, Trace $Q < +\infty$. *Then for arbitrary* $t > 0$ *and sequence* (x_m) *of elements in* H *there exists* $\varphi \in UC_b(H)$ *such that the function* $P_t\varphi$ *is not Lipschitz in any neigbourhood of any point* x_m, $m \in \mathbb{N}$.

Proof. It follows from the previous considerations that for arbitrary v and arbitrary Borel bounded function φ and arbitrary $x \in H$,

$$\partial_v P_t\varphi(x) = \frac{1}{\sqrt{t}} \int_H \varphi(x+y)\langle Q^{-1/2}, (tQ)^{-1/2}y\rangle N_{tQ}(dy) . \qquad (3.56)$$

Let (v_n) be a sequence of vectors such that: $v_n \in \text{Im}\, Q^{1/2}$, $|v_n| = 1$ and $\lim_n |Q^{-1/2}v_n| = +\infty$. For each $n, m \in \mathbb{N}$ define the following linear functionals $F_{n,m}$ on UC_b ,

$$F_{n,m}(\varphi) = \frac{1}{\sqrt{t}} \int_H \varphi(x+y)\langle Q^{-1/2}v_n, (tQ)^{-1/2}y\rangle N_{tQ}(dy) . \qquad (3.57)$$

Then for the norm $\|F_{n,m}\|$ of $F_{n,m}$,

$$\|F_{n,m}\| = \frac{1}{\sqrt{t}} \int_H |\langle Q^{-1/2}v_n, (tQ)^{-1/2}y\rangle| N_{tQ}(dy) . \qquad (3.58)$$

Since, for each $h \in H$, $y \to \langle h, Q^{-1/2}y\rangle$ is a Gaussian random variable on $(H, B(H), N_Q)$ with mean 0 and the second moment $|h|^2$, one easily gets that

$$\|F_{n,m}\| = \frac{1}{\sqrt{t}} \sqrt{\frac{2}{\pi}} |Q^{-1/2}v_n| . \qquad (3.59)$$

Therefore ,for arbitrary m, $\lim_n \|F_{n,m}\| = +\infty$. Consequently by the Banach-Steinhaus principle of the condensation of singularities, there exists $\varphi \in UC_b$ such that for each m,

$$\sup_n |\partial_{v_n} P_t\varphi(x_m)| = \sup_n |F_{n,m}\varphi| = +\infty . \qquad (3.60)$$

Assume that for some m the function $P_t\varphi$ is Lipschitz in a neighbourhood of x_m. Then there exists $\delta > 0$ such that for $\psi = P_t\varphi$

$$|\psi(x_m + y) - \psi(x_m)| \leq C|y| \quad \text{if} \quad |y| \leq \delta. \tag{3.61}$$

Let $\sigma_n > 0$ be such that $\sigma_n < \delta$ and

$$\frac{|\psi(x_m + \sigma_n v_n) - \psi(x_m)|}{|\sigma_n|} \geq \frac{1}{2}|\partial_{v_n}\psi(x_m)|. \tag{3.62}$$

Then

$$\frac{1}{2}|\partial_{v_n}\psi(x_m))| \leq \frac{C|\sigma_n v_n|}{\sigma_n} \leq C, \quad n = 1, 2, \ldots,$$

a contradiction with 3.60.   □

There are many reasons to *conjecture* that for arbitrary $t > 0$, $x \in H$ there exist initial functions $\varphi \in UC_b(H)$ such that the generalized solution u of the heat equation in the form 3.48 is not satisfied at (t, x). The following theorem is a partial confirmation of the conjecture.

Denote by $C_{fi}^\infty(H)$ the set of all cylindrical functions φ of the form:

$$\varphi(y) = \psi(\langle y, e_1\rangle, \ldots, \langle y, e_n\rangle), \quad y \in H, \tag{3.63}$$

where ψ is any bounded C^∞ function of n variables, $n = 1, 2, \ldots$ and (e_k) is the basis determined by Q. It is clear that $C_{fi}^\infty(H) \subset UC_b(H)$. By direct calculations, or by the finite dimensional theory, $D_Q^2 u(t, x)$, $t > 0$, $x \in H$ is a trace class operator. Let us fix $t > 0$ and $x \in H$ and define on functions 3.63 a linear functional F:

$$F(\varphi) = \text{Tr } D_Q^2 P_t\varphi(x), \quad \varphi \in C_{fi}^\infty(H). \tag{3.64}$$

Theorem 3.13. *The functional F, given by 3.64, can not be extended from $C_{fi}^\infty(H)$ to a continuous functional on $UC_b(H)$.*

Proof. By Proposition 3.8, setting, for simplicity of the notation, $x = 0, t = 1$, one gets that

$$
\begin{aligned}
F(\varphi) &= \text{Tr } D_Q^2 u(1, 0) \\
&= \int_H \sum_{j=1}^n \langle y, e_j\rangle \frac{\partial \psi}{\partial x_j}(\langle y, e_1\rangle, \ldots, \langle y, e_n\rangle) N_Q(dy).
\end{aligned}
$$

Integrating by parts in R^n:

$$
\begin{aligned}
F(\varphi) &= \text{Tr } D_Q^2 u(1, 0) \\
&= \frac{1}{t} \int_H \psi(\langle y, e_1\rangle, \ldots, \langle y, e_n\rangle) \left[\sum_{j=1}^n \left(\frac{\langle y, e_j\rangle^2}{\lambda_j} - 1\right)\right] N_Q(dy).
\end{aligned}
$$

It is therefore clear that the supremum of $F(\varphi)$, with respect to all functions $\varphi \in C_{fi}^\infty(H)$ having the supremum not greater than 1 is equal to,

$$\sup_n \int_H \left| \sum_{j=1}^n \left(\frac{\langle y, e_j \rangle^2}{\lambda_j} - 1 \right) \right| N_Q(dy).$$

Let $\eta_j(y) = \frac{\langle y, e_j \rangle^2}{\lambda_j} - 1$, $j = 1, 2, \ldots, y \in H$. Then the sequence (η_j) consists of independent random variables or $(H, B(H), N_Q)$, identically distributed, with mean value 0 and the finite second moment. It therefore easily follows, for instance by the central limit theorem that,

$$\lim_{n \to +\infty} \int_H |\eta_1 + \ldots + \eta_n| N_Q(dy) = +\infty.$$

This proves the result. □

It is well known that if $\dim H < +\infty$ then the heat semigroups is continuous on $(0, +\infty)$ in the operator norm and is even analytic. It follows from the following theorem that neither of those properties hold if $\dim H = +\infty$. The following theorem is taken from [Gu] and the proof from [NeZa].

Theorem 3.14. If $\dim H = +\infty$ and $\ker Q = \{0\}$ then the heat semigroup is not continuous on $(0, +\infty)$ in the operator norm.

Proof. If $t \neq s$ the operator $(\frac{s}{t-1})I$ is not Hilbert–Schmidt and by the Feldman–Hajek Theorem, the measures $N(x, tQ)$ and $N(x, sQ)$ are singular. Consequently

$$
\begin{aligned}
\|P_t - P_s\| &= \sup_{\substack{\varphi \in UC_b(H) \\ \|\varphi\|_0 \leq 1}} |P_t\varphi - P_s\varphi| \\
&= \sup_{\substack{\varphi \in UC_b(H) \\ \|\varphi\|_0 \leq 1}} \left| \int_H \varphi(y) N(x, tQ)(dy) - \int_H \varphi(y) N(x, sQ)(dy) \right| \\
&= \mathrm{var}(N(x, tQ) - N(x, sQ)) = 2. \square
\end{aligned}
$$

Remark 3.15. It follows from the semigroup theory that analytic semigroups are operator norm continuous on $(0, +\infty)$ consequently the heat semigroup on an infinite dimensional Hilbert space is not analytic, see [DaPrZa2].

4. Transition semigroups

Transition semigroups, in spaces of continuous functions and in a space of square integrable functions are introduced. Their strong continuity is discussed.

4.1 Transition semigroups in the space of continuous functions

As we have already seen semigroups of operators appear naturally in connection with the heat equation. It turns out that the same is true with more general Kolmogorov equations. The corresponding semigroups are called *transition semigroups*. In the first section we investigate their time continuity in the spaces of continuous functions. Although we are mainly interested in the properties of the heat semigroups, several results will be proved in a much more general situation, without an additional effort. In particular some results will be valid for an important class of the *translation invariant* transition semigroups, solving the most general Kolmogorov equation with *constant coefficients*.

Let (M, d) be a metric, separable and complete space. Any function $p(t, x, \Gamma)$, $t \geq 0$, $x \in M$ $\Gamma \in B(M)$, with the following properties:

i) for arbitrary $\Gamma \in B(M)$, $p(\cdot, \cdot, \Gamma)$ is a Borel function from $[0, +\infty) \times M$ into $[0, 1]$,

ii) for arbitrary $(t, x) \in [0, +\infty] \times M$, $p(t, x, \cdot \Gamma)$ is a probability measure on $B(M)$,

iii) for arbitrary $x \in M$, $p(0, x, \cdot) = \delta_{\{x\}}$,

iv) $p(t + s, x, \Gamma) = \int_M p(t, x, dy) p(s, y, \Gamma)$, $t, s \geq 0$, $x \in M$, $\Gamma \in B(M)$,

v) $p(t, x, \cdot) \to \delta_{\{x\}}$ weakly as $t \to 0$,

is called a *transition function* on M.

If p is a transition function then the formula:

$$P_t \varphi(x) = \int_M p(t, x, dy) \varphi(y) dy \, , x \in M \, , \tag{4.1}$$

defines a semigroup of operators on the space $E = B_b(M)$ of bounded Borel functions on M. Semigroups of this form are called *transition semigroups*. For example if M is a Hilbert space H then

$$p(t, x, \Gamma) = N(x, tQ)(\Gamma), \quad t \geq 0, x \in H, \Gamma \in B(H)$$

is a transition function on H determining the heat semigroup. It is a special case of *translation invariant* transition functions for which, by definition,

$$p(t, x, \Gamma) = p(t, x + y, \Gamma + y) \text{ forall } x, y \in H, \Gamma \in B(\mathcal{H}) \, . \tag{4.2}$$

Transition semigroups defined by translation invariant transition functions are called *translation invariant* transition semigroups on the Hilbert space H. We will recall their description here as several results valid for the heat semigroups can be easily extended to them. A family of probability measures μ_t, $t \geq 0$, on a separable Hilbert space H, is called *infinite divisible* if

$$\mu_0 = \delta_{\{0\}} \quad \text{and} \quad \mu_t \to \mu_0 \text{ weakly as } t \downarrow 0 \tag{4.3}$$

$$\mu_{t+s} = \mu_t * \mu_s, \quad \text{for all } t, s \geq 0, \tag{4.4}$$

where $*$ denotes the convolution operation. Then the formula

$$p(t, x, \Gamma) = \mu_t\{\Gamma - x\}, \quad (t, x, \Gamma) \in R_+ \times H \times B(H), \quad (4.5)$$

defines a translation invariant transition function on H. It is clear that also conversely, if p is a translation invariant transition function on H, then the family (μ_t) given by ,

$$\mu_t(\Gamma) = p(t, 0, \Gamma) \, t \geq 0 \, \Gamma \in \mathcal{B}(\mathcal{H}), \quad (4.6)$$

is infinite divisible. Infinite divisible families are completely described by the so called *Levy-Khinchin* formula for the characteristic functions $\hat{\mu}_t(\lambda), \lambda \in H$ of $\mu_t, t \geq 0$. Namely

$$\hat{\mu}_t(\lambda) = \int_H e^{i\langle \lambda, x \rangle} \mu_t(dx) = e^{t\psi(\lambda)}, \quad t \geq 0, \lambda \in H, \quad (4.7)$$

where

$$\psi(\lambda) = i\langle a, \lambda \rangle - \frac{1}{2}\langle Q\lambda, \lambda \rangle + \int_{|x|>1} (e^{i\langle \lambda, x \rangle} - 1)\nu(dx)$$
$$+ \int_{|x|\leq 1} [e^{i\langle \lambda, x \rangle} - 1 - i\langle \lambda.x \rangle]\nu(dx), \quad \lambda \in H.$$

In the formula $a \in H$, Q is a nonnegative, trace class operator on H and ν is a measure concentrated on $H \setminus \{0\}$ such that

$$\nu\{x : |x| > 1\} < +\infty, \quad \int_{|x|\leq 1} |x|^2\nu(dx) < +\infty, \quad (4.8)$$

see [GiSk] and [Pa].

The following theorem, as the other results from the present section, are taken from [TeZa].

Theorem 4.1. *Any translation invariant transition semigroup is strongly continuous on* $UC_b(H)$.

Proof. Assume that $\varphi \in UC_b(H)$, $t > 0$. Then

$$|P_t\varphi(x) - P_t f(z)| = \left| \int_H [\varphi(x + y) - \varphi(z + y)]\mu_t(dy) \right|$$
$$\leq \int_H [\varphi(x + y) - \varphi(z + y)]\mu_t(dy).$$

For $\varepsilon > 0$ there exists $\delta > 0$ such that if $|x - x'| < \delta$ then $|f(x) - f(x')| < \varepsilon$. Thus if $|x - z| < \delta$ then also $|(x + y) - (z + y)| < \delta$ and $|\varphi(x + y) - \varphi(z + y)| < \varepsilon$. This gives $|P_t f(x) - P_t f(z)| < \varepsilon$. In a similar way, $|P_t\varphi(x) - \varphi(x)| = |\int_H [\varphi(x + y) - \varphi(z + y)]\mu_t(dy)|$. For $\varepsilon > 0$ there exists $\delta > 0$ such that $|\varphi(x + y) - \varphi(x)| < \varepsilon$ if $|y| < \delta$ and x arbitrary. Therefore

$$|P_t\varphi(x) - \varphi(x)| \leq \varepsilon \int_{|y|\leq\delta} \mu_t(dy) + 2\|\varphi_0\| \int_{|y|>\delta} \mu_t(dy).$$

Since $\mu_t \to \delta_{\{0\}}$ weakly as $t \downarrow 0$, $P_t\varphi \to \varphi$ uniformly.   □

Only in special cases transition semigroups are strongly continuous on the larger space of all bounded and continuous functions on M. In particular heat semigroups are not strongly continuous on $C_b(H)$, both in finite and infinite dimensional Hilbert spaces H, with the exception of the trivial case when $Q = 0$. This is why the heat semigroup is studied, in the notes, on $UC_b(H)$ only.

Theorem 4.2. *Let (P_t) be a transition semigroup on a metric space M. If there exists a sequence (x_n), of elements in M, which is not totally bounded and such that for each $n \in \mathbb{N}$ and all sufficiently small $t > 0$,*

$$p(t, x_n, E \setminus \{x_n\}) = 1 \tag{4.9}$$

then for arbitrary $\alpha \in (0,1)$, there exists $\varphi \in C_b(M)$ and a sequence $(t_n) \downarrow 0$ such that

$$\sup_{x \in E} |P_{t_n}\varphi(x) - \varphi(x)| \geq \alpha, \quad \text{for all } n \in \mathbb{N}$$

Proof. One can assume, without any loss of generality, that there exists $R > 0$ such that $d(x_n, x_m) > R$, for all $n, m \in \mathbb{N}$. Fix $\varepsilon > 0$. Since p is a transition function we can choose, by property v), a decreasing sequence $(t_n) \downarrow 0$ such that

$$p\left(t, x_n, \left\{x: d(x, x_n) \leq \frac{R}{2}\right\}\right) > 1 - \varepsilon, \quad t \in (0, t_n].$$

Moreover by 4.9, for all $n \in \mathbb{N}$, there exists $r_n \in (0, \frac{R}{2})$ such that

$$p(t_n, x_n, \{x: d(x, x_n) \leq r_n\}) < \varepsilon.$$

Let now φ_n be a continuous function on M with values in $[0, 1]$ such that $\varphi_n(x_n) = 1$ and $\varphi_n(x) = 0$ for all $x \in M$ such that $d(x_n, x) > r_n$. Define

$$\varphi(x) = \sum_{n=1}^{+\infty} \varphi_n(x), \quad x \in M.$$

It is clear that $\varphi \in C_b(M)$, $\varphi(x_n) = 1$ and $\varphi(x) \in [0, 1]$ for all $x \in M$. Moreover

$$
\begin{aligned}
P_{t_n}\varphi(x_n) &= P_{t_n}\varphi_n(x_n) + \sum_{m \neq n} P_{t_n}\varphi_n(x_n) \\
&\leq p(t_n, x_n, \{x: d(x, x_n) \leq r_n\}) + P\left(t_n, x_n, \left\{x: d(x, x_n) > \frac{R}{2}\right\}\right) \\
&< 2\varepsilon, \quad n \in \mathbb{N}
\end{aligned}
$$

and this completes the proof.   □

Remark 4.3. We conjecture that a translation invariant semigroup is strongly continuous on $C_b(H)$ if an only if in the Levy-Khinchin representation : $a = 0, Q = 0$ and the measure ν is finite. That those conditions are sufficient for strong continuity directly follows from an explicit expression for the corresponding infinite divisible family (μ_t).

Also the space $UC_b(H)$ is too large for the strong continuity of several transition semigroups which will be studied in the notes. In the next chapter we will investigate the so called Ornstein-Uhlenbeck semigroups for which the transition function p is of the term:

$$p(t, x, \Gamma) = N(S(t)x, Q_t)(\Gamma), \quad (t, x, \Gamma) \in R_+ \times H \times B(H). \quad (4.10)$$

In 4.10 , $S(t)$, $t \geq 0$, is a C_0-semigroup on H generated by an operator A and Q_t is a nonnegative operator of the form

$$Q_t = \int_0^t S(\sigma)QS^*(\alpha)d\sigma, \quad t \geq 0 \quad (4.11)$$

where Q is a symmetric, bounded nonnegative operator. We assume that

$$\operatorname{Tr} Q_t < +\infty \quad \text{for all } t > 0. \quad (4.12)$$

Only if 4.12 holds the definition 4.10 is meaningful. If $A = 0$ then the Ornstein- Uhlenbeck semigroup is identical with the heat semigroup.

Theorem 4.4. *Assume that (P_t) is an Ornstein -Uhlenbeck semigroup. Then,*

i) *For arbitrary $\varphi \in UC_b(H)$,*

$$P_t\varphi(x) \to \varphi(x) \quad (4.13)$$

uniformly with respect to x in compact sets,

ii) *For arbitrary $\varphi \in UC_b(H)$, 4.13 holds uniformly on bounded sets if and only if the generator A is bounded*

iii) *For arbitrary $\varphi \in UC_b(H)$, 4.13 holds uniformly on H if and only if $A = 0$.*

Proof. One can easily show that,

$$|P_t\varphi(x) - \varphi(S(t)x)| \to 0$$

uniformly with respect to x in the whole H. Therefore one can always reduce the proof to the case of $Q = 0$.

i) If K is compact then $\sup_{x \in K} |S(t)x - x| \to 0$ at $s \downarrow 0$. Thus if $\varphi \in UC_b(H)$ then

$$\sup_{x \in K} |P_t\varphi(x) - \varphi(x)| = \sup_{x \in K} |\varphi(S(t)x) - \varphi(x)| \to 0,$$

as required

ii) It is well known that the generator A is bounded if and only if the convergence $S(t) \to I$ as $t \downarrow 0$ is uniform on bounded sets. Since the convergence is always uniform on compact sets, one can find, if A is unbounded, a number $\alpha > 0$, a sequence $t_n \downarrow 0$ and a sequence $\{x_n\}$ such that, $|x_n| = 1$, $|x_n - x_m| \geq \alpha$, and $|S(t_n)x_n - x_n| \geq \alpha$, $n, m \in \mathbb{N}$, $n \neq m$. Replacing t_n by properly chosen smaller, positive number, we can assume that

$$|S(t_n)x_n - x_n| = \frac{\alpha}{2} \quad \text{and} \quad |S(t_n)x_n - x_m| \geq \frac{\alpha}{2},$$

$n, m \in \mathbb{N}$, $n \neq m$, and $(t_n) \downarrow 0$. Let $\tilde{\varphi} \in UC_b(H)$ be such that $\tilde{\varphi}(0) = 1$ and $\tilde{\varphi}(x) = 0$ if $|x| \geq \frac{\alpha}{2}$. Define $\varphi_n(x) = \tilde{\varphi}(x - x_n)$, $n \in \mathbb{N}$, $x \in H$ and conclude as in the proof of the previous theorem

iii) By ii) we can assume that A is bounded and $A \neq 0$. We can find again $(t_n) \downarrow 0$ such that $S(t_n) \neq I$. Since operators $S(t_n)$ are invertible there exists a sequence (x_n) such that $|x_n - x_m| > 1$, $|S(t_n)x_n - x_m| > 1$ for all $n, m \in \mathbb{N}$, $n \neq m$. The proof can be now completed as in ii).   □

4.2 Transition semigroups in spaces of square summable functions

It is well known, see [Yo], that the heat semigroup P_t on R^d can be uniquely extended from the set of all bounded, Borel functions with compact supports, to the Hilbert space $L^2(R^d)$ of all square summable functions, equipped with the norm:

$$\|\varphi\|_2 = \left(\int_{R^d} |\varphi(x)|^2 dx \right)^{1/2}.$$

The extended semigroup $(\widetilde{P}_t)$ is strongly continuous and consists of contraction operators:

$$\|\widetilde{P}_t \varphi\|_2 \leq \|\varphi\|_2, \quad t \geq 0, \ \varphi \in L^2(R^d).$$

The space $L^2(R^d)$ does not have a natural counterpart if R^d is replaced by an infinite dimensional Hilbert space H, as there are no natural generalizations of the Lebesgue measure to H.

In this section, see [Za3], we introduce a set of measures μ such that transition semigroups, and the heat semigroup in particular, have strongly continuous extensions to the space $\mathcal{H} = L^2(H, \mu)$ of Borel functions φ such that

$$\|\varphi\|_{\mathcal{H}}^2 = \int_H |\varphi(x)|^2 \mu(dx) < +\infty,$$

with the Hilbertian norm $\| \cdot \|_{\mathcal{H}}$.

Let M be a separable metric space equipped with the Borel σ-field $\mathcal{B}(M)$. A measure μ on $\mathcal{B}(M)$ is said to be *locally finite* if there exists an increasing sequence of open sets (U_n) such that,

$$\mu(U_n) < +\infty, \quad n \in \mathbb{N} \text{ and } \bigcup_{n=1}^{+\infty} U_n = M.$$

Let $\omega \geq 0$ be a nonnegative number. A locally finite measure μ on M is said to ω- *excessive*, with respect to a transition semigroup (P_t), if and only if

$$\int_M p(t, x, \Gamma)\mu(dx) \leq e^{\omega t}\mu(\Gamma), \quad \text{forall } t \geq 0, \ \Gamma \in \mathcal{B}(M).$$

Here p stands for the transition function determining (P_t). The 0-excessive measures are called shortly *excessive*.

Theorem 4.5. *If a measure μ is ω-excessive for a transition semigroup (P_t) then the semigroup (P_t) can be uniquely extended from $B_b(M) \cap L^2(M, \mu)$ to a strongly continuous semigroups $(\widetilde{P}_t)$ on $L^2(M, \mu)$. Moreover*

$$\|\widetilde{P}_t\|_{L^2(M,\mu)} \leq e^{\frac{1}{2}\omega t}, \quad t \geq 0 \tag{4.14}$$

Proof. Denote $\mathcal{H} = L^2(M, \mu)$ and assume that $\varphi \in B_b(M) \cap \mathcal{H}$. Then

$$\|P_t\varphi\|_{\mathcal{H}}^2 = \int_M \left| \int_M p(t, x, dy)\varphi(y) \right|^2 \mu(dx).$$

By the Schwartz inequality,

$$\left| \int_M p(t, x, dy)\varphi(y) \right|^2 \leq \left(\int_M p(t, x, dy)1^2 \right)\left(\int_M p(t, x, dy)|\varphi(y)|^2 \right), \quad x \in M.$$

Consequently

$$\|P_t\varphi\|_{\mathcal{H}}^2 \leq \int_M \left[\int_M p(t, x, dy)|\varphi(y)|^2 \right]\mu(dx) \tag{4.15}$$

$$\leq e^{\omega t} \int_M |\varphi(y)|^2 \mu(dy).$$

The second estimate obviously holds if the function $|\varphi|^2$ is replaced by an indicator function of any Borel subset. Thus, by a standard limiting procedure, that estimate holds for an arbitrary nonnegative measurable function and in conclusion for $|\varphi|^2$. It follows easily from 4.15 that the extension $(\widetilde{P}_t)$ exists and that 4.14 holds. It remains to show that,

$$\lim_{t \to 0} \|\widetilde{P}_t\varphi - \varphi\|_{\mathcal{H}} = 0, \quad \text{for all } \varphi \in \mathcal{H}. \tag{4.16}$$

Assume first that $\varphi \in \mathcal{H}$ is a bounded continuous function such that for some $n \in \mathbb{N}$, $\varphi = 0$ on U_n^c. Then

$$\|P_t\varphi - \varphi\|_{\mathcal{H}}^2 = \int_M |P_t\varphi(x) - \varphi(x)|^2\mu(dx) \tag{4.17}$$

$$= -2\int_M P_t\varphi(x)\varphi(x)\mu(dx) + \int_M |P_t\varphi(x)|^2\mu(dx) + |\varphi|_{\mathcal{H}}^2.$$

However

$$\int_M P_t\varphi(x)\varphi(x)\mu(dx) = \int_{U_n} P_t\varphi(x)\varphi(x)\mu(dx),$$

and by the Lebesgue dominated convergence theorem

$$\lim_{t\to 0}\int_M P_t\varphi(x)\varphi(x)\mu(dx) = \int_{U_n} |\varphi(x)|^2\mu(dx)$$

$$= \|\varphi\|_{\mathcal{H}}^2.$$

Moreover by 4.14,

$$\varlimsup_{t\to 0}\|P_t\varphi\|_{\mathcal{H}}^2 \le \|\varphi\|_{\mathcal{H}}^2,$$

and consequently

$$\varlimsup_{t\to 0}\|P_t\varphi - \varphi\|_{\mathcal{H}}^2 \le -2\|\varphi\|_{\mathcal{H}}^2 + \|\varphi\|_{\mathcal{H}}^2 + \|\varphi\|_{\mathcal{H}}^2.$$

Since the functions φ with the imposed properties form a dense set in $\mathcal{H}$ the proof of the theorem is complete.   □

A locally finite measure μ is said to be *invariant* for (P_t) if

$$\int_M p(t, x, \Gamma)\mu(dx) = \mu(\Gamma), \quad t \ge 0, \ \Gamma \in \mathcal{B}(M).$$

It is clear that invariant measures are excessive. They exist for many transition semigroups and the properties of the extended transition semigroups were an object of numerous studies, see e.g. [MaRa], [Chow], [DaPrZa3], [ChojGo2] and [AhFuZa]. The Lebesgue measure on R^d is invariant for the classical heat semigroup. There are no however invariant measures for nondegenerate heat semigroups on infinite dimensional Hilbert spaces. It will easily follow from the next theorem that there exist *many* excessive measures for them.

Theorem 4.6. *For arbitrary locally finite measure ν and a number $\omega > 0$, the measure μ:*

$$\mu(\Gamma) = \int_0^{+\infty}\int_M e^{-\omega t}p(t, x, \Gamma)dt\nu(dx), \quad \Gamma \in \mathcal{B}(M) \tag{4.18}$$

is ω-excessive for (P_t). If the measure μ, defined by 4.18 with $\omega = 0$, is locally finite, then it is excessive.

Proof. Let $P_t^*\nu$ be a measure defined as follows:

$$P_t^*\nu(\Gamma) = \int_M p(t, x, \Gamma)\nu(dx), \quad t \geq 0, \ \Gamma \in \mathcal{B}(M).$$

Then $P_{t+s}^*\nu = P_t^*(P_s\nu)$, $t, s \geq 0$ and the measure μ given by 4.18 can be written more compactly:

$$\mu = \int_0^{+\infty} e^{-\omega t} P_t^*\nu dt, \quad (\omega \geq 0). \tag{4.19}$$

From 4.19

$$\begin{aligned}
P_s^*\mu &= \int_0^{+\infty} e^{-\omega t} P_s^*(P_t^*\nu)dt = \int_0^{+\infty} e^{-\omega t} P_{s+t}^*\nu dt \\
&= e^{\omega s} \int_0^{+\infty} e^{-\omega r} P_r^*\nu dr \leq e^{\omega s}\mu.
\end{aligned}$$

This proves the result.   □

Theorem 4.7. *If*

$$\mu = \int_0^{+\infty} N(0, tQ)dt,$$

then the heat semigroup has a strongly continuous extension to $L^2(H, \mu)$. The extended semigroup consists of contraction operators.

Proof. It is easy to check, see e.g. [Gr], that measure μ is finite on bounded sets. By Theorem 4.6, with $\nu = \delta_{\{0\}}$, the measure μ is excessive and by Theorem 4.5 the conclusion follows.   □

Remark 4.8. In the case, important for the Malliavin calculus, when $H = L^2(0, 1)$ and the covariance operator Q has the kernel $q(s, t) = s \wedge t$, $t, s \in (0, 1)$, a different weight,

$$\mu = \int_0^1 N(0, tQ)dt,$$

was proposed in paper [NuUs]. The proof that the transition semigroup has an extension to a C_0- semigroup on $L^2(H, \mu)$, was in [NuUs] more involved and the contraction property was missing.

5. Heat equation with a first order term

Heat equation with a first order perturbation having linear but unbounded coefficients, is studied. Regular and irregular initial functions are considered. The perturbed equation may have much stronger regularizing properties than the unperturbed one.

5.1 Introduction

Let A be the infinitesimal generator of a C_0-semigroups $S(t)$, $t \geq 0$ on a separable Hilbert space H. In the present chapter we will study the following parabolic equation on H.

$$\frac{\partial u}{\partial t}(t,x) = \frac{1}{2}\operatorname{Tr}[D_Q^2 u(t,x)] + \langle Ax, D_x u(t,x)\rangle , \tag{5.1}$$

$$u(0,x) = \varphi(x), \quad t > 0, \ x \in H. \tag{5.2}$$

It is a Kolmogorov equation corresponding to the solution of the following linear stochastic equation, see the Preliminaries,

$$dX = AX dt + dW(t), \quad X(0) = x, \tag{5.3}$$

where W is a Wiener process on H with the covariance operator Q. Let us recall that solutions to this equation are given by the formula:

$$X(t,x) = S(t)x + \int_0^t S(t-s)\,dW(s), \quad t \geq 0. \tag{5.4}$$

The corresponding transition semigroup, called *Ornstein-Uhlenbeck*, is of the form:

$$P_t\varphi(x) = \int_H \varphi(S(t)x + y)N_{Q_t}(dy), \quad x \in H, \ t \geq 0 , \tag{5.5}$$

where

$$Q_t = \int_0^t S(\sigma)QS^*(\sigma)d\sigma, \quad \operatorname{Tr} Q_t < +\infty, \quad t \geq 0 . \tag{5.6}$$

In particular $\operatorname{Tr} Q_t < +\infty$ if Q is a trace class operator.

We will find conditions on the initial function φ and on the operators A, Q , under which the function u:

$$u(t,x) = P_t\varphi(x) = \int_H \varphi(S(t)x + y)N_{Q_t}(dy), \quad x \in H, \ t \geq 0 \tag{5.7}$$

is a solution to $5.1 - 5.2$. We will not use any result on the stochastic equation 5.3 or on the stochastic formula 5.3. Note that even if $u(t,\cdot)$ is a smooth function, the right hand side of 5.1–5.2 has a meaning only if $x \in D(A)$. Thus if the operator A is unbounded, the Kolmogorov equation 5.1 has a coefficient which is not everywhere defined. This fact introduces additional difficulties in the study of $5.1 - 5.2$.

5.2 Regular initial functions

We start with rather special initial functions. Note that if the first derivative $D_x u(t, x)$ belongs to $D(A^*)$, then equation 5.1–5.2 can be rewritten as follows:

$$\frac{\partial u}{\partial t}(t, x) = \frac{1}{2}\text{Tr}[D_Q^2 u(t, x)] + \langle x, A^* D_x u(t, x)\rangle, \qquad (5.8)$$

$$u(0, x) = \varphi(x), \qquad t > 0, \ x \in H. \qquad (5.9)$$

We have also written the second order term in the more convenient Gross form.

A continuous function $u(t, x), t \geq 0, x \in H$, is called a *classical solution* to 5.1–5.2 if $u(0, x) = \varphi(x), x \in H$, and 5.8 holds for all $t \geq 0, x \in H$. In particular the first derivative of the initial function φ has values in $D(A^*)$. Let us recall that the *generalized solution* to 5.8–5.9 is of the form:

$$u(t, x) = \int_H \varphi(S(t)x + y)N_{Q_t}(dy), \qquad t \geq 0, \ x \in H. \qquad (5.10)$$

Theorem 5.1. *Assume that* $\text{Tr}\, Q < +\infty$, $\varphi \in UC_b^2(H)$, $D_x\varphi(x) \in D(A^*)$ *for all* $x \in H$ *and the function* $A^* D_x\varphi$ *is continuous and bounded. Then the generalized solution to 5.8–5.9 is the classical one.*

Proof. By the mean value theorem

$$\varphi(S(t)x + y) = \varphi(S(t)x) + \langle D_x\varphi(S(t)x), y\rangle \qquad (5.11)$$
$$+ \frac{1}{2}\langle D_x^2\varphi(S(t)x + \sigma_t(y)y)y, y\rangle$$

where $\sigma_t(\cdot)$ is a Borel function from H into $[0, 1]$. Therefore

$$u(t, x) = \varphi(S(t)x) + \frac{1}{2}\int_H \langle D_x^2\varphi(S(t)x + \sigma_t(y)y)y, y\rangle N_{Q_t}(dy), \qquad (5.12)$$

and

$$\frac{u(t, x) - u(0, x)}{t} - \frac{1}{2}\text{Tr}\, Q D_x^2 u(0, x) - \frac{\varphi(S(t)x) - \varphi(x)}{t} \qquad (5.13)$$

$$= \frac{1}{2}\left[\frac{1}{t}\int_H \langle D_x^2\varphi(S(t)x + \sigma_t(y)y)y, y\rangle N_{Q_t}(dy) - \int_H \langle D_x^2\varphi(x)y, y\rangle N_Q(dy)\right].$$

We show first that

$$\frac{\varphi(S(t)x) - \varphi(x)}{t} \to \langle x, A^* D_x\varphi(x)\rangle, \qquad \text{as } t \downarrow 0. \qquad (5.14)$$

Again, by the mean value theorem:

$$\varphi(S(t)x) - \varphi(x) = \langle D_x\varphi(x + \xi_t(S(t)x - x)), S(t)x - x\rangle$$

where $\xi_t \in [0, 1]$. However, for arbitrary $x \in H$

$$S(t)x - x = A\left(\int_0^t S(s)x\,ds\right),$$

and

$$
\begin{aligned}
\frac{\varphi(S(t)x) - \varphi(x)}{t} &= \frac{1}{t}\left\langle D_x\varphi(x + \xi_t(S(t)x - x)), A\int_0^t S(s)x\,ds\right\rangle \\
&= \left\langle A^*D_x\varphi(x + \xi_t(S(t)x - x)), \frac{1}{t}\int_0^t S(s)x\,ds\right\rangle.
\end{aligned}
$$

Since $S(t)x - x \to 0$, $\frac{1}{t}\int_0^t S(s)x\,ds \to x$ as $t \to 0$, the relation 5.14 holds. To prove that the right-hand side of 5.13 tends to zero as $t \downarrow 0$ we need the following lemma.

Lemma 5.2. *If* $\operatorname{Tr} Q < +\infty$ *then* $\frac{1}{t}Q_t \to Q$ *in the trace norm, as* $t \to 0$.

Proof. Note that

$$\frac{1}{t}Q_t - Q = \frac{1}{t}\int_0^t [S(u)QS^*(u) - Q]\,du, \quad t > 0,$$

and it is enough to show that

$$\|S(u)QS^*(u) - Q\|_1 \to 0 \quad \text{as } u \to 0.$$

However

$$
\begin{aligned}
\|S(u)QS^*(u) - Q\|_1 &= \|(S(u)Q^{1/2})(Q^{1/2}S^*(u)) - Q^{1/2}Q^{1/2}\|_1 \\
&\leq \|(S(u)Q^{1/2} - Q^{1/2})Q^{1/2}S^*(u)\|_1 \\
&\quad + \|Q^{1/2}(Q^{1/2}S^*(u) - Q^{1/2})\|_1 \\
&\leq \|(S(u) - I)Q^{1/2}\|_{HS}\|Q^{1/2}S^*(u)\|_{HS} \\
&\quad + \|Q^{1/2}\|_{HS}\|Q^{1/2}(S^*(u) - I)\|_{HS}.
\end{aligned}
$$

We will prove that

$$\|(S(u) - I)Q^{1/2}\|_{HS} \to 0, \quad \|Q^{1/2}(S^*(u) - I)\|_{HS} \to 0, \text{ as } u \downarrow 0. \quad (5.15)$$

Denote by (λ_i, e_i) the eigensequence corresponding to Q, then

$$\|(S(u) - I)Q^{1/2}\|_{HS}^2 = \sum_{i=1}^{+\infty} \lambda_i\|S(u)e_i - e_i\|^2.$$

Since $\sum_{i=1}^{+\infty} \lambda_i < +\infty$, $S(u)e_i \to e_i$ as $u \downarrow 0$ for all $i \in \mathbb{N}$, and $\|Q^{1/2}(S^*(u) - I)\|_{HS} = \|(S(u) - I)Q^{1/2}\|_{HS}$, 5.15 holds and the proof of the lemma is complete. $\qquad\square$

Let I_t denote the right-hand side of 5.13 then,

$$2I_t = \frac{1}{t} \int_H [\langle D_x^2\varphi(S(t)x + \sigma_t(y)y)y, y\rangle - \langle D_x^2\varphi(S(t)x)y, y\rangle)]N_{Q_t}(dy)$$
$$+ \frac{1}{t} \operatorname{Tr} Q_t D_x^2\varphi(S(t)x) - \operatorname{Tr} Q D_x^2\varphi(x)$$
$$= I_t^1 + I_t^2.$$

However

$$|I_t^1| \leq \frac{1}{t} \int_H |D_x^2\varphi(S(t)x + \sigma_t(y)y) - D_x^2\varphi(S(t)x)| \, |y|^2 N_{Q_t}(dy)$$
$$\leq \frac{1}{t} \left(\int_H |D_x^2\varphi(S(t)x + \sigma_t(y)y) - D_x^2\varphi(S(t)x|^2 N_{Q_t}(dy) \right)^{1/2} \left(\int_H |y|^4 N_{Q_t}(dy) \right)^{1/2}.$$

By elementary properties of Gaussian measures on Hilbert spaces, see ([DaPrZa2], p. 57), for a constant $c_1 > 0$,

$$\frac{1}{t} \left(\int_H |y|^4 N_{Q_t}(dy) \right)^{1/2} \leq c_1 \operatorname{Tr} Q_t,$$

and Lemma 5.2 implies that for a constant $c_2 > 0$,

$$\frac{1}{t} \operatorname{Tr} Q_t \leq c_2, \quad t \in (0, 1). \tag{5.16}$$

Moreover $N_{Q_t} \Rightarrow \delta_{\{0\}}$ weakly as $t \downarrow 0$, and $D_x^2\varphi(z)$, $z \in H$ is bounded and uniformly continuous, therefore, taking into account 5.16, $|I_t^1| \to 0$ as $t \downarrow 0$ uniformly with respect to $x \in H$. In addition,

$$|I_t^2| \leq \left\| \left(\frac{1}{t} Q_t \right) D_x^2\varphi(S(t)x) - Q D_x^2\varphi(x) \right\|_1$$
$$\leq \left\| \frac{1}{t} Q_t - Q \right\|_1 \|D_x^2\varphi(S(t)x) - D_x^2\varphi(x)\|$$
$$+ \|Q\|_1 \|D_x^2\varphi(S(t)x) - D_x^2\varphi(x)\| \to 0 \quad \text{as } t \downarrow 0,$$

uniformly on compact sets of $x \in H$. This way it was shown that,

$$\frac{\partial^+ u(0, x)}{\partial t} = \frac{1}{2} \operatorname{Tr} Q D_x^2 u(0, x) + \langle x, A^* D_x u(0, x) \rangle,$$

for arbitrary $x \in H$.

It is easy to see that if a function φ satisfies assumptions of Theorem 5.1 then also $u(t, \cdot)$ satisfies the same assumptions for all $t \geq 0$ and consequently,

$$\frac{\partial^+ u(t,x)}{\partial t} = \frac{1}{2} \operatorname{Tr} Q D_x^2 u(t,x) + \langle x, A^* D_x u(t,x) \rangle$$

for all $t \geq 0$, $x \in H$ and the proof of theorem can be completed in the same way as that of Proposition 3.5 in §3.2.   □

Remark 5.3. If $A^* D\varphi$ is a bounded function, *uniformly* continuous on bounded sets then it follows, from the proof of the theorem, that the time derivative $\frac{\partial}{\partial t} u(t,x)$ is uniformly continuous with respect to (t,x) from compact subsets of $(0, +\infty) \times H$.

Although classical solutions are very special we will show now that arbitrary generalized solution is a limit of the classical ones.

Theorem 5.4. *For arbitrary* $\varphi \in UC_b(H)$ *there exists a bounded sequence of initial functions* (φ_n) *from* $UC_b^2(H)$, *converging pointwise to* φ, *for which the Kolmogorov equation 5.8–5.9 has classical solutions* u_n *converging pointwise to the generalized solution* u.

Proof. A function $\varphi : H \to R$ is called *cylindrical* if there exist $n \in \mathbb{N}$ and vectors $a_1, \ldots, a_n \in H$ such that

$$\varphi(x) = \varphi\left(\sum_{i=1}^n \langle x, a_i \rangle a_i\right), \quad x \in H,$$

Then and only then there exists $\psi : R^n \to R$ such that

$$\varphi(x) = \psi(\langle x, a_1 \rangle, \ldots, \langle x, a_n \rangle), x \in H. \tag{5.17}$$

A cylindrical function φ is called *special* if in the representation 5.17, $a_i \in D(A^*)$, $i = 1, \ldots, n$ and $\psi \in UC_b^2(R^n)$. Then, by direct calculations,

$$D\varphi(x) = \sum_{i=1}^n \frac{\partial \psi}{\partial \xi_i}(\langle x, a_1 \rangle, \ldots, \langle x, a_n \rangle) a_i,$$

$$D^2\varphi(x) = \sum_{i,j=1}^n \frac{\partial \psi}{\partial \xi_i \partial \xi_j}(\langle x, a_1 \rangle, \ldots, \langle x, a_n \rangle) a_i \otimes a_j,$$

$$A^* D\varphi(x) = \sum_{i=1}^n \frac{\partial \psi}{\partial \xi_i}(\langle x, a_1 \rangle, \ldots, \langle x, a_n \rangle) A^* a_i, x \in H.$$

Therefore φ satisfies assumptions of Theorem 5.1 and the corresponding generalized solution is the classical one.

To finish the proof it is enough to show that for arbitrary $\varphi \in UC_b(H)$ there exists a sequence (φ_n) of special cylindrical functions such that,

i) $\|\varphi_n\|_0 \leq \|\varphi\|_0, \quad n \in \mathbb{N}$
ii) $\lim_n \varphi_n(x) = \varphi(x) = \varphi(x), \quad x \in H.$

Since the set $D(A^*)$ is linear and dense in H, there exists an orthonormal and complete sequence (e_i) composed of elements from $D(A^*)$. Define

$$\widetilde{\psi}_n(\xi) = \varphi(\xi_1 e_1 + \ldots + \xi_n e_n), \quad \xi = (\xi_1, \ldots, \xi_n) \in R^n.$$

For each $n \in \mathbb{N}$ there exists $t_n > 0$ such that $\sup_{\xi \in R^n} |\widetilde{\psi}_n(\xi) - \psi_n(\xi)| \leq \frac{1}{n}$, where

$$\psi_n(\xi) = \frac{1}{\sqrt{(2\pi t_n)^n}} \int_{R^n} \widetilde{\psi}_n(\xi) e^{-\frac{|\xi - \eta|^2}{2t_n}} \, d\eta, \quad \xi \in R^n,$$

and $\psi_n \in UC_b^\infty(R^n)$, $n \in \mathbb{N}$. Moreover, for $x \in H$,

$$
\begin{aligned}
|\varphi(x) \quad &- \quad \psi_n(\langle x, e_1 \rangle, \ldots, \langle x, e_n \rangle)| \\
&\leq \quad |\varphi(x) - \varphi(\langle x, e_1 \rangle e_1 + \ldots + \langle x, e_n \rangle e_n)| \\
&\quad + |(\widetilde{\psi}_n - \psi)(\langle x, e_1 \rangle, \ldots, \langle x, e_n \rangle)| \\
&\leq \quad |\varphi(x) - \varphi(\langle x, e_1 \rangle e_1 + \ldots + \langle x, e_n \rangle e_n)| + \frac{1}{n} \to 0 \text{ as } n \to +\infty. \square
\end{aligned}
$$

5.3 General initial functions

This section is concerned with a class of Kolmogorov equations of the type 5.1 – 5.2 for which generalized solutions are smooth functions in the space variable for each $t > 0$ and for *each* bounded Borel initial function φ. We follow basically [DaPrZa1].

We will need the following hypothesis, called *range condition*, see [Za1],

$$S(t)(H) \subset Q_t^{1/2}(H), \quad \text{for all } t > 0. \tag{5.18}$$

If 5.18 holds we define

$$\Lambda_t = Q_t^{-\frac{1}{2}} S(t), \quad t > 0 \tag{5.19}$$

where $Q_t^{-\frac{1}{2}}$ denotes the pseudoinverse of $Q_t^{1/2}$. It follows from the closed graph theorem that Λ_t, $t > 0$, is a bounded operator. The range condition is closely related to the regularity of the generalized solution as the following theorem shows, see [DaPrZa1].

Theorem 5.5. *The hypothesis 5.18 holds if and only if for arbitrary $\varphi \in B_b(H)$ and arbitrary $t > 0$, $P_t \varphi \in C_b^\infty(H)$.*

Proof. Assume that 5.18 holds, $t > 0$, $\varphi \in B_b(H)$ and $x \in H$. Since $S(t)x \in Q_t^{1/2}(H)$ the measures $N(S(t)x, Q_t)$, $N(0, Q_t)$ are equivalent and the corresponding density $\varrho_t(x, \cdot)$ is given by the Cameron–Martin formula:

$$\frac{dN(S(t)x, Q_t)}{dN(0, Q_t)}(y) = \varrho_t(x, y), \quad y \in H,$$

where
$$\varrho_t(x,y) = e^{\langle \Lambda_t x, Q_t^{-1/2} y \rangle - \frac{1}{2}|\Lambda_t x|^2}, \quad y \in H.$$

Therefore
$$P_t\varphi(x) = \int_H \varphi(y) e^{\langle \Lambda_t x, Q_t^{-1/2} y \rangle - \frac{1}{2}|\Lambda_t x|^2} N_{Q_t}(dy).$$

In exactly the same way as in the proof of Theorem 3.6, one shows that $P_t\varphi(x)$ is differentiable an arbitrary number of times. In particular, for $g, h \in H$,

$$\langle DP_t\varphi(x), g \rangle = \int_H \langle \Lambda_t g, Q_t^{-1/2} y \rangle \varphi(S(t)x + y) N_{Q_t}(dy), \qquad (5.20)$$

$$\langle D^2 P_t\varphi(x)h, g \rangle = \int_H [\langle \Lambda_t g, Q_t^{-1/2} y \rangle \langle \Lambda_t h, Q_t^{-1/2} y \rangle - \langle \Lambda_t g, \Lambda_t h \rangle] (5.21)$$
$$\cdot \varphi(S(t)x + y) N_{Q_t}(dy).$$

Assume now that for arbitrary $\varphi \in B_b(H)$, $P_t\varphi$ is a continuous function but nevertheless for some $x_0 \in H$, $S(t)x_0 \notin Q_t^{1/2}(H)$. Then, for all $n \in \mathbb{N}$, measures $N(S(t)\frac{x_0}{n}, Q_t)$, $N(0, Q_t)$ are singular. Consequently for arbitrary $n \in \mathbb{N}$ there exists a Borel set $K_n \subset H$ such that

$$N\left(S(t)\frac{x_0}{n}, Q_t\right)(K_n) = 0, \quad N(0, Q_t)(K_n) = 1,$$

and if $K = \bigcap_{n=1}^{+\infty} K_n$, then

$$N\left(S(t)\frac{x_0}{n}, Q_t\right)(K) = 0, \quad N(0, Q_t(K)) = 1.$$

If $\varphi = \chi_K$, then $P_t\varphi(\frac{x_0}{n}) = 0$, $n \in \mathbb{N}$ and $P_t\varphi(0) = 1$. Therefore the function $P_t\varphi$ is not continuous at 0. $\qquad \square$

It is our aim to find conditions, in addition to 5.6 and 5.18, which imply that generalized solutions satisfy the parabolic equation for arbitrary $\varphi \in UC_b(H)$.

We start from rewriting formulae 5.20 and 5.21 in a more compact way. The precise meaning of the integrals in the following proposition are given in the proof.

Proposition 5.6. *If $\varphi \in B_b(H)$ and conditions 5.6, 5.18 hold then*

$$DP_t\varphi(x) = \Lambda_t^* \left[\int_H Q_t^{-1/2} y \varphi(S(t)x + y) N_{Q_t}(dy) \right], \qquad (5.22)$$

$$D^2 P_t\varphi(x) = \Lambda_t^* \left[\int_H (Q_t^{-\frac{1}{2}} y \otimes Q_t^{-\frac{1}{2}} y - I) \varphi(S(t)x + y) N_{Q_t}(dy) \right] \Lambda_t, \quad (5.23)$$

where the integral in 5.22 is an element in H and the integral in 5.23 is a Hilbert–Schmidt operator on H.

Proof. For arbitrary $g \in H$, and an arbitrary orthonormal, complete basis (e_k) in H,

$$\langle g, Q_t^{-1/2} y \rangle = \sum_{k=1}^{+\infty} \langle g, e_k \rangle \langle e_k, Q_t^{-1/2} y \rangle, \quad y \in H,$$

where $\xi_k(\cdot) = \langle e_k, Q_t^{-1/2}(\cdot) \rangle$, $k = 1, 2, \ldots$ is an orthonormal sequence on $L^2(H, \mathcal{B}(H), N_{Q_t})$. Let $\zeta(y) = \varphi(S(t)x + y)$, $y \in H$. Then

$$\int_H \langle g, Q_t^{-1/2} y \rangle \zeta(y) N_{Q_t}(dy) = \int_H \left\langle g, \sum_{k=1}^{+\infty} e_k \xi_k(y) \right\rangle \zeta(y) N_{Q_t}(dy)$$

$$= \left\langle g, \sum_{k=1}^{+\infty} e_k ((\xi_k, \zeta)) \right\rangle,$$

where $((\cdot, \cdot))$ denotes the scalar product in $L^2(H, \mathcal{B}(H), N_{Q_t})$. However, denoting by $\|| \cdot \||$ the norm on $L^2(H, \mathcal{B}(H), N_{Q_t})$,

$$\sum_k |((\xi_k, \zeta))|^2 \le \|| \zeta \||^2 = \int_H \varphi^2(S(t)x + y) N_{Q_t}(dy) \le \|\varphi\|_0^2,$$

so the series $\sum_{k=1}^{+\infty} e_k ((\xi_k, \zeta))$ converges in H and it is natural to define,

$$\int_H Q_t^{-1/2} y \varphi(S(t)x + y) N_{Q_t}(dy) = \sum_{k=1}^{+\infty} e_k ((\xi_k, \zeta)) .$$

Moreover,

$$\left| \int_H Q_t^{-1/2} y \varphi(S(t)x + y) N_{Q_t}(dy) \right| \le \|\varphi\|_0. \tag{5.24}$$

It is well known that operators $e_k \otimes e_j$, $k, j \in \mathbb{N}$, form an orthonormal and complete basis in the space of Hilbert–Schmidt operators on H. Let $g, h \in H$, then

$$\int_H [\langle g, Q_t^{-1/2} y \rangle \langle h, Q_t^{-1/y} y \rangle - \langle g, h \rangle] \varphi(S(t)x + y) N_{Q_t}(dy)$$

$$= \int_H \left[\left\langle g, \sum_{k=1}^{+\infty} e_k \xi_k(y) \right\rangle \left\langle h, \sum_{j=1}^{+\infty} e_j \xi_j(y) \right\rangle - \langle g, h \rangle \right] \zeta(y) N_{Q_t}(dy)$$

$$= \left\langle g, \int_H \left\{ \left[\sum_{k,j=1}^{+\infty} (e_k \otimes e_j) \xi_k(y) \xi_j(y) \right] - I \right\} \zeta(y) N_{Q_t}(dy) h \right\rangle$$

$$= \left\langle g, \left[\sum_{k \ne j}^{+\infty} (e_k \otimes e_j)((\xi_k \xi_j, \zeta)) + \sum_{k=1}^{+\infty} e_k \otimes e_k ((\xi_k^2 - 1, \zeta)) \right] h \right\rangle .$$

It is therefore natural to set,

$$\int_H (Q_t^{-\frac{1}{2}} y \otimes Q_t^{-\frac{1}{2}} y - I)\varphi(S(t)x + y)N_{Q_t}(dy) \tag{5.25}$$

$$= \sum_{k \neq j}^{+\infty} (e_k \otimes e_j)((\xi_k \xi_j, \zeta)) + \sum_{k=1}^{+\infty} e_k \otimes e_k ((\xi_k^2 - 1, \zeta))$$

However, compare the proof of Theorem 3.7,

$$\sum_{k \neq j}^{+\infty} |((\xi_k \xi_j, \zeta))|^2 + \sum_{k=1}^{+\infty} |((\zeta_k^2 - 1, \zeta))|^2, \le 2\|\zeta\|^2 \le 2\|\varphi\|_0^2$$

and the integral 5.25 defines a Hilbert–Schmidt operator and

$$\left\| \int_H (Q_t^{-1/2} y \otimes Q_t^{-1/2} y - I)\varphi(S(t)x + y)N_{Q_t}(dy) \right\|_{HS} \le \sqrt{2}\|\varphi\|_0. \square \tag{5.26}$$

As a corollary we have the following result with a new estimate 5.28.

Theorem 5.7. *Assume that $\varphi \in B_b(H)$ and conditions 5.6, 5.18 are satisfied.*

i) *If the operator $\Lambda_t A$ has a continuous extension to H then $DP_t\varphi(x) \in D(A^*)$ for all $x \in H$ and*

$$\|A^* DP_t\varphi\|_0 \le \|\Lambda_t A\| \, \|\varphi\|_0 \tag{5.27}$$

ii) *If the operator Λ_t belongs to $l^4(H)$, then $D^2 P_t\varphi(x)$ is trace class operator and*

$$\| \operatorname{Tr} D^2 P_t\varphi\|_0 \le 4\|\varphi\|_0\|\Lambda_t\|_4^2 \tag{5.28}$$

Proof. The estimate 5.27 follows from 5.24 and 5.22. Similarly, from 5.23 and 5.26 one has that

$$D^2 P_t\varphi(x) = \Lambda_t^* G \Lambda_t \tag{5.29}$$

where G is a Hilbert–Schmidt operator with the estimate 5.26. Consequently, by the lemma which follows,

$$| \operatorname{Tr} \Lambda_t^* G \Lambda_t| \le \|\Lambda_t^* G \Lambda_t\|_1 \le 2^{\frac{3}{2}}\|G\|_2\|\Lambda_t\|_4^2 \, .$$

Taking into account the estimate 5.26, the proof is complete. $\square$

We formulate and prove the lemma which has just been used.

Lemma 5.8. *Assume that G is a self-adjoint Hilbert–Schmidt operator and $R \in l^1(H)$. Then*

$$R^* G R \in l^4(H) \tag{5.30}$$

and

$$\|R^* G R\|_1 \le 2^{3/2}\|G\|_2 \cdot \|R\|_4^2 \tag{5.31}$$

Proof. Assume first that G is also nonnegative definite. Then

$$\|R^*GR\|_1 = \|G^{1/2}R\|_2^2,$$

and by 2.4,

$$\|G^{1/2}R\|_2 \le 2^{1/2}\|G^{1/2}\|_4 \cdot \|R\|_4.$$

Since $\|G^{1/2}\|_4 = \|G\|_2^{1/2}$,

$$\|R^*GR\|_1 \le 2\|G^{1/2}\|_4^2\|R\|_4^2 \tag{5.32}$$
$$\le 2\|G\|_2\|R\|_4^2.$$

In the general case the operator G has a representation:

$$G = G_1 - G_2,$$

where G_1 and G_2 are respectively the positive and negative parts of G. Moreover,

$$\|R^*GR\|_1 \le \|R^*G_1R\|_1 + \|R^*G_2R\|_1,$$

and by 5.32

$$\|R^*GR\|_1 \le 2[\|G_1\|_2 + \|G_2\|]\|R\|_4^2.$$

But $\|G\|_2 = (\|G_1\|_2^2 + \|G_2\|^2)^{1/2}$ and therefore

$$\|G_1\|_2 + \|G_2\|_2 \le \sqrt{2}\|G\|_2.$$

Finally

$$\|R^*GR\|_1 \le 2^{3/2}\|G\|_2\|R\|_4^2.$$

Theorem 5.9. *Assume in addition to conditions 5.6, 5.18, that the operators $\Lambda_t A$, $t > 0$ have continuous extensions to H, form a strongly continuous function of $t > 0$, and the operators $\Lambda_t Q^{\frac{1}{2}} t > 0$, are in $l^4(H)$. Then for arbitrary $\varphi \in UC_b(H)$, the generalized solution is uniformly continuous on $[0, +\infty[\times H$ and satisfies 5.8 for all $t > 0$ and $x \in H$.*

Proof. It follows from the demonstrations of Proposition 5.6 and of Theorem 5.7 that

$$A^*DP_t\varphi(x) = (\Lambda_t A)^* \left[\int_H Q_t^{-1/2} y\varphi(S(t)x + y)N_{Q_t}(dy) \right], \tag{5.33}$$

$$\operatorname{Tr} Q^{\frac{1}{2}} D^2 P_t\varphi(x)Q^{\frac{1}{2}} = \tag{5.34}$$

$$\operatorname{Tr} Q^{\frac{1}{2}} \Lambda_t^* \left[\int_H (Q_t^{-1/2}y \otimes Q_t^{-1/2}y - I)\varphi(S(t)x + y)N_{Q_t}(dy) \right] \Lambda_t Q^{\frac{1}{2}}$$

Therefore, under the assumptions of the theorem, functions

$$A^*DP_t\varphi(x), \operatorname{Tr} Q^{-1/2}D^2 P_t\varphi(x)Q^{-1/2}, t > 0, x \in H$$

are continuous in both variables. Let $\varepsilon(A)$ denote the linear subspace of $UC_{bc}(H) = UC_b(H) \oplus iUC_b(H)$ spanned by functions ψ_h, $h \in D(A^*)$, where

$$\psi_h(x) = e^{i\langle h, x\rangle}.$$

By direct calculations,

$$
\begin{aligned}
u_h(t, x) &= P_t\psi_h(x) = e^{i\langle x, S^*(t)h\rangle - 1/2\langle Q_t h, h\rangle} \\
\frac{\partial}{\partial t}u_h(t, x) &= u_h(t, x)\left[i\langle x, S^*(t)A^*h\rangle - \frac{1}{2}\langle QS^*(t)h, S^*(t)h\rangle\right], \\
\langle x, A^*Du_n(t, x)\rangle &= u_n(t, x)i\langle x, A^*S^*(t)h\rangle, \\
\frac{1}{2}\operatorname{Tr} QD^2u_n(t, x) &= -\frac{1}{2}u_n(t, x)\langle QS^*(t)h, S^*(t)h\rangle, \quad t > 0,
\end{aligned}
$$

and therefore the theorem holds for $\varphi \in \varepsilon(A)$. It is easy to show that for arbitrary $\varphi \in UC_b(H)$ there exists a sequence (φ_n) uniformly bounded such that $\varphi_n \in UC_{bc}(H)$, $n \in \mathbb{N}$ and $\varphi_n \to \varphi$ uniformly on compact sets. Let u_n and u be the corresponding generalized solutions to 5.8 and $x \in H$ a fixed element. It follows from 5.33, 5.34 that

$$A^*Du_n(\cdot, x) \to A^*Du(\cdot, x), \operatorname{Tr} D_Q^2u_n(\cdot, x) \to \operatorname{Tr} D_Q^2u(\cdot, x),$$

uniformly on compact subsets of $]0, +\infty[$, as $n \to +\infty$. Since functions u_n satisfy 5.8 on $]0, +\infty[\times H$ therefore the sequence $\frac{\partial}{\partial t}u_n(\cdot, x)$ converges uniformly on compact subsets of $]0, +\infty[\times H$. Consequently the time derivative $\frac{\partial}{\partial t}u(\cdot, x)$ exists and is continuous on $]0, +\infty[$. By the obvious limit argument the function u satisfies 5.8. □

Remark 5.10. Instead of using exponential functions one could take in the proof cylindrical functions φ of the form:

$$\varphi(x) = \psi(\langle x, h_1\rangle, \ldots, \langle x, h_n\rangle), \quad x \in H,$$

where $h_1, \ldots, h_n \in D(A^*)$, $\psi \in UC_b(R^n)$, $n \in \mathbb{N}$. Direct computations imply that also for them the problem 5.8–5.9 is solvable.

We finish this section by a result due to G.Da Prato, see [DaPr2], obtained inedependently in [ChojGo1], which complements Theorem 5.7 . Its starting point is a remark that for all $t > 0$, and $\phi \in B_b(H)$,

$$P_t\phi(x) = P_{t/2}(P_{t/2}(\phi))(x) \quad x \in H.$$

If the condition 5.18 holds then, by Theorem 5.5, the function $P_{t/2}(\phi)$ is continuously differentiable with bounded first derivative and therefore, with a slight abouse of notation, $DP_t\phi = P_{t/2}(S^*(t/2)DP_{t/2}(\phi))$. Let us recall that if $S(t)(H) \subset D(A)$ $t > 0$ then the semigroup S is called *differentiable*. The following proof was communicated to us by B. Goldys.

Theorem 5.11. *Assume that conditions 5.6, 5.18 hold. Then,* *(i)* $DP_t\phi(x) \in \text{dom}\,(A^*)$ *for all* $t > 0$, $x \in H$ *and all* $\phi \in B_b(H)$ *if and only if the semigroup* $S(t)$ *is differentiable.*

(ii) *The operator* $D^2 P_t \phi(x)$ *is of trace class for all* $t > 0$, $x \in H$ *and* $\phi \in B_b(H)$.

Proof. Let $B_b\,(H, H)$ and $B_b\,(H, L_{HS}(H))$, denote the spaces of all measurable, bounded mappings from H into H and from H into $L_{HS}(H)$ respectively. Define, for $x \in H$, $t \geq 0$,

$$N_1(t)\phi(x) = \int_H S^*(t)\phi\,(S(t)x + y)\,\mu_t(dy), \quad \phi \in B_b\,(H, H)\,,$$

and

$$N_2(t) = \int_H S^*(t)\phi\,(S(t)x + y)\,S(t)\mu_t(dy), \quad \phi \in B_b\,(H, L_{HS}(H))\,.$$

Starting from appropriate cylindrical functions it is easy to prove the following lemma

Lemma 5.12. *The families* $(N_1(t))$ *and* $(N_2(t))$ *form semigroups of linear operators on* $B_b\,(H, H)$ *and on* $B_b\,(H, L_{HS}(H))$. *Moreover,*

$$DP_t\phi(x) = N_1(t)D\phi(x), \quad \phi \in C_b^1(H), \tag{5.35}$$

and

$$D^2 P_t\phi(x) = N_2(t)D^2\phi(x), \quad \phi \in C_b^2(H). \tag{5.36}$$

To prove part *(i)* of the theorem assume that the semigroup $S(t)$ is differentiable. Then $A^* N_1(t)$ is well defined for every $x \in H$. It follows from 5.18 , Theorem 5.5 and 5.35 that

$$A^* DP_t\phi(x) = A^* N_1\left(\frac{t}{2}\right) DP_{t/2}\phi(x),$$

and sufficiency follows. Assume that $DP_t\phi(x) \in \text{Dom}\,(A^*)$ and take function

$$\phi(x) = f\,(\langle x, h \rangle)\,, \quad x \in H,$$

where $h \in H$ and $f \in C_b^1\,(R)$. Then

$$DP_t\phi(x) = \left(\int_H f'\,(S(t)x + y)\,\mu_t(dy)\right) S^*(t)h.$$

If we choose f in such a way that the integral does not vanish then the necessity follows.

It remains to prove *(ii)*. If 5.18 holds then, by Theorem 2.3 from the Preliminaries, for a constant $C > 0$,

$$\|S(t)^*h\|^2 \leq \|Q_t^{1/2}h\|^{1/2}\; h \in H\,.$$

Since Q_t is a trace class operator, the operator $S(t)$ is Hilbert-Schmidt for all $t > 0$ and hence, by the semigroup property, it is also of trace class . It follows that $N_2(t)\phi(x)$, $\phi \in C_b(H, L_{HS}(H))$ is of trace class for every $t > 0$. By 5.36

$$D^2 P_t \phi(x) = N_2\left(\frac{t}{2}\right) D^2 P_{t/2}\phi(x)$$

and (ii) follows.   □

5.4 Range condition and examples

In this section we investigate the range condition 5.18 in more detail and present some examples.

5.4.1 Range condition. The range condition has a useful control theoretic interpretation, see [Za2]. Consider a linear control system:

$$\frac{dy}{dt}(t) = Ay + Q^{1/2}u, \quad y(0) = x. \tag{5.37}$$

For arbitrary function $u(\cdot) \in L^2([0,T]; H)$, $x \in H$, denote by $y^{x,u(\cdot)}(t)$, $t \in [0,T]$ the solution to 5.37. System 5.37 is said to be *null controllable in time* $T > 0$ if for arbitrary x, there exists $u \in L^2([0,T]; H)$ such that $y^{x,u(\cdot)}(T) = 0$.

Theorem 5.13. *The range condition 5.18 holds for a given $T > 0$ if and only if system 5.37 is null controllable in time $T > 0$.*

Proof. Define on $L^2([0,T], H)$ a linear operator L_T with values in H by the formula:

$$L_T u = \int_0^T S(T-s)Q^{1/2}u(s)\,ds, \quad u \in L^2([0,T], H).$$

Then its adjoint L_T^* acts from H into $L^2([0,T], H)$ and is of the form:

$$L_T^* h(s) = Q^{1/2} S^*(T-s)h, \quad s \in [0,T].$$

Since $y(T) = S(T)x + L_T u$, the system 5.37 is null controllable in time T if and only if

$$\text{Range } S(T) \subseteq \text{Range } L_T. \tag{5.38}$$

It is therefore enough to show that Range L_T = Range $Q_T^{1/2}$. However, by the Theorem 2.3 from the Preliminaries, the identity of images is equivalent to the existence of positive constants c, C, such that

$$c|Q_T^{1/2}h| \leq |L_T^* h| \leq C|Q_T^{1/2}h|, \quad \text{for all } h \in H. \tag{5.39}$$

It follows from the formulae for L_T^* and Q_T that:

$$|L_T^* h|^2 = \int_0^T |Q^{1/2}S^*(s)h|^2\,ds = \langle Q_T h, h \rangle = |Q_T^{1/2}h|^2$$

so 5.39 holds with $c = C = 1$.   □

The operator Λ_t defined by 5.19 has also control theoretic interpretation, see [Za2], and [DaPrZa2].

Proposition 5.14. *For arbitrary $T > 0$ and $x \in H$,*

$$|\Lambda_T x|^2 = \inf \left\{ \int_0^T |u(s)|^2 \, ds; \ S(T)x + L_T u = 0 \right\}. \qquad (5.40)$$

Therefore the number $|\Lambda_T x|^2$ can be interpreted as the minimal energy which the system 5.37 requires to drive the state x to the origin 0.

We deduce now some consequences of the general characterizations.

Proposition 5.15. *If for all $x \in H$ and some $T > 0$,*

$$\int_0^T |Q^{-1/2} S(s) x|^2 \, ds < +\infty$$

then 5.37 is null controllable in time T and

$$\|\Lambda_T\| \leq \frac{1}{T} \left(\int_0^T \|Q^{-1/2} S(s)\|^2 \, ds \right)^{1/2} \qquad (5.41)$$

Proof. Define $u(s) = -\frac{1}{t} Q^{-1/2} S(s) x$, $s \in [0, T]$. Then

$$
\begin{aligned}
y^{x, u(\cdot)}(T) &= S(T)x + \int_0^T S(T-s) Q^{1/2} \left[-\frac{1}{T} Q^{-1/2} S(s) x \right] ds \\
&= S(T)x - \frac{1}{T} \int_0^T S(T-s) S(s) x \, ds = 0.
\end{aligned}
$$

Thus 5.37 holds and

$$|\Lambda_T x|^2 \leq \int_0^T \left| \frac{1}{T} Q^{-1/2} S(s) x \right|^2 ds \leq \frac{|x|^2}{T^2} \int_0^T \|Q^{-1/2} S(s)\|^2 \, ds,$$

so 5.41 holds as well.

Corollary 5.16. *If $Q = I$ then for arbitrary $T > 0$ there exists $M > 0$ such that*

$$\|\Lambda_t\| \leq \frac{M}{\sqrt{t}}, \qquad t \in [0, T]$$

Corollary 5.17. *If for some $\beta \in [0, 1)$, $M >$ and all $t > 0$,*

$$\|Q^{-1/2} S(t)\| \leq \frac{M}{t^{\frac{\beta}{2}}}, \qquad t > 0, \qquad (5.42)$$

then

$$\|\Lambda_t\| \leq \frac{M}{\sqrt{1-\beta}} \frac{1}{\sqrt{t^{1+\beta}}}, \qquad t > 0. \qquad (5.43)$$

Assumptions of Corollary 5.17 are satisfied if $Q = I$, and then $\beta = 0$, or if the operator A generates an analytic semigroup, see e.g. [DaPrZa2], such that $\mathrm{Im}\, Q^{1/2} \supset D(-A)^{\frac{\beta}{2}}$.

5.4.2 Examples. We illustrate the general theory by considering specific applications.

Example 5.18. Let (e_n) be a complete and orthonormal basis in H. Define operators A and Q by the relations:

$$Ae_n = -\alpha_n e_n, \quad Qe_n = \gamma_n e_n, \quad n = 1, \ldots \quad (5.44)$$

where $\alpha_n, \gamma_n > 0$, $n \in \mathbb{N}$. The operators $S(t)$, Q_t, $\Lambda_t = Q_t^{-1/2} S(t)$, $\Lambda_t Q^{1/2}$ and $\Lambda_t A$ have exactly the same system of eigenvectors as A and Q and the corresponding sequences of eigenvalues are as follows:

$$(e^{-t\alpha_n}), \quad \left(\frac{1}{2}\frac{\gamma_n}{\alpha_n}(1 - e^{-2\alpha_n t})\right), \quad \left(\sqrt{\frac{2\alpha_n}{\gamma_n(e^{2\alpha_n t} - 1)}}\right), \quad \left(\sqrt{\frac{2\alpha_n}{e^{2\alpha_n t} - 1}}\right).$$

and

$$\left(\sqrt{\frac{2\alpha_n^3}{\gamma_n(e^{2\alpha_n t} - 1)}}\right).$$

Consequently, if the positive sequence (α_n) is separated from 0, then

$$\operatorname{Tr} Q_t < +\infty \quad \text{iff} \quad \sum_{n=1}^{+\infty} \frac{\gamma_n}{\alpha_n} < +\infty$$

and the range condition, as well as the requirement $\Lambda_t Q^{1/2} \in l^4(H)$, hold under rather mild assumptions on the sequences (α_n) and (γ_n). It is interesting to note that the condition

$$\Lambda_t Q^{1/2} \in l^4(H)$$

involves only the operator A and not the covariance operator Q.

Let in particular

$$\alpha_n = n^\alpha, \gamma_n = n^{-\gamma}, \alpha, \gamma > 0.$$

Then,

$$\operatorname{Tr} Q_t < +\infty \text{ iff } \alpha + \gamma > 1.$$

If,

$$\alpha + \gamma > 1,$$

then, for all $t > 0$, the range condition holds, the operator $\Lambda_t Q^{1/2}$ belongs to $l^4(H)$ and the operator $\Lambda_t A$ is bounded. Consequently in that situation the results on the parabolic equation proved in the present chapter are applicable.

We finally consider specific operators Q and A, commonly used in applications.

Example 5.19. Let $H = L^2(0,1)$ and $S(t)$, $t \geq 0$ be the heat semigroup corresponding to the Dirichlet boundary condition. Thus $A = \frac{d^2}{d\xi^2}$ and the domain $D(A)$ consists of all absolutely continuous functions x such that $x(0) = x(1) = 0$, $\frac{d}{d\xi}x$ is also absolutely continuous with $\frac{d^2}{d\xi^2}x$ in H. Let Q be the following operator:

$$Qx(\xi) = \int_{-\pi}^{\pi} q(\xi,\eta)x(\eta)\,d\eta, \quad \xi \in (0,1), \ x \in H,$$

where

$$q(\xi,\eta) = \min(\xi,\eta) - \xi\eta, \quad \xi,\eta \in (0,1).$$

Eigenfunctions and eigenvalues corresponding to Q are as follows:

$$e_n(\xi) = \sqrt{2}\sin(n\pi\xi)\,, \xi \in (0,1)\,, \gamma_n = (n\pi)^{-2}\ n = 1,\dots.$$

Since the eigenfunctions form an orthonormal and complete basis in H and the eigenvalues are positive the operator Q is positive as well. Moreover:

$$Ae_n(\xi) = -(n\pi)^2 e_n(\xi)\,, \xi \in (0,1)n = 1,\dots.$$

and

$$\alpha_n = (n\pi)^2, \quad n = 1,\dots.$$

Therefore this specific example is of the previously considered form.

In a similar way one can treat the case when the covariance operator is determined by the Brownian sheet, see § 2.3.

Example 5.20. Let $H = L^2(0,1)$ and $S(t)$, $t \geq 0$ be the heat semigroup corresponding to the mixed boundary conditions : $A = \frac{d^2}{d\xi^2}$ and the domain $D(A)$ consists of all absolutely continuous functions x such that $x(0) = \frac{d}{d\xi}x(1) = 0$, $\frac{d}{d\xi}x$ is also absolutely continuous with $\frac{d^2}{d\xi^2}x$ in H. The operator Q :

$$Qx(\xi) = \int_{-\pi}^{\pi} q(\xi,\eta)x(\eta)\,d\eta, \quad \xi \in (0,1), \ x \in H,$$

has now a simpler kernel,

$$q(\xi,\eta) = \min(\xi,\eta), \quad \xi,\eta \in (0,1).$$

Eigenfunctions and eigenvalues corresponding to Q are as follows:

$$e_n(\xi) = \sqrt{2}\sin((n + \frac{1}{2})\pi\xi)\,, \xi \in (0,1)\,, \gamma_n = ((n + \frac{1}{2})\pi)^{-2}\ n = 1,\dots.$$

Again the eigenfunctions form an orthonormal and complete basis in H and the eigenvalues are positive and consequently the operator Q is positive . Moreover:

$$Ae_n(\xi) = -((n + \frac{1}{2}\pi))^2 e_n(\xi), \xi \in (0,1), \ n = 1, \ldots .$$

and

$$\alpha_n = ((n + \frac{1}{2})\pi)^2, \ n = 1, \ldots .$$

Thus also this example is of the previously considered form. Note that in both cases $A = -Q^{-1}$.

6. General parabolic equations. Regularity

Regularity of the solutions of general second order parabolic equations on Hilbert spaces is sudied. Nonsmooth initial functions are also considered using the Bismut-Elworthy-Xe formula.

6.1 Convolution type and evaluation maps

The present chapter is concerned with the Kolmogorov equation

$$
\begin{aligned}
\frac{\partial u}{\partial t}(t,x) &= \frac{1}{2}\operatorname{Tr}[(G(x)Q^{1/2})^* D^2 u(t,x)(G(x)Q^{1/2})] \qquad (6.1)\\
&\quad + \langle Ax + F(x), Du(t,x)\rangle, \quad t \geq 0, \ x \in D(A)\\
u(0,x) &= \varphi(x), \quad x \in H.
\end{aligned}
$$

The generalized solution to 6.1 is defined by the usual formula:

$$
\begin{aligned}
u(t,x) &= \mathbb{E}(\varphi(X(t,x))),\\
&= P_t\varphi(x), \quad t \geq 0, \ x \in H
\end{aligned}
$$

where the process $X(t,x)$ $t \geq 0$, $x \in H$, satisfies the stochastic equation,

$$
\begin{aligned}
dX &= (AX + F(X))\,dt + G(X(t))\,dW(t), \qquad (6.2)\\
X(0) &= x \in H, \qquad (6.3)
\end{aligned}
$$

on H.

We first prove results on existence of solutions to 6.2 and on their dependence on initial data. To do so it will be convenient to treat 6.2, as a fixed point problem:

$$X = K(x, X), \qquad (6.4)$$

where $K(x, .)$ is a convolution like transformation defined by the right hand side of 2.28. We will use a functional analytic approach. The approach is based on implicit function theorem, postponed to the Appendix, and on properties of deterministic and stochastic convolutions, gathered in the present section. Denote by $H^p([0,T])$ the space of all progressively measurable H-valued process ψ defined on $[0,T]$, $T > 0$, $p \geq 2$, equipped with the norm:

$$\|\psi\|_{H^p[0,T]} = \sup_{t \le T}(\mathbb{E}\|\psi(t)\|_H^p)^{1/p}. \tag{6.5}$$

Similarly denote by $\mathcal{H}^p([0,T])$ the space of all progressively measurable $\mathcal{H}$-valued processes Ψ, see §§ 2.3.2, defined on $[0,T]$, equipped with the norm:

$$\|\Psi\|_{\mathcal{H}^p[0,T]} = \sup_{t \le T}(\mathbb{E}\|\Psi(t)\|_\mathcal{H}^p)^{1/p}. \tag{6.6}$$

The normed spaces $H^p([0,T])$, $\mathcal{H}^p([0,T])$ are Banach spaces denoted shortly by H^p and $\mathcal{H}^p$. We will need the following equivalent norms:

$$\|\psi\|_{p,\lambda,T} = \sup_{t \le T} e^{-\lambda t}(\mathbb{E}(\|\psi(t)\|_H)^p)^{1/p} \tag{6.7}$$

$$\||\Psi\||_{p,\lambda,T} = \sup_{t \le T} e^{-\lambda t}(\mathbb{E}\|\Psi(t)\|_\mathcal{H}^p)^{1/p}. \tag{6.8}$$

Assume now that $S(t)$, $t \ge 0$ is a C_0-semigroup of bounded linear operators on H and define for $\psi \in H^p([0,T])$, $\Psi \in \mathcal{H}^p([0,T])$ two *convolution type* mappings:

$$\mathcal{T}_0(\psi)(t) = \int_0^t S(t-s)\psi(s)\,ds, \quad t \in [0,T], \tag{6.9}$$

$$\mathcal{T}_1(\Psi)(t) = \int_0^t S(t-s)\Psi(s)\,dW(s), \quad t \in [0,T]. \tag{6.10}$$

Let $M_T = \sup_{t \le T}\|S(t)\|_{L(H,H)}$.
We have the following easy consequence of the Burkholder-Davis-Gundy inequalities 2.23.

Theorem 6.1. *For arbitrary $p \ge 2$, $T > 0$, the formulae 6.9 and 6.10, define linear operators from $H^p[0,T]$ into $H^p[0,T]$ and from $\mathcal{H}^p[0,T]$ into $H^p[0,T]$ respectively. In addition, for arbitrary $\alpha \in]0,1[$, there exists $\lambda > 0$ such that,*

$$\|\mathcal{T}_0(\psi)\|_{p,\lambda,T} \le \alpha\|\psi\|_{p,\lambda,T}, \tag{6.11}$$

$$\|\mathcal{T}_1(\Psi)\|_{p,\lambda,T} \le \alpha\||\Psi\||_{p,\lambda,T}. \tag{6.12}$$

Proof. We will show only how to select $\lambda > 0$ to fulfill 6.12. By the very definition:

$$\mathbb{E}\|\mathcal{T}_1(\Psi)(t)\|_H^p = \mathbb{E}\|\int_0^t S(t-s)\Psi(s)\,dW(s)\|_H^p$$

$$\le M_T^p c_p \mathbb{E}\left(\int_0^t \|\Psi(s)\|_\mathcal{H}^2\,ds\right)^{p/2}$$

$$\le M_T^p c_p t^{p/2-1}\mathbb{E}\left(\int_0^t \|\Psi(s)\|_\mathcal{H}^p\,ds\right)$$

$$\le M_T^p c_p t^{p/2-1}\int_0^t e^{\lambda t p}\mathbb{E}(e^{-\lambda s}\|\Psi(s)\|)_\mathcal{H}^p\,ds$$

$$\leq M_T^p c_p t^{p/2-1} \left(\int_0^t e^{\lambda t p}\, ds \right) (\|||\Psi|||_{p,\lambda,T}^p)$$

$$\leq M_T^p c_p t^{p/2-1} \frac{e^{\lambda t p}}{\lambda p} \|||\Psi|||_{p,\lambda,T}^p.$$

Therefore,

$$\|\mathcal{T}_1(\Psi)\|_{p,\lambda,T}^p \leq \frac{1}{\lambda} M_T^p c_p \frac{T^{p/2-1}}{p} \|||\Psi|||_{p,\lambda,T}^p$$

and it is enough to choose

$$\lambda < \frac{M_T^p c_p T^{p/2-1}}{p}. \quad \Box \qquad (6.13)$$

Assume now that $F : H \to H$, $G : H \to \mathcal{H}$ are given transformations. They induce mappings $\mathcal{F}$ and $\mathcal{G}$ on stochastic processes according to the formulae:

$$\mathcal{F}(\psi)(t) = F(\psi(t)), \quad \psi \in H^p[0,T], \ t \in [0,T], \qquad (6.14)$$
$$\mathcal{G}(\psi)(t) = G(\psi(t)), \quad \psi \in H^p[0,T], \ t \in [0,T]. \qquad (6.15)$$

They are often called *evaluation maps*.

Theorem 6.2. *If $F : H \to H$, $G : H \to \mathcal{H}$ are Lipschitz mappings with the Lipschitz constant $\gamma > 0$ then also mappings $\mathcal{F} : H^p([0,T]) \to H^p([0,T])$, $\mathcal{G} : H^p([0,T]) \to \mathcal{H}^p([0,T])$ are Lipschitz with the same constants in the norms $\|\cdot\|_{p,T,\lambda}$, $\|||\cdot|||_{p,T,\lambda}$ for any $\lambda > 0$.*

Proof. We consider for instance transformation $\mathcal{G}$. Let $\psi_1, \psi_2 \in H^p([0,T])$ then

$$e^{\lambda t p} \mathbb{E} \|\mathcal{G}(\psi_1)(t) \ - \ \mathcal{G}(\psi_2)(t)\|_{\mathcal{H}}^p$$
$$= e^{\lambda t p} \mathbb{E} \|G(\psi_1(t)) - G(\psi_2(t))\|_{\mathcal{H}}^p$$
$$\leq \gamma^p e^{\lambda t p} \mathbb{E} \|\psi_1(t) - \psi_2(t)\|_H^p$$
$$\leq \gamma^p \|\psi_1 - \psi_2\|_{p,\lambda,T}^p.$$

Consequently

$$\|||\mathcal{G}(\psi_1) - \mathcal{G}(\psi_2)|||_{p,\lambda,T}^p \leq \gamma^p \|\psi_1 - \psi_2\|_{p,\lambda,T}^p$$

as required. $\Box$

Theorem 6.3. *Assume that Lipschitz mappings $F : H \to H$, $G : H \to \mathcal{H}$ have directional derivatives*

$$\partial^k F(x; y_1, \ldots, y_k), \quad \partial^k G(x; y_1, \ldots, y_k), \quad x \in H, \ y_1, \ldots, y_k \in H \quad (6.16)$$

for $k = 1, \ldots, n$, continuous in all variables and such that for $k = 1, \ldots, n$

$$\sup_{x \in H} \left[\sup_{\|y_i\| \le 1, \, i=1,\dots,k} \|\partial^k F(x; y_1, \dots, y_k)\|_H \right] < +\infty, \qquad (6.17)$$

$$\sup_{x \in H} \left[\sup_{\|y_i\| \le 1, \, i=1,\dots,k} \|\partial^k G(x; y_1, \dots, y_k)\|_{\mathcal{H}} \right] < +\infty. \qquad (6.18)$$

Then the transformations $\mathcal{F} : H^p([0,T]) \to H^p([0,T])$, $\mathcal{G} : H^p([0,T]) \to \mathcal{H}^p([0,T])$, *have directional derivatives*

$$\partial^k \mathcal{F}(X; Y_1, \dots, Y_k), \quad \partial^k \mathcal{G}(X; Y_1, \dots, Y_k), \quad k = 0, 1, \dots, n \qquad (6.19)$$

for each $X \in H^p([0,T])$ *and* $Y_1, \dots, Y_k \in H^{kp}([0,T])$. *The derivatives are continuous from* $H^p \times \underbrace{H^{kp} \times \dots \times H^{kp}}_{k-times}$ *into* H^p, *and* $\mathcal{H}^p$ *respectively and*

$$\sup_{X \in H^p} \left[\sup_{\|Y_i\|_{H^p} \le 1, \, i=1,\dots,k} \|\partial^k \mathcal{F}(X; Y_1, \dots, Y_k)\|_{H^p} \right] < +\infty, \quad (6.20)$$

$$\sup_{X \in H^p} \left[\sup_{\|Y_i\|_{H^{kp}} \le 1, \, i=1,\dots,k} \|\partial^k \mathcal{G}(X; Y_1, \dots, Y_k)\|_{\mathcal{H}^p} \right] < +\infty. \quad (6.21)$$

Moreover

$$\partial^k \mathcal{F}(X; Y_1, \dots, Y_k)(t) = \partial^k F(X(t); Y_1(t), \dots, Y_k(t)), \quad t \in [0, T], \quad (6.22)$$
$$\partial^k \mathcal{G}(X; Y_1, \dots, Y_k)(t) = \partial^k G(X(t); Y_1(t), \dots, Y_2(t)), \quad t \in [0, T]. \quad (6.23)$$

Proof. We will consider for instance the operator $\mathcal{F}$. Note that if the formula 6.22 holds then it defines a transformation with the properties specified in the theorem. The theorem is true if $n = 1$. If it is true for some k then for arbitrary $X \in H^p$, $Y_1, \dots, Y_{k+1} \in H^{(k+1)p}$, $\sigma > 0$:

$$\frac{1}{\sigma}[\partial^k \mathcal{F}(X(t) + \sigma Y_{k+1}(t); Y_1(t), \dots, Y_k(t)) - \partial^k \mathcal{F}(X; Y_1, \dots, Y_k)(t)]$$

$$= \frac{1}{\sigma}[\partial^k F(X(t) + \sigma Y_{k+1}(t); Y_1(t), \dots, Y_k(t)) - \partial^k F(X(t); Y_1(t), \dots, Y_k(t))]$$

$$= \int_0^1 \partial^k F(X(t) + \sigma s Y_{k+1}(t); Y_1(t), \dots, Y_k(t), Y_{k+1}(t)) \, ds.$$

Therefore,

$$\left\| \frac{1}{\sigma}[\partial^k \mathcal{F}(X + \sigma Y_{k+1}; Y_1, \dots, Y_k) - \partial^k \mathcal{F}(X; Y_1, \dots, Y_k)] \right.$$

$$\left. - \partial^{k+1} F(X(\cdot); Y_1(\cdot), \dots, Y_k(\cdot), Y_{k+1}(\cdot)) \right\|_{\mathcal{H}^p}^p$$

$$\le \mathbb{E}\left[\int_0^1 |\partial^{k+1} F(X(t) + \sigma s Y_{k+1}(t); Y_1(t), \dots, Y_k(t)) \right.$$

$$\left. - \partial^{k+1} F(X(t); Y_1(t), \dots, Y_k(t)) \right]$$

$$\to 0 \quad \text{because } \sigma \downarrow 0 \quad \square$$

Let $\varphi : H \to R^1$ be a given function. If ξ is an H-valued random variable defined on $(\Omega, \mathcal{F}, \mathbb{P})$ then $\Phi(\xi)$ given by the evaluation formula:

$$\Phi(\xi)(\omega) = \varphi(\xi(\omega)), \quad \omega \in \Omega$$

is a real-valued random variable provided φ is a Borel function.

The following classical result can be proved in a similar way as the previous theorems and therefore its proof will be omitted.

Theorem 6.4. i) *If φ is a bounded and continuous function then Φ is a continuous map from $L^p(\Omega; H)$ into $L^q(\Omega; H)$ for any $p, q \geq 1$.*

ii) *If φ is bounded and continuous together with its first derivative then for arbitrary $p > 2$ and $q \in [1, \frac{p}{p-q}]$, Φ is Fréchet differentiable from $L^p(\Omega; H)$ into $L^q(\Omega; H)$.*

iii) *If φ is bounded and continuous together with its first and second derivatives then for arbitrary $p > 4$ and $q \in [1, \frac{p}{p-2}]$, Φ is twice Fréchet differentiable from $L^p(\Omega; H)$ into $L^q(\Omega; H)$.*

6.2 Solutions of stochastic equations

We go back to the equation 6.2 and using results of the §6.1 and of the Appendix we investigate existence of solutions to 6.2 and the character of their dependence on initial conditions $x \in H$.

Theorem 6.5. *Assume that $F : H \to H$, $G : H \to \mathcal{H}$ are Lipschitz mappings. Then for each $p \geq 2$, $T > 0$ and each $x \in H$ the equation 6.2 has a unique solution $X(\cdot, x)$ in $H^p([0, T])$. Moreover the mapping $x \to X(\cdot, x)$ from H into $H^p([0, T])$ is Lipschitz.*

Proof. According to the definitions one has to show that the integral equation 2.28 has a unique solution in $H^p([0, T])$. For each $x \in H$ and $Y \in H^p([0, T])$ define a mapping $\mathcal{K}$:

$$\mathcal{K}(x, Y)(t) = S(t)x + \int_0^t S(t - s)F(Y(s))\, ds \qquad (6.24)$$

$$+ \int_0^t S(t - s)G(Y(s))\, dW(s), \quad t \in [0, T],$$

from $H \times H^p([0, T])$ into $H^p([0, T])$. By results of §6.1 the mapping $\mathcal{K}$ is well defined. It is linear in the variable x and is the sum of linear, integral mappings $\mathcal{T}_0$ and $\mathcal{T}_1$ composed with evaluation maps $\mathcal{F}$ and $\mathcal{G}$. By Theorem 6.1 and Theorem 6.2 we can apply Theorem 10.1 from the Appendix with the function H replaced by $\mathcal{K}$, the space Λ replaced by H and the space E replaced by $H^p([0, T])$. This way we get existence of a solution $X(\cdot, x)$ which depends continuously on x. The mapping $x \to X(\cdot, x)$ is in fact Lipschitz and this follows from 10.3. $\square$

Theorem 6.6. *In addition to assumptions of Theorem 6.5 assume that* $F \in C_b^2(H, H)$, $G \in C_b^2(H, \mathcal{H})$. *Then for each* $p \geq 2$ *and* $T > 0$ *the mapping* $x \to X(\cdot, x)$, *from* H *into* $H^p([0, T])$ *is 2-times continuously differentiable in* x. *Moreover for any* $y, z \in H$ *the processes* $\zeta^y(t) = \partial_x X(t, x; y)$, $\eta^{y,z}(t) = \partial_x^2 X(t, x; y, z)$, $t \in [0, T]$ *are the unique solutions of the following equations:*

$$d\zeta^y = (A\zeta^y + \partial_x F(X; \zeta^h)) \, dt + \partial_x G(X; \zeta^h) \, dW(t) \qquad (6.25)$$

$$\zeta^y(0) = y,$$

$$d\eta^{y,z} = (A\eta^{y,z} + \partial_x F(X; \eta^{y,z})) \, dt + \partial_x G(X; \eta^{y,z}) \, dW(t) \qquad (6.26)$$

$$+ \partial_x^2(X; \zeta^y, \zeta^z) \, dt + \partial_x^2 G(X; \eta^{y,z}, \eta^{y,z}) \, dW(t).$$

Proof. To prove 6.25 we apply Theorem 10.2 from the Appendix. Its assumptions are satisfied by Theorem 6.3 and Theorem 6.1. Note that equation 6.25 in equivalent to the integral one:

$$\zeta^y(t) = S(t)y + \int_0^t S(t - s) \partial_x F(X(s, x); \zeta^y(s)) \, ds \qquad (6.27)$$

$$+ \int_0^t S(t - s) \partial_x G(X(s, x); \zeta^y(s)) \, dW(s), \quad t \in [0, T]$$

which has a unique solution by the contraction mapping principle. It is equivalent to equation 10.4.

To derive 6.26 we apply Theorem 10.4 from the Appendix, taking $E = H^p([0, T])$ and $E_0 = H^{2p}([0, T])$. By Theorem 6.3 and Theorem 6.1 assumptions of Theorem 10.4 from the Appendix are satisfied. Moreover the equation 6.26 is equivalent to 10.6.   □

6.3 Space and time regularity of generalized solutions

As a first consequence of the previous results we have a theorem on differentiability of the generalized solution.

Theorem 6.7. *Assume that mappings:* $\varphi \in H \to R^1$, $F : H \to H$, $G : H \to \mathcal{H}$ *belong respectively to* $C_b^2(H; R^1)$, $C_b^2(H, H)$ *and* $C_b^2(H, \mathcal{H})$. *Then for arbitrary* $t \geq 0$ *the function* $u(t, \cdot)$ *is twice continuously differentiable.*

Proof. The function $x \to u(t, x)$ can be regards as a composition of the mappings from H into $L^p(\Omega; H)$ and from $L^p(\Omega; H)$ into $L^q(\Omega; R^1)$, given by $x \to X(t, x)$, $\xi \to \varphi(\xi)$ and of the linear, integration operator $\eta \to \mathbb{E}(\eta)$ from $L^q(\Omega; R^1)$ into R^1. Therefore it is enough to apply Theorem 6.6 and Theorem 6.4.   □

We will show now that the generalized solution u is continuous with respect to time and that the same is true for its first and second space derivatives. To do so we consider subspaces $H_c^p[0, T]$ and $\mathcal{H}_c^p[0, T]$ of $H^p[0, T]$ and $\mathcal{H}^p[0, T]$

consisting of those processes ψ and Ψ which are continuous in $t \in [0,T]$. Thus $\psi \in H_c^p[0,T]$ if $\psi \in H^p[0,T]$ and for each $t_0 \in [0,T]$,

$$\lim_{t \to t_0} \mathbb{E}|\psi(t) - \psi(t_0)|_H^p = 0$$

In a similar way $\Psi \in \mathcal{H}_c^p[0,T]$ if $\Psi \in \mathcal{H}^p[0,T]$ and for each $t_0 \in [0,T]$,

$$\lim_{t \to [0,T]} \mathbb{E}\|\Psi(t) - \Psi(t_0)\|_{\mathcal{H}}^p = 0$$

The spaces $H_c^p[0,T]$, $\mathcal{H}_c^p[0,T]$ are linear, closed subspaces of $H^p[0,T]$, and $\mathcal{H}^p[0,T]$ respectively and therefore are Banach spaces as well. We have also the following important proposition:

Proposition 6.8. *The formulae 6.9 and 6.10 define continuous, linear mappings from $H^p[0,T]$ and from $\mathcal{H}^p[0,T]$ into $H_c^p[0,T]$ respectively.*

Proof. We prove the result for $\mathcal{T}_1$ only; the proof for $\mathcal{T}_0$ is similar. Let $0 \leq s < t \leq T$ and $h = t - s$. Then

$$\mathbb{E}|\mathcal{T}_1\Psi(s) - \mathcal{T}_1\Psi(t)|^p$$

$$\leq 2^{p-1}\mathbb{E}\left|\int_0^s S(s-\sigma)\Psi(\sigma)\,dW(\sigma) - S(h)\int_0^s S(s-\sigma)\Psi(\sigma)\,dW(\sigma)\right|^p$$

$$+2^{p-1}\mathbb{E}\left|\int_s^{s+h} S(s+h-\sigma)\Psi(\sigma)\,dW(\sigma)\right|^p$$

$$\leq 2^{p-1}M_T^p s^{p/2-1}\left(\mathbb{E}\int_0^s \|(S(h)-I)\Psi(\sigma)\|_{\mathcal{H}}^p\,d\sigma\right)$$

$$+2^{p-1}M_T^p h^{p/2-1}\mathbb{E}\left(\int_0^h \|\Psi(\sigma)\|_{\mathcal{H}}^p\,d\sigma\right) = I_1 + I_2$$

Since $\|(S(h)-I)\Psi(\sigma)\|_{\mathcal{H}}^p \leq (M+1)^p\|\Psi(\sigma)\|_{\mathcal{H}}^p$, $\|(S(h)-I)\Psi(\sigma)\|_{\mathcal{H}}^p \to 0$ as $h \to 0$, $\mathbb{P}$-almost surely and $\mathbb{E}\int_0^T \|\Psi(\sigma)\|_{\mathcal{H}}^p\,d\sigma < +\infty$, the first integral I_1 converges to 0 as $h \to 0$, (uniformly in $t, s \in [0,T]$). It is clear that also $I_2 \to 0$ as $h \to 0$ and the result follows. □

Taking into account Proposition 6.8 and the proofs of Theorem 6.6 and Theorem 6.7 we arrive at following time-regularity result:

Theorem 6.9. *Under the assumptions of Theorem 6.7 the generalized solution $u(t,x)$, $t \in [0,T]$, $x \in H$ is continuous with respect to both variables together with its first and second x-derivatives.*

6.4 Strong Feller property

We continue our study of the regularity properties of the generalized solution u determined by the stochastic equation,

$$dx = (Ax + F(X)) dt + G(X(t)) dW(t), \qquad (6.28)$$
$$X(0) = x,$$

discussed in the previous sections. Let P_t be the corresponding transition semigroup:

$$P_t\varphi(x) = \mathbb{E}(\varphi(X(t,x))),$$
$$= u(t,x), \quad t \geq 0, \ x \in H$$

We consider irregular initial functions φ. For applications it is important to know when the transition semigroup (P_t) is *strong Feller* in the sense that for arbitrary $t > 0$, $\varphi \in B_b(H)$, $P_t\varphi \in C_b(H)$. Necessary and sufficient conditions for the strong Feller property in the case $F = 0$ and $G = I$ were given in Theorem 5.5, §5.3. It is easy to check that if , in particular, $Q = I$, then the condition 5.18 holds and the corresponding Ornstein–Uhlenbeck semigroup is strong Feller. In this section we consider rather general F and G, assume that $Q = I$ and follow [PeZa].

Theorem 6.10. *Assume that $F : H \to H$, $G : H \to L(H)$ are Lipschitz transformations and that for all $x \in H$, $G(x)$ is an invertible mapping such that for some $K > 0$,*

$$\|G^{-1}(x)\| \leq K, \quad x \in H. \qquad (6.29)$$

Assume in addition that for all $t > 0$, $S(t) \in L_{HS}(H,H)$,

$$\int_0^t \|S(\sigma)\|_{HS}^2 d\sigma < +\infty. \qquad (6.30)$$

Then for arbitrary $T > 0$ there exists a constant $C_T > 0$ such that for all $\varphi \in B_b(H)$ and all $t \in [0,T]$,

$$|P_t\varphi(x) - P_t\varphi(y)| \leq \frac{C_T}{\sqrt{t}} \|\varphi\|_0 |x - y|, \quad x, y \in H. \qquad (6.31)$$

Proof. We start from proving 6.31 in the case of smooth F, G and φ.

Lemma 6.11. *Assume that the mappings F, G and the function φ have bounded and uniformly continuous derivatives up to the second order. Then for each $t > 0$, $P_t\varphi(\cdot) \in UC_b^2(H)$ and*

$$\varphi(X(t,x)) = P_t\varphi(x) \qquad (6.32)$$
$$+ \int_0^t \langle DP_{t-s}\varphi(X(t,x)), G(X(s,x)) \, dW(S), \quad \mathbb{P}\text{-}a.s.$$

Proof. Sketch. Let $\{e_n\}$ be a complete orthonormal basis in H. For each n, let $X_n(\cdot) = X_n(\cdot, x)$ be the solution to the equation

$$
\begin{aligned}
dX_n &= (A_n X_n + F(X_n))\,dt + G(X_n)Q_n^{1/2}\,dW, \qquad (6.33)\\
X_n(0) &= x,
\end{aligned}
$$

where $A_n = nA(n - A)^{-1}$ is the Yosida approximation of A and Q_n is the orthogonal projection of H onto $\lim\{e_1,\dots,e_n\}$, $n \in \mathbb{N}$.

It can be shown that the function

$$
u_n(t, x) = \mathbb{E}(\varphi(X_n(t, x))),\, (t, x) \in [0, +\infty) \times H \qquad (6.34)
$$

is the classical solution of the Kolmogorov equation

$$
\begin{aligned}
\frac{\partial}{\partial t} u_n(t, x) &= \frac{1}{2}\,\mathrm{Tr}[G(x)D^2 u_n(t, x)G^*(x)] \qquad (6.35)\\
&\quad + \langle A_n x + F(x), Du_n(t, x)\rangle,\\
u_n(0, x) &= \varphi(x), \qquad t \ge 0,\ x \in H \qquad (6.36)
\end{aligned}
$$

Applying ItÔ's formula to the process $\psi(s) = u_n(t - s, X_n(s, x))$, $s \in [0, t]$, $x \in H$ one obtains $\mathbb{P}$-a.s.,

$$
\varphi(X_n(t, x)) = u_n(t, x) + \int_0^t \langle Du_n(t-s, X_n(s, x))G(X_n(x, s))\,dW(s)\rangle. \quad (6.37)
$$

Note that

$$
\begin{aligned}
u(s, x) &= \mathbb{E}(\varphi(X_n(s, x))),\\
Du_n(s, x) &= \mathbb{E}[(DX_n(s, x))^* D\varphi(X_n(s, x))], \qquad (6.38)\\
u(s, x) &= \mathbb{E}(\varphi(X(s, x))),\\
Du(s, x) &= \mathbb{E}[(DX(s, x))^* D\varphi(X(s, x))]. \qquad (6.39)
\end{aligned}
$$

Applying Theorem 10.6 and Theorem 10.7, and taking into account that the approximation procedure affects only the semigroup and the Wiener process, one can pass to the limit in 6.38 to get 6.39 and in 6.37 to arrive at 6.32. $\quad\square$

We derive now the so called Bismut-Elworthy-Xe formula, see [DaPrElZa].

Lemma 6.12. *Under the assumptions of Lemma 6.11, the directional derivative $\langle DP_t\varphi(x), h\rangle$ is given by the formula,*

$$
\langle DP_t\varphi(x), h\rangle \qquad (6.40)
$$

$$
= \frac{1}{t}\mathbb{E}\left\{ \varphi(X(t, x))\int_0^t \langle B^{-1}(X(s, x))(DX(s, x)h),\, dW(s)\rangle \right\}.
$$

Proof. Fix $h \in H$ and define $u(t, x) = P_t\varphi(x)$, $t \geq 0$, $x \in H$. Multiplying the both sides of 6.32 by

$$\int_0^t \langle B^{-1}(X(s, x))[DX(s, x)h], dW(s) \rangle,$$

and taking expectations one gets,

$$\mathbb{E}\left(\varphi(X(t, x)) \int_0^t \langle G^{-1}(X(s, x))(DX(s, x)h), dW(s) \rangle \right)$$

$$= \mathbb{E}\left[\int_0^t \langle G^*(X(s, x))DP_{t-s}\varphi(X(s, x)), G^{-1}(X(s, x))(DX(s, x)h) \rangle\, ds \right],$$

$$= \mathbb{E}\left[\int_0^t \langle DP_{t-s}\varphi(X(s, x)), DX(s, x)h \rangle\, ds \right],$$

$$= \int_0^t \langle D[\mathbb{E}(P_{t-s}\varphi(X(s, x)))], h \rangle\, ds,$$

$$= \int_0^t \langle D(P_s P_{t-s}\varphi)(x), h \rangle\, ds = t\langle DP_t\varphi(x), h \rangle. \square$$

Lemma 6.13. *Under the assumptions of Lemma 6.11, the estimate 6.31 holds true.*

Proof. Fix $T > 0$, then by Lemma 6.12,

$$|\langle DP_t\varphi(x), h \rangle|^2 \leq \frac{1}{t^2}\|\varphi\|_0^2 \mathbb{E}\left\{ \int_0^t |G^{-1}(X(s, x)h|^2\, ds \right\} \qquad (6.41)$$

$$\leq \frac{K^2}{t^2}\|\varphi\|_0^2 \mathbb{E}\left\{ \int_0^t |DX(s, x)h|^2\, ds \right\},$$

and we need an estimate on the process $Y(t) = DX(t, x)h$, $t \geq 0$. By Theorem 6.6 in §6.2, the process Y is the mild solution of the equation:

$$dY(t) = [AY(t) + DF(X(t); Y(t))]\, dt + DG(X(t); Y(t))\, dW(s) \quad (6.42)$$
$$Y(0) = h$$

The equation 6.42 is linear, with random coefficients, and equivalent to the integral equation:

$$Y(t) = S(t)h + \int_0^t S(t - s)DF(X(S); Y(s))\, ds$$

$$+ \int_0^t S(t - s)DG(X(S); Y(s)\, dW(s)$$

By our assumptions, for arbitrary $T > 0$, there exists $M > 0$, such that:

$$|DF(x;y)| \leq M|y|,$$
$$\|S(\sigma)DG(x;y)\|_{HD} \leq M|y|, \quad x,y \in H$$

Applying the contraction mapping principle in $\mathcal{H}^2[0,T]$ one easily obtains that 6.42 has a unique solution satisfying

$$\mathbb{E}|Y(t)|^2 \leq C|h|^2, \quad t \in [0,T], \tag{6.43}$$

if T is sufficiently small. Reiterating the procedure one gets 6.43 for arbitrary $T > 0$, $h \in H$. Applying 6.43 to 6.41 one has that

$$|\langle DP_t\varphi(x), h\rangle|^2 \leq \frac{K^2}{t}C\|\varphi\|_0^2|h|^2, \quad t \in [0,T].$$

Let x, y be arbitrary elements in H. Then by the mean value theorem, for $\sigma(y) \in [0,T]$:

$$P_t\varphi(x) - P_t\varphi(y) = \langle DP_t\varphi(x + \sigma(y)(x-y)), x-y\rangle.$$

Therefore

$$|P_t\varphi(x) - P_t\varphi(y)| \leq \frac{K\sqrt{C}}{\sqrt{t}}\|\varphi\|_0|x-y|,$$

as required.   □

Lemma 6.14. *If 6.31 holds for arbitrary* $\varphi \in UC_b^2(H)$, *then it holds for all* $\varphi \in B_b(H)$.

Proof. If $\varphi \in UC_b(H)$, then there exists a sequence (φ_n) of functions from $UC_b^2(H)$ such that $\lim_n \varphi_n(x) = \varphi(x)$, $\lim_n \|\varphi_n\|_0 = \|\varphi\|$, and therefore 6.31 holds for all $\varphi \in UC_b(H)$. From elementary properties of measures on metric spaces one has the following estimate of the variation of the signed measure $P(t,x,\cdot) - P(t,y,\cdot)$:

$$\mathrm{Var}(P(t,x,\cdot) - P(t,y,\cdot)) = \sup_{\substack{\|\varphi\|_0 \leq 1 \\ \varphi \in UC_b(H)}} |P_t\varphi(x) - P_t\varphi(y)|$$

$$\leq \frac{C_T}{\sqrt{t}}|x-y|, \quad x,y \in H.$$

But then for $\varphi \in B_b(H)$:

$$|P_t\varphi(x) - P_t\varphi(y)| \leq \left|\int_H (z)[P(t,x,dz) - P(t,y,dz)]\right|$$

$$\leq \|\varphi\|_0 \mathrm{Var}(P(t,x,\cdot) - P(t,y,\cdot))$$

$$\leq \|\varphi\|_0 \frac{C_T}{\sqrt{t}}|x-y|. \square$$

It follows from Lemma 6.14 that to prove the theorem one can assume that $\varphi \in UC_b^2(H)$. It is possible, see [PeZa], to construct approximations F_n of F and G_n of G with the following properties:

i) F_n, G_n, $n \in \mathbb{N}$ are twice Fréchet differentiable with bounded and continuous derivatives,

ii) F_n, G_n, $n \in \mathbb{N}$ satisfy the Lipschitz condition,

iii) Operators G_n, $n \in \mathbb{N}$ are invertible and

$$\varlimsup_{n \to +\infty} [\sup_{x \in H} \|G_n^{-1}(x)\|] \le \sup_{x \in H} \|G^{-1}(x)\|,$$

iv) $\quad \lim_{n \to +\infty} |F_n(x) - F(x)| = 0, \quad \lim_{n \to +\infty} \|G_n(x) - G(x)\| = 0, \quad x \in H$

To complete the proof of the theorem denote by $X_n(\cdot, x)$, the solution of the equation

$$\begin{aligned} dX_n &= (AX_n + F_n(X_n))\, dt + G_n(X_n)\, dW(t) \\ X_n(0) &= x \end{aligned}$$

and let P_t^n, $t \ge 0$ be the corresponding semigroup. Fix $T > 0$, $\varphi \in UC_b^2(H)$. By Lemma 6.13

$$|P_t^n\varphi(x) - P_t^n\varphi(y)| \le \frac{C_T}{\sqrt{t}}\|\varphi\|_0|x - y|, \quad x, y \in H,\ t \in [0, T].$$

However for fixed $t > 0$ and $x \in H$ there exists a subsequence $(X_{n_k}(t, x))$ such that $X_{n_k}(t, x) \to X(t, x)$, $\mathbb{P}$-a.s. as $k \to +\infty$. Since φ is bounded and continuous function we have

$$P_t^{n_k}\varphi(x) = \mathbb{E}(\varphi(X_{n_k}(t, x))) \to \mathbb{E}(\varphi(X(t, x))) = P_t\varphi(x),$$

as $k \to +\infty$ and therefore

$$|P_t\varphi(x) - P_t\varphi(y)| \le \frac{C_T}{\sqrt{t}}\|\varphi\|_0|x - y|, \quad x, y \in H,\ t \in [0, T]$$

as required. □

Example 6.15. Assume that $H = L^2(0, \pi)$ and $A = \frac{d^2}{d\xi^2}$ with the Dirichlet boundary conditions. Eigenvalues and eigenfunctions of A are $(-n^2)$ and $(\sqrt{\frac{2}{\pi}} \sin n\xi)$ respectively. Consequently

$$\|S(t)\|_{HS}^2 = \sum_{n=1}^{+\infty} e^{-2tn^2},$$

$$\int_0^t \|S(s)\|_{HS}^2\, ds = \sum_{n=1}^{+\infty} \frac{1 - e^{-2tn^2}}{2n^2} < +\infty, \quad t > 0.$$

Let $F(x)(\xi) = f(X(\xi))$, $\xi \in (0, \pi)$, $x \in H$ where f is a given Lipschitz function. If in addition $B \equiv I$ then all conditions of the theorem are satisfied and the transition semigroup corresponding to the stochastic parabolic equation,

$$\frac{\partial X}{\partial t}(t,\xi) = \frac{\partial^2 X}{\partial \xi^2}(t,\xi) + f(X(t,\xi)) + \frac{\partial W}{\partial t}(t,\xi)$$

$$X(0,\xi) = x(\xi), \quad \xi \in (0,\pi), \ t > 0,$$

transforms bounded functions onto continuous and the estimate 6.31 holds.

7. General parabolic equations. Uniqueness.

The problem of uniqueness of the solutions to the Kolmogorov equation is discussed. ItÔ formula and regularization schemes for stochastic evolution equations are used.

7.1 Uniqueness for the heat equation

First we discuss uniqueness in the simplest case of the infinite dimensional heat equation.

Theorem 7.1. *Assume that functions* $\varphi(x)$, $u(t,x)$, $D_t u(t,x)$, $D_x u(t,x)$, $D_x^2 u(t,x)$, $t > 0$, $x \in H$, *are uniformly continuous on closed, bounded subsets of* $]0,+\infty[\times H$ *and that equation:*

$$\frac{\partial u}{\partial t}(t,x) = \frac{1}{2}\operatorname{Tr} D_Q^2 u(t,x), \quad t > 0 \ x \in H,$$

$$\lim_{t\downarrow 0} u(t,x) = \varphi(x), \quad x \in H,$$

holds. If, in addition, the solution u *is continuous on* $[0,+\infty) \times H$ *and for arbitrary* $T > 0$ *there exists* $M > 0$ *such that,*

$$|u(t,x)| + \|D_x u(t,x)\| \leq M e^{M\|x\|}, \quad (t,x) \in [0,T] \times H,$$

then

$$u(t,x) = \mathbb{E}(\varphi(x + W(t))), \quad t \geq 0, \ x \in E.$$

Proof. Fix $t > t_0 > 0$, $x \in H$ and define,

$$\psi(s) = u(t - s, x + W(s)), \quad s \in [0, t_0].$$

Due to the assumptions imposed on u one can apply ItÔ's formula to ψ:

$$\psi(t_0) = \psi(0) \tag{7.1}$$

$$+ \int_0^{t_0} \left[\frac{1}{2}\operatorname{Tr} D_Q^2 u(t - s, x + W(s)) - \frac{\partial u}{\partial t}(t - s, x + W(s)) \right] ds$$

$$+ \int_0^{t_0} \langle D_x u(t - s, x + W(s)), dW(s) \rangle$$

But u satisfies the heat equation and therefore,

$$u(t - t_0, x + W(t_0)) = u(t, x) + \int_0^{t_0} \langle D_x u(t - s, x + W(s), dW(s)) \rangle. \quad (7.2)$$

However,

$$\mathbb{E} \int_0^{t_0} \|D_x u(t - s, x + W(s)\|^2 ds \leq M^2 e^{2\|x\|} \int_0^{t_0} ds < +\infty.$$

Applying expectation operator to the both sides of 7.2 one arrives at

$$\mathbb{E}u(t - t_0, x + w(t_0)) = u(t, x).$$

Since $|u(t - t_0, x + w(t_0))| \leq M e^{\|x\|} e^{\sup_{s \leq t} \|W(s)\|}$
and, by Fernique's theorem, see [DaPrZa2],

$$\mathbb{E}(e^{\sup_{s \leq t} \|W(s)\|}) < +\infty,$$

the Lebesgue theorem implies

$$u(t, x) = \lim_{t_0 \downarrow 0} \mathbb{E}u(t - t_0, x + w(t_0)) = \mathbb{E}(\varphi(x + w(t))). \quad \square$$

7.2 Uniqueness in the general case

We pass now to the general case and consider Kolmogorov equations of the form:

$$\frac{\partial u}{\partial t}(t, x) = \frac{1}{2} \text{Tr}[(G(x)Q^{1/2})D^2 u(t, x)(G(x)Q^{1/2})^*] \quad (7.3)$$
$$+ \langle Ax + F(x), Du(t, x), \quad t > 0, \ x \in D(A), \rangle$$
$$u(0, x) = \varphi(x), \quad x \in H.$$

As before A denotes the infinitesimal generator of a C_0-semigroup $S(t)$, $t \geq 0$ on H, Q is a selfadjoint bounded, nonnegative operator on H, and $F : H \to H$ and $G : H \to \mathcal{H}$ are Lipschitz mappings. Equation 7.3 corresponds to the stochastic equation

$$dX(t) = (AX(t) + F(X(t)) dt + G(X(t)) dW(t) \quad (7.4)$$
$$X(0) = x \in H, \quad t > 0$$

which has a unique solution $X(\cdot, x)$ in $H^p([0, T])$, for any $T > 0$ and $p \geq 2$.

Our aim is to prove the following theorem.

Theorem 7.2. *Assume that F and G are Lipschitz mappings and $u(t, x)$, $t \geq 0$, $x \in H$ is a bounded, continuous function such that for each $t > 0$, the first and second space derivatives $Du(t, x)$, $D^2 u(t, x)$, $x \in H$, exist are bounded on $(0, +\infty) \times H$ and uniformly continuous on bounded subsets of $[\varepsilon, +\infty) \times H$, for any $\varepsilon > 0$. If u satisfies 7.3 then*

$$u(t, x) = \mathbb{E}(\varphi(X(t, x))), \quad t \geq 0, \ x \in H, \quad (7.5)$$

where X is the solution to 7.4.

For the proof we will need some preparatory work. We can not repeat the proof from the previous section because of two reasons. To apply the Itô formula to the process:

$$u(t - s, X(s, x)), \quad s \in [0, t_0], \ t_0 < t ,$$

the function $u(t, x)$, $t > 0$, $x \in H$ should have continuous the first time derivative, which exists, in general, only if $x \in D(A)$. Moreover the solution X of 7.4 is not given in the integral form:

$$X(t) = x + \int_0^t (AX(S) + F(X(S))) \, ds \tag{7.6}$$
$$+ \int_0^t G(X(S)) \, dW(S), \quad t \geq 0,$$

required by the formulation of the Itô formula, but it satisfies a convolution type version of the stochastic equation, see the §§ 2.3.3. To overcome the difficulties we will approximate both the function u and the process X in such a way that Itô's formula will be applicable and by passing to the limit in the formulae we will arrive at 7.5.

We first need the concept of strong solutions of the stochastic equations. If a solution to 7.4 takes values in $D(A)$,

$$\int_0^t |AX(S)| \, ds < +\infty, \text{ for all } t \geq 0, \mathbb{P} - a.s.,$$

and 7.6 holds, then X is called a *strong solution* to 7.5.

Since $S(t)$, $t \geq 0$ is a C_0-semigroup, there exist constants $M > 0$, $\omega \in R^1$ such that,

$$|S(t)| \leq M e^{\omega t}, \quad t \geq 0 \tag{7.7}$$

and, for all $\lambda > \omega$, the formula

$$R_\lambda x = \int_0^{+\infty} e^{-\lambda t} S(t) x \, dt, \quad x \in H \tag{7.8}$$

defines a bounded linear operator. It is well known and is easy to see that,

$$\|\lambda R_\lambda\| \leq \frac{\lambda}{\lambda - \omega} \quad \text{and} \quad \lambda R_\lambda x \to x \text{ as } \lambda \to +\infty.$$

We fix $\lambda > \omega$ and denote by T_λ the operator λR_λ. Then T_λ is a linear, bounded operator from H into the space $D(A)$, equipped with the graph norm. For $\lambda > \omega$ define

$$F_\lambda(x) = T_\lambda F(x), \quad G_\lambda(x) = T_\lambda G(x), \quad x \in H.$$

It is easy to see that if F and G are Lipschitz mappings with respect to the spaces H, U then F_λ and G_λ are Lipschitz with respect to $D(A)$, U.

Consequently, for each initial $T_\lambda x$, $x \in H$, there exists a unique solution X_λ of the equation

$$dX_\lambda = (\widehat{A} X_\lambda + F_\lambda(X_\lambda)) \, dt + G_\lambda(X_\lambda) \, dW \qquad (7.9)$$
$$X_\lambda(0) = T_\lambda x$$

where $\widehat{A}$ is the restriction of A to $D(A)$. The operator $\widehat{A}$ generates the same semigroup $S(t)$, $t \geq 0$ but restricted $D(A)$.

Lemma 7.3. *If F and G are Lipschitz mappings from H into H and H into $\mathcal{H}$ respectively then, for arbitrary $\lambda > \omega$, $p \geq 2$, $T > 0$, the equation 7.9 has a unique solution X_λ in $H^p([0,T], D(A))$ which is a strong solution of 7.4 with F, G and x are replaced by F_λ, G_λ and $T_\lambda x$. Moreover if $\lambda \to +\infty$, $X_\lambda \to X$ in $H^p([0,T], H)$.*

Proof. That the solution X_λ exists has been already shown. The convergence property follows from Theorem 10.6 ii). Since X_λ is a solution to 7.9 it is also a solution to 7.4 with F, G and x replaced by F_λ, G_λ and $T_\lambda x$. By the very definition, for all $t > 0$, $\int_0^t |A X_\lambda(S)| \, ds < +\infty$, $\mathbb{P}$-a.s. and this easily implies that X_λ is the required strong solution.   □

We pass now to the proof of Theorem 7.2

Proof. Assume that u satisfies equation 7.3 and has the properties formulated in Theorem 7.2. Fix any $t_0 \in (0, t)$, $\lambda > 0$ and define

$$\psi(s) = u_\lambda(t - s, X_\lambda(s)), \text{ where } u_\lambda(s, x) = u(s, T_\lambda x), \quad s \in [0, t_0]. \quad (7.10)$$

Assumptions of the Itô formula are satisfied and

$$dψ(s) = \left[-\frac{\partial u_\lambda}{\partial t}(t - s, X_\lambda(s)) + \mathcal{L}_\lambda u_\lambda(t - s, \cdot)(X_\lambda(s)) \right] ds \quad (7.11)$$
$$+ \langle D_x u_\lambda(t - s, X_\lambda(s)), G_\lambda(X_\lambda(s)) \, dW(S) \rangle, \quad s \in [0, t_0].$$

Here

$$\mathcal{L}_\lambda \varphi(x) = \frac{1}{2} \operatorname{Tr}(G_\lambda(x) Q^{1/2})^* D_x^2 \varphi(x)(G_\lambda(x) Q^{1/2})$$
$$+ \langle Ax + F_\lambda(x), D_x \varphi(x) \rangle, \quad x \in D(A),$$

is the value of the operator determined by the process X_λ, on the function φ. Note that

$$\frac{\partial u_\lambda}{\partial t}(s, x) = \frac{\partial u}{\partial t}(s, T_\lambda x), \quad D_x u_\lambda(s, x) = T_\lambda^* D_x u(s, T_\lambda x),$$
$$D_x^2 u_\lambda(s, x) = T_\lambda^* D_x^2 u(s, T_\lambda x) T_\lambda, \quad s > 0, \ x \in H.$$

Therefore

$$\mathcal{L}_\lambda u_\lambda(s,\cdot)(x) = \frac{1}{2}\operatorname{Tr}[(\mathcal{T}_\lambda G_\lambda(x)Q^{1/2})^* D_x^2 u(s,\mathcal{T}_\lambda x)](\mathcal{T}_\lambda G_\lambda(x)Q^{1/2})$$
$$+\langle A\mathcal{T}_\lambda x + \mathcal{T}_\lambda F_\lambda(x), D_x u(S,\mathcal{T}_\lambda x)\rangle,$$

and

$$-\frac{\partial u_\lambda}{\partial t}(s,x) + \mathcal{L}_\lambda u_\lambda(s,\cdot)(x) = -\frac{\partial u}{\partial t}(s,\mathcal{T}_\lambda x) + \mathcal{L}_\lambda u_\lambda(s,\cdot)(x)$$
$$= -\frac{\partial u}{\partial t}(s,\mathcal{T}_\lambda x) + \mathcal{L}u(s,\cdot)(\mathcal{T}_\lambda x) + \mathcal{L}_\lambda u_\lambda(s,\cdot)(x) - \mathcal{L}u(s,\cdot)(\mathcal{T}_\lambda x)$$
$$= \mathcal{L}_\lambda u_\lambda(s,\cdot)(x) - \mathcal{L}u(s,\cdot)(\mathcal{T}_\lambda x)$$
$$= \frac{1}{2}\operatorname{Tr}\{[(\mathcal{T}_\lambda G_\lambda(x)Q^{1/2})(\mathcal{T}_\lambda G_\lambda(x)Q^{1/2})^*$$
$$-(G(\mathcal{T}_\lambda x)Q^{1/2})(G(\mathcal{T}_\lambda x)Q^{1/2})^*]D_x^2 u(s,\mathcal{T}_\lambda x)\}$$
$$+\langle \mathcal{T}_\lambda F_\lambda(x) - F(\mathcal{T}_\lambda x), D_x u(s,\mathcal{T}_\lambda x)\rangle, \quad s > 0, \ x \in H,$$

where $\mathcal{L}$ is the operator determined by the process X. Since $\mathbb{E}(\psi(t_0)) = \mathbb{E}(u_\lambda(t - t_0, X_\lambda(t_0))) = \mathbb{E}(u(t - t_0, \mathcal{T}_\lambda X_\lambda(t_0)))$, $\mathbb{E}(\psi(0)) = u(t, \mathcal{T}_\lambda x)$, one gets from 7.11 that

$$\mathbb{E}(u(t - t_0, \mathcal{T}_\lambda X_\lambda(t_0))) \tag{7.12}$$
$$= u(t,\mathcal{T}_\lambda x) + \frac{1}{2}\mathbb{E}\int_0^{t_0}\operatorname{Tr}\{[(\mathcal{T}_\lambda G_\lambda(X_\lambda(S))Q^{1/2}))(\mathcal{T}_\lambda G_\lambda(X_\lambda(S))Q^{1/2})^*$$
$$-(G(\mathcal{T}_\lambda X_\lambda(S))Q^{1/2})(G(\mathcal{T}_\lambda X_\lambda(S))Q^{1/2})^*]D_x^2 u(t - s, \mathcal{T}_\lambda X_\lambda(s))\}\,ds$$
$$+\mathbb{E}\int_0^{t_0}\langle \mathcal{T}_\lambda F_\lambda(X_\lambda(s)) - F(\mathcal{T}_\lambda X_\lambda(s)), D_x u(t - s, \mathcal{T}_\lambda X_\lambda(s))\rangle ds$$
$$= u(t, \mathcal{T}_\lambda(x)) + \frac{1}{2}I_\lambda^1(t_0) + I_\lambda^2(t_0).$$

Taking into account that the process X_λ has continuous trajectories one gets, by the Lebesgue dominated convergence theorem, that

$$\mathbb{E}(u(t,\mathcal{T}_\lambda X_\lambda(t)) = u(t,\mathcal{T}_\lambda x) + \frac{1}{2}I_\lambda^1(t) + I_\lambda^2(t).$$

It is therefore enough to show that $I_\lambda^1(t) \to 0$, $I_\lambda^2(t) \to 0$ as $\lambda \to +\infty$. We will prove for instance that $\lim_{\lambda \to +\infty} I_\lambda^1(t) = 0$. Note that if B and C are Hilbert–Schmidt operators then

$$|\operatorname{Tr} BB^* - \operatorname{Tr} CC^*| \ \leq \ \|BB^* - CC^*\|_1 \tag{7.13}$$
$$\leq \ \|(B - C)B^*\|_1 + \|C(B - C)^*\|_1$$
$$\leq \ \|B - C\|_{HS}(\|B\|_{HS} + \|C\|_{HS})$$

Let $M = \sup\{\|D_x^2 u(s,x)\|; \ s > 0, x \in H\}$. By 7.13,

$$I_\lambda^1(t) \leq M\mathbb{E}\int_0^t \|(\mathcal{T}_\lambda G_\lambda(X_\lambda(S))$$
$$-G(\mathcal{T}_\lambda X_\lambda(S)))Q^{1/2}\|_{HS}(\|\mathcal{T}_\lambda G_\lambda(X_\lambda(S))Q^{1/2}\|_{HS}$$
$$+\|G(\mathcal{T}_\lambda X_\lambda(S))Q^{1/2}\|_{HS})ds$$
$$\leq M\left(\mathbb{E}\int_0^T \|(\mathcal{T}_\lambda G_\lambda(X_\lambda(S)) - G(\mathcal{T}_\lambda X_\lambda(S)))Q^{1/2}\|_{HS}^2 ds\right)^{1/2}$$
$$\cdot\left(\mathbb{E}\int_0^T [\|\mathcal{T}_\lambda G_\lambda(X_\lambda(S))Q^{1/2}\|_{HS} + \|G(\mathcal{T}_\lambda X_\lambda(S))Q^{1/2}\|_{HS}]^2 ds\right)^{1/2}$$

Taking into account that G is Lipschitz form H into $\mathcal{H}$ and that $X_\lambda \to X$ as $\lambda \to +\infty$ in $H^2[0, t; H]$, one sees that $I_\lambda^1(t) \to 0$ as $\lambda \to +\infty$. $\square$

8. Parabolic equations in open sets

Heat equation with a linear first order term is considered in open subsets of a Hilbert space. Boundary conditions are of the Dirichlet type. Regularity of the generalized solution is studied.

8.1 Introduction

Let $\mathcal{O}$ be an open subset of a separable Hilbert space H. The present chapter is devoted to the Kolmogorov equation in the set $\mathcal{O}$, with the Dirichlet boundary condition:

$$\frac{\partial u}{\partial t}(t, x) = \frac{1}{2}\operatorname{Tr} D_Q^2 u(t, x) + \langle Ax, Du(t, x)\rangle, \quad x \in \mathcal{O}, \ t > 0, \quad (8.1)$$
$$u(0, x) = \varphi(x), \ x \in \mathcal{O}, \quad u(t, y) = 0, \ y \in \partial\mathcal{O}, \ t > 0. \quad (8.2)$$

If $\mathcal{O} = H$ then the problem 8.1–8.2 is identical with the one considered in § 5. Its generalized solution was given by the formula:

$$u(t, x) = P_t\varphi(x) = \mathbb{E}(\varphi(X^x(t))), \quad t \geq 0, \ x \in H,$$

with X^x the solution to the stochastic Ornstein-Uhlenbeck equation:

$$dX = AXdt + dW(t), X(0) = x. \quad (8.3)$$

The generalised solution to 8.1–8.2 is defined by a modified formula:

$$u(t, x) = \mathbb{E}(\varphi(X^x(t)); \tau_{\mathcal{O}}^x > t), \quad x \in \mathcal{O} \quad (8.4)$$

where

$$\tau_{\mathcal{O}}^x = \inf\{t > 0; \ X^x(t) \in \mathcal{O}^c\}, \quad (8.5)$$

is the so called *exit time* of the process X^x from $\mathcal{O}$. As a counterpart of the semigroup (P_t), it is convenient to introduce a family of linear operators (R_t), corresponding to the Dirichlet problem in $\mathcal{O}$, and acting on Borel functions defined on $\mathcal{O}$ in the following way

$$R_t\varphi(x) = \mathbb{E}(\varphi(X^x(t)); \tau_{\mathcal{O}}^x > t), \quad x \in \mathcal{O}. \tag{8.6}$$

One can show that the family (R_t) forms a semigroup: $R_{t+s} = R_t R_s$ which will be called the *restricted semigroup*.

We will prove that , under weak requirements , the function $R_t\varphi$ is as regular in $\mathcal{O}$, for each $t > 0$, as the function $P_t\varphi$ in H. We limit our presentation to the differentiability of the generalized solution in the space variable and follow the paper [DaPrGoZa]. It is an open problem to characterize those initial functions for which the generalized solution is the classical one.

8.2 Main theorem

Assume that $\gamma(t)$, $t > 0$, is a decreasing positive function such that

$$\|D^2 P_t\varphi(x)\| \le \gamma(t)\|\varphi\|_0, \quad \text{for all } \varphi \in B_b(H), \ t > 0, \ x \in H \tag{8.7}$$

The main result of the present section is the following theorem taken from [DaPrGoZa].

Theorem 8.1. *Assume that 8.7 holds and that for a decreasing sequence* $(s_n) \to 0$, $s_1 = t$ *and an open set* $U \subset \mathcal{O}$:

$$\sum_n \left[\sup_{x \in U} \mathbb{P}^{1/2}(\tau_{\mathcal{O}}^x \le s_{n-1})\right] \gamma^{1/2}(s_n) < +\infty. \tag{8.8}$$

Then, for arbitrary $\varphi \in B_b(\mathcal{O})$, *the function* $R_t\varphi$ *is continuously differentiable in* U.

The proof will be based on several results of independent interest. The following proposition, due to Dynkin, see [Dy], is valid for general transition semigroups. If $\varphi \in B_b(H)$ then $\check{R}_t\varphi$ denotes a function on H given by the formula:

$$\check{R}_t\varphi(x) = \begin{cases} R_t\varphi(x) & \text{if } x \in \mathcal{O} \\ 0 & \text{if } x \in \mathcal{O}^c. \end{cases}$$

Proposition 8.2. *For arbitrary* $t > 0$, $s \in [0, t]$, $x \in \mathcal{O}$ *and* $\varphi \in B_b(\mathcal{O})$:

$$|R_t\varphi(x) - P_s(\check{R}_{t-s}\psi)(x)| \le \|\varphi\|_0 \mathbb{P}(\tau_{\mathcal{O}}^x \le s).$$

Proof.

$$
\begin{aligned}
R_t\varphi(x) &= \mathbb{E}(\varphi(X^x(t)); \; X^x(r) \in \mathcal{O} \quad \text{for all } r \in [0,t]) \\
&= \mathbb{E}(\varphi(X^x(t)); \; X^x(r) \in \mathcal{O} \quad \text{for all } r \in [s,t]) \\
&= \mathbb{E}(\varphi(X^x(t)); \; X^x(r) \in \mathcal{O} \quad \text{for some } r \in [0,s] \\
&\qquad\qquad\qquad\quad \text{and } X^x(r) \in \mathcal{O} \quad \text{for all } r \in [s,t]) \\
&= I_1 + I_2.
\end{aligned}
$$

By the Markov property, see [DaPrZa2],

$$
\begin{aligned}
I_1 &= \mathbb{E}(\varphi(X^x(t)); \; X^x(r) \in \mathcal{O} \text{ for all } r \in [s,t]) \\
&= \mathbb{E}(R_{t-s}\varphi(X^x(s)); \; X^x(s) \in \mathcal{O}) = P_s(\check{R}_{t-s}\varphi)(x).
\end{aligned}
$$

Since

$$
\begin{aligned}
|I_2| &\leq |\varphi|_0 \mathbb{P}(X^x(r) \in \mathcal{O}^c \text{ for some } r \in [0,s] \text{ and } X^x(r) \in \mathcal{O} \text{ for all } r \in [s,t]) \\
&\leq |\varphi|_0 \mathbb{P}(X^x(r) \in \mathcal{O}^c \text{ for some } r \in [0,s]) \\
&\leq |\varphi|_0 \mathbb{P}(\tau_{\mathcal{O}}^x \leq s),
\end{aligned}
$$

the proof is complete. □

Proposition 8.3. *Assume that for $\varphi \in B_b(H)$, $P_t\varphi$ is a continuous function and that*

$$
\mathbb{P}(\tau_{\mathcal{O}}^x \leq s) \to 0 \text{ as } s \to 0 \text{ uniformly with respect to } x \text{ from compact sets.}
$$

Then for arbitrary $\varphi \in B_b(\mathcal{O})$ and arbitrary $t > 0$, $R_t\varphi$ is a continuous function.

Proof. The result follows immediately from Proposition 8.2.

We need the following interpolatory result.

Proposition 8.4. *Assume that g is of class C^2 in an open set U and let $d(x) = distance(x, U^c) \wedge 2$, $x \in U$. Then*

$$
|Dg(x)| \leq \frac{4}{d(x)} \sqrt{|g|_U} \sqrt{|g|_U + \|D^2g\|_U}, \quad x \in U,
$$

where $|\cdot|_U$ and $\|\cdot\|_U$ denote the suprema over the set U.

Proof. Assume that the closure of the open ball,

$$
B(x,r) = \{y \in H; \; \|y - x\| < r\},
$$

is contained in U and $|h| = r$. By the mean value theorem, for each $t \in [0,1]$ there exists $z \in B(x,r)$ such that

$$
g(x+th) = g(x) + t\langle Dg(x), h\rangle + \frac{1}{2}t^2\langle D^2g(z)h, h\rangle. \tag{8.9}
$$

From 8.9:

$$\left|\left\langle Dg(x), \frac{h}{|h|}\right\rangle\right| \leq \frac{2\|g\|_U}{t|h|} + \frac{1}{2}t|h|\|D^2g\|_U$$

$$\leq \left(\frac{2\|g\|_U}{r}\right)\frac{1}{t} + t\left(\frac{1}{2}r\|D^2g\|_U\right),$$

and therefore, for all $t \in (0,1)$

$$|Dg(x)| \leq \left(\frac{1\|g\|_U}{r}\right)\frac{1}{t} + t\left(\frac{1}{2}r\|D^2g\|_U\right)$$

$$\leq \alpha\frac{1}{t} + t\beta.$$

Setting $t = \sqrt{\frac{\alpha}{\alpha+\beta}}$ are gets that $\alpha\frac{1}{t} + t\beta \leq 2\sqrt{\alpha(\alpha+\beta)}$. Consequently

$$|Dg(x)| \leq 2\sqrt{\frac{2\|g\|_U}{r}}\sqrt{\frac{2\|g\|_U}{r} + \frac{1}{2}r\|D^2g\|_U}$$

$$\leq \frac{4}{r}\sqrt{\|g\|_U}\sqrt{\|g\|_U + \frac{1}{4}r^2\|D^2g\|_U}$$

Since r was an arbitrary, positive number smaller than $d(x)$, the estimate follows.   □

Proof. We pass to the proof of the main theorem. Note that

$$R_t\varphi(x) = P_{S_1}(\check{R}_{t-S_1}\varphi)(x) + \sum_{n=1}^{+\infty}(P_{s_{n+1}}\check{R}_{t-s_{n+1}}\varphi)(x) - P_{S_n}(\check{R}_{t-s_{n+1}}\varphi)(x))$$

$$= \psi_1(x) + \sum_{n=1}^{+\infty}\psi_{n+1}(x) - \psi_n(x), \quad x \in \mathcal{O}.$$

By 8.7

$$\|D^2\psi_{n+1}(x) - D^2\psi_n(x)\| \leq 2\gamma(s_n)|\varphi|_0, \quad x \in H,$$

and by the lemma:

$$|D\psi_{n+1}(x) - D\psi_n(x)|$$
$$\leq \frac{4}{d(x)}\sqrt{|\psi_{n+1} - \psi_n|_U}\sqrt{|\psi_{n+1} - \psi_n| + \|D^2\psi_{n+1} - D^2\psi_n\|}.$$

However

$$|\psi_{n+1}(x) - \psi_n(x)| \leq 2\mathbb{P}(\tau_{\mathcal{O}}^x \leq s_n),$$

and the result follows.   □

8.3 Estimates of the exit probabilities

To apply Theorem 8.1 to specific examples one needs good estimates on the function γ, and on the *exit probabilities* :

$$\mathbb{P}(\tau_{\mathcal{O}}^x \leq s), \quad s > 0.$$

Estimates on the function γ can be deduced from results on the Ornstein-Uhlenbeck transition semigroups in § 5. In particular taking into account formula 8.8 for $D^2 P_t \varphi(x)$ and the definition 5.19 of the function Λ_t one finds that for interesting examples one has that:

$$0 \leq \gamma(s) \leq \frac{M}{s^\delta} \tag{8.10}$$

for some positive δ. In particular if $Q = I$, then $\delta = 1$, see § 5.

The following theorem plays an essential role in the estimates of the exit probabilities. It is an improved version of a result form [DaPrGoZa] with a simpler proof based on the arguments of [Pe] and on the properties of Gaussian measures.

Theorem 8.5. *Assume that for some* $0 < \alpha < 1, T > 0,$

$$\int_0^T t^{-\alpha} \|S(t) Q^{1/2}\|_{HS}^2 \, dt < +\infty. \tag{8.11}$$

Then the process $Z(t) = \int_0^t S(t-s) \, dW(s), t \in [0,T],$ *where* W *is a* Q*-Wiener process, has continuous version and there exist positive constants* $C_1 > 0,$ $C_2 > 0$ *such that for all* $s \in (0,T]$ *and all* $r > 0,$

$$\mathbb{P}(\sup_{t \leq s} \|Z(t)\| \geq r) \leq C_1 e^{-C_2 \frac{r^2}{s^\alpha}}. \tag{8.12}$$

Proof. We will use the following factorization:

$$Z(t) = \frac{\sin \beta \pi}{\pi} \int_0^t (t-s)^{\beta-1} S(t-s) Y(s) ds, \quad t \geq 0, \tag{8.13}$$

where

$$Y(s) = \int_0^t (s-\sigma)^{-\beta} S(s-\sigma) dW(\sigma), \quad s \geq 0 \tag{8.14}$$

see [DaPrZa2] and [DaPrZa4], valid for $\beta \in (0, \frac{1}{2})$. Define:

$$\beta = \frac{\alpha}{2}, \quad n_0 = \left[\frac{1}{\alpha}\right] + 1, \quad q_0 = 2n_0, \quad p_0 = \frac{q_0}{q_0 - 1}.$$

Then $p_0^{-1} + q_0^{-1} = 1$. Moreover if $n \geq n_0, n \in \mathbb{N}$ and $q = 2n, p = q(q-1)^{-1}$ then by the Hölder inequality:

$$\|Z(t)\| \leq \frac{\sin\beta\pi}{\pi} \int_0^t (t-s)^{\beta-1}\|S(t-s)\| \, \|Y(s)\| ds$$

$$\leq \frac{\sin\beta\pi}{\pi} \left(\int_0^t (t-s)^{(\beta-1)p}\|S(t-s)\|^p ds \right)^{1/p} \left(\int_0^t \|Y(s)\|^q ds \right)^{1/q}.$$

Therefore, for all $u \in (0,T]$:

$$\sup_{t\leq u} \|Z(t)\| \leq cu^{\frac{2(\beta-1)p+1}{p}} \left(\int_0^u \|Y(s)\|^q ds \right)^{2/q}$$

where $c = \left(\frac{\sin\beta\pi}{\pi}\right)^2 M^2$, $M = \sup(\|S(t)\|, \, t \leq T)$. For arbitrary $\lambda > 0$, $n \geq n_0$, $u \in [0,T]$,

$$\frac{\sup_{t\leq u}\|Z(t)\|^2}{c\lambda u^\alpha} \leq \left(\frac{1}{u}\int_0^u \left(\frac{\|Y(s)\|^2}{\lambda} \right)^n ds \right)^{1/n},$$

$$\mathbb{E}\left(\frac{\sup_{t\leq u}\|Z(t)\|^2}{c\lambda u^\alpha} \right)^n \leq \frac{1}{u}\int_0^u \mathbb{E}\left(\frac{\|Y(s)\|^2}{\lambda} \right)^n ds.$$

Consequently:

$$\mathbb{E}\left(\sum_{n\geq n_0} \frac{1}{n!} \left(\frac{\sup_{t\leq u}\|Z(t)\|^2}{c\lambda u^\alpha} \right)^n \right) \leq \frac{1}{u}\int_0^u \mathbb{E}(e^{\frac{\|Y(s)\|^2}{\lambda}})ds. \qquad (8.15)$$

By Theorem 2.7 from the Preliminaries, if $\mathbb{E}\|Y(T)\|^2 \leq \frac{\lambda}{2}$ and $s \in (0,T]$,

$$\mathbb{E}(e^{\frac{\|Y(s)\|^2}{\lambda}}) \leq \frac{1}{\sqrt{1-\frac{2}{\lambda}\mathbb{E}\|Y(s)\|^2}} \qquad (8.16)$$

$$\leq \frac{1}{\sqrt{1-\frac{2}{\lambda}\mathbb{E}\|Y(T)\|^2}} = C(\lambda).$$

Note that $\mathbb{E}\|Y(T)\|^2 = \int_0^T t^{-\alpha}\|S(t)Q^{1/2}\|_{HS}^2 dt$. Moreover, for $n = 0, 1, \ldots, n_0 - 1$:

$$\mathbb{E}\left(\frac{\sup_{t\leq u}\|Z(t)\|^2}{c\lambda u^\alpha} \right)^n \leq \left(\mathbb{E}\left(\frac{\sup_{t\leq u}\|Z(t)\|^2}{c\lambda u^\alpha} \right)^{n_0} \right)^{\frac{n}{n_0}}$$

and

$$\sum_{n=0}^{n_0-1} \frac{1}{n!}\mathbb{E}\left(\frac{\sup_{t\leq u}\|Z(t)\|^2}{c\lambda u^\alpha} \right)^n \qquad (8.17)$$

$$\leq \sum_{n=0}^{n_0-1} \frac{1}{n!} \left(\left[\mathbb{E}\left(\frac{\sup_{t\leq u}\|Z(t)\|^2}{c\lambda u^\alpha} \right)^{n_0} \right]^{\frac{n}{n_0}} \right)^n$$

$$\leq e^{\left(\mathbb{E}\left(\frac{\sup_{t\leq u}\|Z(t)\|^2}{c\lambda u^\alpha} \right)^{n_0} \right)^{\frac{1}{n_0}}}.$$

Since,

$$\mathbb{E}\left(\frac{\sup_{t\leq u}\|Z(t)\|^2}{c\lambda u^\alpha}\right)^{n_0} \leq n_0!\frac{1}{u}\int_0^u \mathbb{E}\left(\frac{1}{n_0!}\left(\frac{\|Y(s)\|^2}{\lambda}\right)^{n_0}\right)ds$$

$$\leq n_0!\frac{1}{u}\int_0^u \mathbb{E}\left(e^{\frac{\|Y(s)\|^2}{\lambda}}\right)ds$$

$$\leq n_0!C(\lambda)$$

then, taking into account 8.15, 8.16 and 8.17,

$$\mathbb{E}\left(e^{\frac{\sup_{t\leq u}\|Z(t)\|^2}{c\lambda u^\alpha}}\right) \leq C(\lambda) + e^{(n_0!C(\lambda))^{1/n_0}}.$$

Finally, by Chebyshev's inequality:

$$\mathbb{P}(\sup_{t\leq u}\|Z(t)\| \geq r) = \mathbb{P}\left(e^{\frac{\sup_{t\leq u}\|Z(t)\|^2}{c\lambda u^\alpha}} \geq e^{\frac{r^2}{c\lambda u^\alpha}}\right)$$

$$\leq e^{-\frac{r^2}{c\lambda u^\alpha}}\mathbb{E}\left(e^{\frac{\sup_{t\leq u}\|Z(t)\|^2}{c\lambda u^\alpha}}\right)$$

$$\leq e^{-\frac{r^2}{c\lambda u^\alpha}}\left(C(\lambda) + e^{(n_0!C(\lambda))^{1/n_0}}\right)$$

and the proof of the estimate 8.12 is complete. The proof of the continuity of Z is a consequence of the formula 8.13 and is left to the reader, see [DaPrZa2].

$\square$

The following theorem provides the required estimates on the exit probabilities.

Theorem 8.6. *Assume that the condition 8.11 holds. Let $a \in \mathcal{O}$, $r > 0$, and $T > 0$, be such that,*

$$B(a,r) \subset \mathcal{O}, \text{ and } \sup_{t\leq T}|S(t)a - a| \leq \frac{r}{4}.$$

Define $M = \sup(\|S(t)\|; t \leq T)$, and $r_0 = r/4M$. Then there exist positive constants C_1, C_2 such that for $s \in [0,T]$,

$$\sup_{x\in B(a,r_0)} \mathbb{P}(\tau^x_{B(a,r)} \leq s) \leq C_1 e^{-C_2\frac{r^2}{s^\alpha}}. \tag{8.18}$$

Proof. To prove 8.18 notice that for $x \in B(a,r_0)$:

$$\|X^x(t) - a\| = \|S(t)x + Z(t) - a\| \leq \|S(t)(x - a)\| + \|S(t)a - a\| + \|Z(t)\|$$
$$\leq 1/2r + \|Z(t)\|.$$

Therefore, for $x \in B(a,r_0)$ and $s \leq T$:

$$\mathbb{P}(\tau^x_{B(a,r)} \leq s) \leq \mathbb{P}(\sup_{t\leq s}\|X^x(t) - a\| \geq r) \leq \mathbb{P}\left(\sup_{t\leq s}\|Z(t)\| \geq \frac{r}{2}\right),$$

and it is enough to apply 8.12.

$\square$

9. Applications

The theory of linear equations is applied to a class of nonlinear parabolic equations arising in stochastic control of infinite dimensional systems. An application to mathematical finance is given as well.

9.1 HJB equation of stochastic control

Results on Kolmogorov equations, discussed in previous chapters, play an important role in the analysis of a class of nonlinear parabolic equations on Hilbert spaces. They appear in connection with control of stochastic infinite dimensional systems. We start from recalling basic control theoretic concepts.

Consider a family of stochastic equations on a Hilbert space H :

$$
\begin{aligned}
dX &= (AX + F(X, z))dt + dW(t) \\
X(0) &= x
\end{aligned}
\tag{9.1}
$$

parametrized by a parameter z from a metric space Z. Here $A : D(A) \subset H \to H$ is the infinitesimal generator of a C_0-semigroup $S(t)$, $t \geq 0$ on H and $F : H \times Z \to H$ is a fixed mapping. Moreover $W(t)$, $t \geq 0$ is a Wiener process on H, with the covariance operator Q, defined on a probability space $(\Omega, \mathcal{F}, \mathbb{P})$. The probability space $(\Omega, \mathcal{F}, \mathbb{P})$ is equipped with an increasing family of σ-fields of events $\mathcal{F}_t \subset \mathcal{F}$, $t \geq 0$, which may happen up to time $t \geq 0$. A *strategy* $z(t)$, $t \geq 0$ is any progressively measurable Z-valued stochastic process . If a strategy $z(\cdot)$ and the initial state $x \in H$ are given then $X^{z(\cdot)}(t, x)$, $t \geq 0$ denotes the solution to 9.1 in which the parameter z was replaced by the process $z(\cdot)$. One of the main problems in the stochastic control theory is to find a strategy $z(\cdot)$ which, for fixed $T > 0$ and $x \in H$, minimizes the cost functional:

$$
J_T(x, z(\cdot)) = \mathbb{E}\left(\int_0^T f(X^{z(\cdot)}(s, x), z(s)) \, ds + \varphi(X^{z(\cdot)}(T, x)) \right),
\tag{9.2}
$$

In 9.2 , $f : H \times Z \to R^1$, $\varphi : H \to R^1$ are given functions.

There is a standard way to approach the problem of minimizing 9.2 under the *constraints* 9.1. One defines the so called *value function V* by the formula :

$$
V(t, x) = \inf_{z(\cdot)} J_t(x, z(\cdot)), \qquad t \geq 0, \ x \in H.
\tag{9.3}
$$

where the infimum is taken with respect to the all processes $z(\cdot)$. Denote by $\mathcal{L}^z$, $z \in Z$, the differential operator determined by the 9.1 with the parameter z fixed :

$$
\mathcal{L}^z \psi(x) = \frac{1}{2} \operatorname{Tr} Q D^2 \psi(x) + \langle Ax + F(x, z), D\psi(x) \rangle, \qquad x \in H.
\tag{9.4}
$$

One can show, by a heuristic argument, that the function V should satisfy the following equation:

$$\frac{\partial u}{\partial t}(t, x) = \inf_{z \in Z}[\mathcal{L}^z u(t, x) + f(x, z)], \tag{9.5}$$

$$u(0, x) = \varphi(x), \quad x \in H, \ t > 0.$$

The equation 9.4 is called a *Hamilton–Jacobi–Bellman* equation or shortly *HJB* equation.

Assume that the equation 9.5 has a solution u, continuous on $[0, +\infty)$ $\times H$ with continuous first derivative $Du(t, x)$, $t > 0$, $x \in H$. Assume in addition that for arbitrary $t > 0$, $x \in H$ the infimum in 9.5 is attained at $k(t, x) \in Z$. Then, in majority of cases, the solution u of 9.5 can be identified with the value function 9.3 and the optimal strategy $\widehat{z}(\cdot)$ is of the form :

$$\widehat{z}(t) = k(T - t, \widehat{X}(t)), \quad t \in [0, T], \tag{9.6}$$

where $\widehat{X}$ is a process satisfying

$$d\widehat{X}(t) = A\widehat{X}(t) + F(\widehat{X}(t), k(T - t, \widehat{X}(t)))dt + dW(t) \tag{9.7}$$

$$\widehat{X}(0) = x, \quad t \in (0, T], \ x \in H.$$

This means that given the function $k(\cdot, \cdot)$, one solves first the stochastic equation 9.7, then the solution $\widehat{X}$ is an optimal process and corresponds to the strategy $\widehat{z}$.

We will not formulate a precise result on existence of an optimal control but only indicate the main steps of the proof that 9.6 defines the optimal control and that $u = V$.

Let $z(\cdot)$ be any admissible control and $X^{z(\cdot)}(t, x)$, $t \geq 0$, the corresponding solution of 9.1. Applying Itô's formula to $\xi(t) = u(T - t, X^{z(\cdot)}(t, x))$, $t \in [0, T]$ and calculating expectations one gets that

$$\mathbb{E}(u(0, X^{z(\cdot)}(T, x))) = u(T, x) \tag{9.8}$$

$$+ \mathbb{E}\left(\int_0^T \left[-\frac{\partial u}{\partial s}(T - t, X^{z(\cdot)}(t, x)) + \mathcal{L}^{z(t)}u(T - t, X^{z(\cdot)}(t, x))\right]dt\right).$$

Since the function u solves 9.5 and the infimum in 9.5 is attained at $k(t, x) \in Z$, therefore, for all $t > 0$, $x \in H$ and $z \in Z$:

$$-\frac{\partial u}{\partial t}(t, x) + \mathcal{L}^z u(t, x) = [-\mathcal{L}^{k(t,x)}u(t, x) - f(x, k(t, x)) + \mathcal{L}^z u(t, x) \tag{9.9}$$

$$+ f(x, z)] - f(x, z) = h(t, x, z) - f(x, z).$$

The function h defined in 9.9 is called the *discrepancy function* of the control problem. Note that,

$$h(t, x, z) \geq 0, \quad \text{for all } t > 0, \ x \in H, \ z \in Z, \tag{9.10}$$

and
$$h(t, x, z) = 0, \quad \text{if } z = k(t, x) \ t > 0, \ x \in H, \ z \in Z. \tag{9.11}$$

But
$$\mathbb{E}(u(0, X^{z(\cdot)}(T, x))) = \mathbb{E}(\varphi(X^{z(\cdot)}(T, x)),$$

and from 9.8 , 9.9,
$$\mathbb{E}(\varphi(X^{z(\cdot)}(T, x))) = u(T, x) \tag{9.12}$$
$$+\mathbb{E}\left(\int_0^T [h(T - t, X^{z(\cdot)}(t, x), z(t)) - f(X^{z(\cdot)}(t, x), z(t))]dt\right).$$

Rearranging terms in 9.12 one arrives at:

$$J_T(x, z(\cdot)) = u(T, x) + \mathbb{E}\left(\int_0^T h(T - t, X^{z(\cdot)}(t, x), z(t))dt\right). \tag{9.13}$$

It follows now from 9.10 and 9.11 that for any admissible strategy $z(\cdot)$,

$$J_T(x, z(\cdot))) \geq u(T, x)$$

and if $z(t) = k(T - t, X^{z(\cdot)}(t, x))$, then $z(\cdot)$ is optimal. In addition we have the equality
$$u(t, x) = V(t, x), \ t \geq 0, \ x \in H.$$
The main identity 9.13 is usually obtained for suitable regularizations of u and X and then, by a limit argument, for the true solutions.

For the control system 9.1 the HJB equation can be written in a more convenient way. The *Hamiltonian*, related to the problem of minimizing 9.2, is a function $\mathcal{H}(x, p)$, $x \in H$, $p \in H$, given by the formula:

$$\mathcal{H}(x, p) = \inf_{z \in Z}\{\langle F(x, z), p\rangle + f(x, z)\}. \tag{9.14}$$

Let $\mathcal{L}$ be the differential operator determined by the linear stochastic equation

$$dX = AX dt + dW(t), \quad x(0) = x. \tag{9.15}$$

Then, as we already know,

$$\mathcal{L}\psi(x) = \frac{1}{2}\operatorname{Tr} QD^2\psi(x) + \langle Ax, D\psi(x)\rangle, \quad x \in D(A), \tag{9.16}$$

and 9.5 becomes

$$\frac{\partial u}{\partial t}(t, x) = \mathcal{L}u(t, x) + \mathcal{H}(x, Du(t, x)), \tag{9.17}$$
$$u(0, x) = \varphi(x), \quad x \in H, \ t > 0.$$

It is important to notice that the HJB equation 9.17 is a *perturbation* of the linear parabolic equation:

$$\frac{\partial u}{\partial t} = \mathcal{L}u, \quad u(0) = \varphi \tag{9.18}$$

by a nonlinear, first order term $\mathcal{H}(x, Du(t, x))$, $x \in H$, $t > 0$. This remark will be used in the study of the HJB equation in the next section. Notice also that the discrepancy function h is related to the Hamiltonian in the following way:

$$h(t, x, z) = [\langle F(x, z), p \rangle + f(x, z)] - \mathcal{H}(x, p), \tag{9.19}$$

with

$$p = Du(t, x). \tag{9.20}$$

Let us consider some examples.

Example 9.1. Assume that the system 9.1 is linear:

$$dX = (AX + Bz)dt + dW, \quad X(0) = x, \tag{9.21}$$

where B is a bounded linear operator from a Hilbert space Z into H. Let

$$J_T(x, z(\cdot)) = \mathbb{E}\left(\int_0^T [g(X(t, x)) + |z(t)|^2]\, dt + \varphi(X(T, x))\right). \tag{9.22}$$

Then $F(x, z) = Bz$, $f(x, z) = g(x) + |z|^2$, $x \in H$, $z \in Z$, $\mathcal{H}(x, p) = -\frac{1}{4}|B^*p|^2$. The corresponding HBJ is of the form:

$$\frac{\partial u}{\partial t}(t, x) = \frac{1}{2}\operatorname{Tr} QD^2 u(t, x) + \langle Ax, Du(t, x) \rangle - \frac{1}{4}|B^* Du(t, x)|^2, \tag{9.23}$$
$$u(0, x) = \varphi(x), \quad x \in H, \ t > 0.$$

Moreover,

$$k(t, x) = -\frac{1}{2}B^* Du(t, x), \quad t > 0, \ x \in H,$$

and the function h is of the form

$$h(t, x, z) = \left|z + \frac{1}{2}B^* p\right|^2, \quad p = Du(t, x), \ t > 0, \ x \in H.$$

Example 9.2. Let again the control system be 9.21, but the space of the control parameters be the unite ball $Z_1 = \{z \in Z; \ |z| \le 1\}$ and,

$$J_T(x, z(\cdot)) = \mathbb{E}\left(\int_0^T g(X(t, x))\, dt + \varphi(X(T, x))\right).$$

Then $F(x, z) = Bz$, $f(x, z) = g(x)$, $x \in H$, $|z| \le 1$, and $\mathcal{H}(x, p) = -|B^*p|$. Therefore the corresponding HJB equation is of the form:

$$\frac{\partial u}{\partial t}(t, x) = \frac{1}{2}\operatorname{Tr} QD^2 u(t, x) + \langle Ax, Du(t, x) \rangle - |B^* Du(t, x)|, \tag{9.24}$$
$$u(0, x) = \varphi(x), \quad x \in H, \ t > 0.$$

Moreover

$$k(t, x) = -\frac{B^* Du(t, x)}{|Du(t, x)|}, \quad t > 0, \ x \in H.$$

9.2 Solvability of HJB equation

Let (P_t) be the transition semigroup corresponding to the parabolic equation:

$$\frac{\partial u}{\partial t}(t,x) = \frac{1}{2}\operatorname{Tr} QD^2u(t,x) + \langle Ax, Du(t,x)\rangle, \qquad (9.25)$$

$$u(0,x) = \varphi(x), \quad x \in H, \ t > 0.$$

Applying the variation of constant formula to HJB equation 9.17 one arrives at an *integral HJB*:

$$u(t,x) = P_t\varphi(x) + \int_0^t P_{t-s}[\mathcal{H}(\cdot, Du(s,\cdot))](x)\, ds . \qquad (9.26)$$

We are concerned here with the equation 9.26 which is more convenient for the analytical treatment than 9.17. If it has a unique solution u then one can usually show, following the steps indicated in the previous section, that the function u is equal to the value function V.

We will assume that:

(A.1) For some $\gamma \in (0,1)$ and $M > 0$,

$$\|A_t\| \le \frac{M}{t^\gamma}, \qquad t > 0. \qquad (9.27)$$

As far as the Hamiltonian $\mathcal{H}$ is concerned we will require:

(A.2) $\mathcal{H}$ is a continuous function of both variables and for a constant $M > 0$

$$|\mathcal{H}(x,p) - \mathcal{H}(x,q)| \le M|p - q|, \quad x,p,q \in H . \qquad (9.28)$$

(A.3) For every $\psi \in C_b(H)$, $\mathcal{H}(\cdot, \psi(\cdot)) \in C_b(H)$, and the transformation

$$\psi \to \mathcal{H}(\cdot, \psi(\cdot)) \qquad (9.29)$$

maps bounded sets in $C_b(H)$ into bounded sets of $C_b(H)$.

For fixed $T > 0$ denote by $C_0^{1,\gamma}$ the space $C_0^{1,\gamma}([0,T],H)$ of all bounded, continuous functions $u : [0,T] \times H \to R^1$ such that for all $t \in (0,T]$, $u(t,\cdot) \in C_b^1(H)$ and the function:

$$t^\gamma Du(t,x), \quad (t,x) \in \,]0,T] \times H ,$$

is measurable and bounded. The space $C_0^{1,\gamma}$ with the norm:

$$\|u\|_{C_0^{1,\gamma}} = \sup_{t\in[0,T]} \|u(t,\cdot)\|_0 + \sup_{t\in\,]0,T]} t^\gamma \|Du(t,\cdot)\|_0 ,$$

where $\|\cdot\|_0$ denotes the supremum norm on $B_b(H)$, is a Banach space.

We will sketch now the proof of the following existence theorem from [CaDaPr]. The case of a linear Hamiltonian was treated in [DaPrZa1].

Theorem 9.3. *If the assumptions* (A.1)–(A.3) *are satisfied then the equation* 9.26 *has a unique solution* u *in the space* $C_0^{1,\gamma}$.

Proof. For any $u \in C_0^{1,\gamma}$ define,

$$\Gamma(u)(t) = \int_0^t P_{t-s}[\mathcal{H}(\cdot, Du(s, \cdot))] \, ds, \quad t \in [0, T].$$

Then Γ is a transformation which maps $C_0^{1,\gamma}$ into $C_0^{1,\gamma}$. We show that it is a contraction on $C_0^{1,\gamma}$, provided $T > 0$, is sufficiently small. Let $u, v \in C_0^{1,\gamma}$, then

$$\begin{aligned}
\|\Gamma(u)(t) - \Gamma(v)(t)\|_0 &\leq M \int_0^t \|Du(s, \cdot) - Dv(s, \cdot)\|_0 \, ds \\
&\leq M \int_0^t s^{-\gamma} s^\gamma \|Du(s, \cdot) - Dv(s, \cdot)\|_0 \, ds \\
&\leq M \left(\int_0^t s^{-\gamma} \, ds \right) \|u - v\|_{C_0^{1,\gamma}} \leq \frac{MT^{1-\gamma}}{1-\gamma} \|u - v\|_{C_0^{1,\gamma}}.
\end{aligned}$$

Moreover,

$$\begin{aligned}
\|D\Gamma(u)(t) - D\Gamma(v)(t)\|_0 &\leq \int_0^t \|DP_{t-s}[\mathcal{H}(\cdot, Du(s, \cdot)) - \mathcal{H}(\cdot, Dv(s, \cdot))]\|_0 \, ds \\
&\leq M \int_0^t (t-s)^{-\gamma} \|\mathcal{H}(\cdot, Du(s, \cdot)) - \mathcal{H}(\cdot, Dv(s, \cdot))\|_0 \, ds \\
&\leq M^2 \int_0^t (t-s)^{-\gamma} \|Du(s, \cdot) - Dv(s, \cdot)\|_0 \, ds \\
&\leq M^2 \int_0^t (t-s)^{-\gamma} s^{-\gamma} s^\gamma \|Du(s, \cdot) - Dv(s, \cdot)\|_0 \, ds \\
&\leq M^2 \left[\int_0^t (t-s)^{-\gamma} s^{-\gamma} \, ds \right] \|u - v\|_{C_0^{1,\gamma}} \\
&\leq M^2 t^{1-2\gamma} \left(\int_0^t (1-\sigma)^{-\gamma} \sigma^{-\gamma} \, d\sigma \right) \|u - v\|_{C_0^{1,\gamma}} \\
&\leq M^2 K t^{1-2\gamma} \|u - v\|_{C_0^{1,\gamma}},
\end{aligned}$$

for some constant K. Consequently,

$$t^\gamma \|D\Gamma(u)(t) - D\Gamma(v)(t)\|_0 \leq M^2 K t^{1-\gamma} \|u - v\|_{C_0^{1,\gamma}}$$

and,

$$\|\Gamma(u) - \Gamma(v)\|_{C_0^{1,\gamma}} \leq \left[\frac{MT^{1-\gamma}}{1-\gamma} + M^2 K T^{1-\gamma} \right] \|u - v\|_{C_0^{1,\gamma}}.$$

Therefore, if the positive number T is sufficiently small, the mapping Γ is a contraction in $C_0^{1,\gamma}$ and the equation 9.26 has a unique solution in $C_0^{1,\gamma}([0, T], H)$. By the standard step method the theorem is true for arbitrary $T > 0$. $\qquad \square$

9.3 Kolmogorov's equation in mathematical finance

Parabolic equations are traditionally used in the mathematical physics and describe a variety of physical phenomena. They are also important in chemistry and biology to model reaction and diffusion processes and the time evolution of populations. Recently they have appeared, see e.g. [MuRu], in the mathematical finance. We present here equations respectively of one and of infinite number of variables. We start from the so called *Black–Scholes equation* in the one state variable.

Let $X(t)$ denote the price of a share at moment $t \geq 0$. It is assumed that X is a stochastic process defined on a probability space $(\Omega, \mathcal{F}, \mathbb{P})$, equipped with the filtration of events $\mathcal{F}_t$ and that for each $t \geq 0$, the price $X(t)$ is measurable with respect to $\mathcal{F}_t$. The time evolution of the price is often modeled by a stochastic Itô equation.

$$dX(t) = X(t)(rdt + \sigma dW(t)), \quad X(0) = x > 0, \tag{9.30}$$

in which $r > 0$ and $\sigma > 0$ stand respectively for the *interest rate* and for the *volatility* of the share. The parameters r and σ can be identified from the real data with sufficient accuracy. An important property of the solution X is that the discounted price of the share :

$$Z(t) = e^{-rt}X(t), \quad t \geq 0, \tag{9.31}$$

is a *fair* process or, in more technical terms, a *martingale*. This means that for arbitrary $t \geq s \geq 0$ and for arbitrary event $A \in \mathcal{F}_s$:

$$\int_A X(t,\omega)\mathbb{P}(d\omega) = \int_A X(s,\omega)\mathbb{P}(d\omega) \ .$$

Let $\varphi : R_+ \to R$ be a given function. For $t > 0$, $\varphi(X(t))$ can be interpreted as the gain of a holder of an European option with the maturity time t. Let $X^x(t)$ denote the solution to 9.30, with x denoting the price of the share at moment 0. Then the price $V(t, x)$, which should be paid by the buyer of the option to the seller, see [MuRu] , is given by the formula:

$$V(t,x) = \mathbb{E}(e^{-rt}\varphi(X^x(t))), \quad t \geq 0, \ x > 0 \tag{9.32}$$

and solves the following equation

$$\frac{\partial V}{\partial t}(t,x) = \frac{1}{2}\sigma^2 x^2 \frac{\partial^2 V}{\partial x^2}(t,x) + rx\frac{\partial V}{\partial t}(t,x) - rV(t,x), \tag{9.33}$$

$$V(0,x) = \varphi(x), \quad V(t,0) = 0, \ t > 0, \ x > 0.$$

The equation 9.33 is the famous Black-Scholes equation. It can be solved explicitly and the resulting formula is called the *Black–Scholes formula* for the price of the European option, [MuRu].

There exists an infinite dimensional version of the Black–Scholes theory, related to the interest rate processes. It leads to the so called *Heath–Jarrow–Morton* equation, see [HeJaMo] and [MuRu]. Below we basically follow [Mu]. Denote by $B(t, \xi)$, $t \geq 0$, $\xi > 0$ the price at moment t of a bond maturing at time $t + \xi$. At a current time $t > 0$ the function $B(t, \cdot)$ is known but its future values, $B(s, \xi)$, $s > t$, $\xi > 0$ can be only estimated and, in general, are of stochastic nature. The *forward rate function* $X(t, \xi)$, $t \geq 0$, $\xi \geq 0$, related to the price process B, is defined by the relation,

$$B(t, \xi) = e^{-\int_0^\xi X(t, \eta) \, d\eta}, \qquad t \geq 0, \; \xi \geq 0, \tag{9.34}$$

or, equivalently, by the formula:

$$X(t, \xi) = -\frac{\partial}{\partial \xi} (\ln B(t, \xi)), \qquad t \geq 0, \; \xi \geq 0. \tag{9.35}$$

In particular $r(t) = X(t, 0)$, $t \geq 0$ is the value of the *interest rate* at time t. Starting from D.Heath, R.Jarrow and A.Morton model [HeJaMo], M. Musiela in [Mu], proposed to represent the process X as a solution of an evolution equation:

$$dX(t, \xi) = \left(\frac{\partial X}{\partial \xi}(t, \xi) + a(\xi, X(t)) \right) dt + \sum_{i=1}^d \sigma_i(\xi, X(t)) \, dW_i(t) \tag{9.36}$$
$$X(0, \xi) = x(\xi),$$

where $W = (W_1, \ldots, W_d)$ is a $d-$ dimensional Wiener process, a and σ_i are functionals to be identified from the data and $x(\cdot)$ is the forward rate function at the initial moment 0. The economical interpretation of X leads to the requirement that for each $T \geq 0$, the process:

$$e^{-\int_0^t X(s, 0) \, ds} B(t, T - t) = e^{-\int_0^t X(s, 0) \, ds} e^{-\int_0^{T-t} X(t, \eta) \, d\eta}, \qquad 0 \leq t \leq T,$$

is a martingale. One can show, see [HeJaMo] and [Mu] , that this is true if and only if the functional a can be expressed in terms of the *volatility* functionals σ_i as follows:

$$a(\xi, y) = \sum_{i=1}^d \sigma_i(\xi, y) \int_0^\xi \sigma_i(\eta, y) \, d\eta,$$

for all $\xi \geq 0$ and all forward rate functions y. It is natural to assume that forward rates are absolutely continuous functions and take as the underling Hilbert space H the space H_γ^1 consisting of absolutely continuous functions y defined on $[0, +\infty)$, such that

$$\|y\|_\gamma^2 = |y(0)|^2 + \int_0^{+\infty} e^{-\gamma \eta} \left| \frac{dy}{d\eta}(\eta) \right|^2 d\eta < +\infty.$$

Here $\gamma \geq 0$ is a parameter to be chosen. If now φ is a functional on H_γ^1 and $X^x(t)$, $t \geq 0$ is a solution to 9.36, then $\varphi(X^x(t))$ is the gain of a holder of an European option, specified by φ, on interest rate derivative. The corresponding price of the option is given by a formula similar to 9.32:

$$V(t,x) = \mathbb{E}\left(e^{-\int_0^t X^x(s,0)\,ds} \varphi(X^x(t)) \right), \quad t \geq 0, \ x \in H_\gamma^1. \tag{9.37}$$

Thus V should satisfy the following Kolmogorov equation on H_γ^1:

$$\frac{\partial V}{\partial t}(t,x) = \frac{1}{2} \sum_{i=1}^d D_{\sigma_i}^2 V(t,x) + D_{(\frac{dx}{d\xi}+a)} V(t,x) - x(0)V(t,x),$$
$$V(0,x) = \varphi(x), \quad t \geq 0, \ x \in H_\gamma^1,$$

where $D_b V$ and $D_{\sigma_i}^2 V$ denote the first and the second directional derivatives of V in the directions b and σ_i respectively, see [GoMu].

10. Appendix

The Appendix is devoted to existence and regular dependence on a parameter of the solutions to a nonlinear equation on a Banach space. These results were used in the analysis of the stochastic evolution equations.

10.1 Implicit function theorems

Let Λ be an open subset of a Banach space and E a Banach space and H a transformation from $\Lambda \times E$ into E. We will assume that

$$\|H(\lambda,x) - H(\lambda,y)\|_E \leq \alpha \|x-y\|_E, \quad \text{for all } \lambda \in \Lambda, \ x,y \in E \tag{10.1}$$

and consider the equation

$$x = H(\lambda,x), \quad (\lambda,x) \in \Lambda \times E \tag{10.2}$$

Theorem 10.1. *Assume that the transformation H satisfies 10.1 with $\alpha \in [0,1[$. Then for arbitrary $\lambda \in \Lambda$ there exists exactly one solution $x = \varphi(\lambda)$ of the equation 10.2. If in addition H is continuous with respect to the first variable then the function φ is continuous on Λ.*

Proof. The first part of the theorem is a consequence of the contraction mapping theorem. Note that for arbitrary $\lambda \in \Lambda$

$$\varphi(\lambda) = H(\lambda, \varphi(\lambda)).$$

To prove the latter part fix $\lambda_0 \in \Lambda$ then:

$$\varphi(\lambda) - \varphi(\lambda_0) = H(\lambda, \varphi(\lambda)) - H(\lambda_0, \varphi(\lambda_0))$$
$$= [H(\lambda, \varphi(\lambda)) - H(\lambda, \varphi(\lambda_0))] + [H(\lambda, \varphi(\lambda_0)) - H(\lambda_0, \varphi(\lambda_0))].$$

Therefore:

$$\|\varphi(\lambda) - \varphi(\lambda_0)\|_E \leq \alpha \|\varphi(\lambda) - \varphi(\lambda_0)\|_E + \|H(\lambda, \varphi(\lambda_0)) - H(\lambda_0, \varphi(\lambda_0))\|_F,$$

and

$$\|\varphi(\lambda) - \varphi(\lambda_0)\|_E \leq \frac{1}{1-\alpha} \|H(\lambda, \varphi(\lambda_0)) - H(\lambda_0, \varphi(\lambda_0))\|_E, \qquad (10.3)$$

so the result follows. $\square$

If G is a mapping from E into F then its directional derivative at x and in direction y will be denoted by $\partial_x G(x; y)$. This convention, naturally extends to higher and partial derivatives.

Theorem 10.2. *Assume that H is continuous with respect to the first variable and satisfies 10.1 with $\alpha \in [0, 1[$. If, in addition, there exist directional derivatives*

$$\partial_\lambda H(\lambda, x; \mu), \qquad \partial_x H(\lambda, x; y),$$

and are continuous with respect to all variables, then for arbitrary $\lambda, \mu \in \Lambda$ there exists directional derivative $\partial_\lambda \varphi(\lambda; \mu)$,

$$\partial_\lambda \varphi(\lambda; \mu) = [I - \partial_x H(\lambda, \varphi(\lambda))]^{-1} \partial_\lambda H(\lambda, \varphi(\lambda); \mu), \qquad (10.4)$$

and is continuous with respect to all variables.

Proof. Let us fix $\lambda_0, \mu_0 \in \Lambda$ and let $t > 0$ be a sufficiently small number. Then

$$\varphi(\lambda_0 + t\mu_0) - \varphi(\lambda_0) = H(\lambda_0 + t\mu_0, \varphi(\lambda_0 + t\mu_0)) - H(\lambda_0, \varphi(\lambda_0))$$
$$= [H(\lambda_0 + t\mu_0, \varphi(\lambda_0 + t\mu_0)) - H(\lambda_0, \varphi(\lambda_0 + t\mu_0))]$$
$$+ [H(\lambda_0, \varphi(\lambda_0 + t\mu_0)) - H(\lambda_0, \varphi(\lambda_0))]$$
$$= \int_0^1 \partial_\lambda H(\lambda_0 + \sigma t\mu_0, \varphi(\lambda_0 + t\mu_0); t\mu_0) \, d\sigma$$
$$+ \int_0^1 \partial_x H[\lambda_0, \varphi(\lambda_0) + \sigma(\varphi(\lambda_0 + t\mu_0) - \varphi(\lambda_0)); \varphi(\lambda_0 + t\mu_0) - \varphi(\lambda_0)] \, d\sigma.$$

Both integrals have a well defined meaning. Moreover the strong integral

$$\int_0^1 \partial_x H(\lambda_0, a + \sigma b) \, d\sigma$$

exists for arbitrary $a, b \in E$ and its operator norm is not greater than α. Consequently

$$\frac{1}{t}(\varphi(\lambda_0 + t\mu_0) - \varphi(\lambda_0))$$

$$= \left[I - \int_0^1 \partial_x H(\lambda_0, \varphi(\lambda_0) + \sigma(\varphi(\lambda_0 + t\mu_0) - \varphi(\lambda_0))\, d\sigma\right]^{-1} \tag{10.5}$$

$$\int_0^1 \partial_\lambda H(\lambda_0 + \sigma t\mu_0, \varphi(\lambda_0 + t\mu_0); \mu_0)\, d\sigma.$$

To pass in 10.5 to the limit as $t \downarrow 0$ we need the following lemma.   □

Lemma 10.3. *Assume that a sequence (B_n) of bounded linear operators strongly converges to B. If $\|B_n\| \le \alpha$, $\|B\| \le \alpha$ for some $\alpha \in [0, 1[$ and all $n = 1, 2, \dots$ then*

$$(I - B_n)^{-1} \to (I - B)^{-1} \quad \text{strongly as } n \to +\infty$$

Proof. Let

$$x_n \to x \quad \text{and} \quad y_n = (I - B_n)^{-1} x_n, \quad y = (I - B)^{-1} x.$$

Since $y_n - B_n y_n = x_n$, $y - By = x$, therefore $y_n - y = B_n(y_n - y) + (x_n - x) + (B_n y - By)$ and $\|y_n - y\|_E \le \frac{1}{1-\alpha}[\|x_n - x\|_E + \|B_n y - By\|_E]$, so the result follows.   □

By Lemma 1 one can pass to the limit in 10.5 as $t \downarrow 0$ to get the formula 10.4. The continuity statement is also a consequence of Lemma 10.3.   □

Theorem 10.4. *Assume that E_0 is a Banach space continuously embedded into E and that 10.2 holds both in E and E_0 with the same constant $\alpha \in [0, 1[$. Let moreover assumptions of Theorem 10.2 be satisfied in E and E_0. If, in addition, there exist second directional derivatives,*

$$\partial_\lambda \partial_x H, \quad \partial_x \partial_\lambda H, \quad \partial_\lambda^2 H, \quad \partial_x^2 H,$$

in all points and in all directions of $\Lambda \times E_0$ and are continuous with respect to all variables as functions with values in E, then there exists the second directional derivative of φ at any point $\lambda \in \Lambda$ and any direction μ_0, ν_0, continuous with respect to all variables. Moreover,

$$\partial_\lambda^2 \varphi(\lambda_0; \mu_0, \nu_0) = [I - \partial_x H(\lambda_0, \varphi(\lambda_0))]^{-1} \tag{10.6}$$
$$\{\partial_x^2 H(\lambda_0, \varphi(\lambda_0)); \partial_\lambda \varphi(\lambda_0; \mu_0) \partial_\lambda \varphi(\lambda_0; \nu_0) + \partial_\lambda \partial_x H$$
$$(\lambda_0, \varphi(\lambda_0)); \partial_\lambda \varphi(\lambda_0; \mu_0), \nu_0) + \partial_x \partial_\lambda H(\lambda_0, \varphi(\lambda_0)); \mu_0, \partial_\lambda \varphi(\lambda_0; \nu_0))$$
$$+ \partial_\lambda^2 H(\lambda_0, \varphi(\lambda_0)); \mu_0, \nu_0)\}.$$

Remark 10.5. Note that by Theorem 10.1 and Theorem 10.2, functions $\varphi(\lambda)$, $\partial_\lambda \varphi(\lambda; \mu)$, $\lambda, \mu \in \Lambda$ are continuous as E_0-valued functions, therefore, by Lemma 10.3 and assumptions of Theorem 10.4, the formula 10.6 defines a continuous function of all three variables. Formulation of the theorem and its proof follow the Appendix in [DaPrZa4], see also [DaPrZa2].

Proof. We start from the obvious identities

$$\varphi(\lambda) = H(\lambda, \varphi(\lambda)), \quad \lambda \in \Lambda,$$

$$
\begin{aligned}
&\partial_\lambda\varphi(\lambda_0 + t\nu_0; \mu_0) - \partial_\lambda\varphi(\lambda_0; \mu_0) \\
=\ &[\partial_\lambda H(\lambda_0 + t\nu_0, \varphi(\lambda_0 + t\nu_0); \mu_0) + \partial_x H(\lambda_0 + t\nu_0, \varphi(\lambda_0 + t\nu_0); \\
&\quad \partial_\lambda\varphi(\lambda_0 + t\nu_0; \mu_0))] - [\partial_\lambda H(\lambda_0, \varphi(\lambda_0); \mu_0) + \partial_x H(\lambda_0, \varphi(\lambda_0); \partial_\lambda\varphi(\lambda_0; \mu_0)] \\
=\ &[\partial_\lambda H(\lambda_0 + t\nu_0, \varphi(\lambda_0 + t\nu_0); \mu_0) - \partial_\lambda H(\lambda_0, \varphi(\lambda_0); \mu_0)] \\
&+ [\partial_x H(\lambda_0 + t\nu_0, \varphi(\lambda_0 + \nu_0); \partial_\lambda\varphi(\lambda_0 + t\nu_0; \mu_0)) \\
&- \partial_x H(\lambda_0, \varphi(\lambda_0); \partial_\lambda\varphi(\lambda_0 + t\nu_0; \mu_0))] \\
&+ \partial_x H(\lambda_0, \varphi(\lambda_0); \partial_\lambda\varphi(\lambda_0 + t\nu_0; \mu_0) - \partial_\lambda\varphi(\lambda_0; \mu_0)) \\
=\ &J_1(t) + J_2(t) + \partial_x H(\lambda_0, \varphi(\lambda_0); \partial_\lambda\varphi(\lambda_0 + t\nu_0; \mu_0) - \partial_\lambda\varphi(\lambda_0; \mu_0)).
\end{aligned}
$$

They imply that

$$
\begin{aligned}
&\frac{1}{t}[\partial_\lambda\varphi(\lambda_0 + t\nu_0; \mu_0) - \partial_\lambda\varphi(\lambda_0; \mu_0)] \\
&\qquad\qquad = [I - \partial_x H(\lambda_0, \varphi(\lambda_0))]^{-1}\left[\frac{1}{t}J_1(t) + \frac{1}{t}J_2(t)\right]
\end{aligned}
$$

Since

$$
\begin{aligned}
\frac{1}{t}J_1(t) =\ &\frac{1}{t}[\partial_\lambda H(\lambda_0 + t\nu_0, \varphi(\lambda_0 + t\nu_0); \mu_0) - \partial_\lambda H(\lambda_0, \varphi(\lambda_0 + t\nu_0); \mu_0)] \\
&+ \frac{1}{t}[\partial_\lambda H(\lambda_0, \varphi(\lambda_0 + t\nu_0); \mu_0) - \partial_\lambda H(\lambda_0, \varphi(\lambda_0); \mu_0)]
\end{aligned}
$$

therefore, by the integral representations (as in the proof of Theorem 10.2), one has that:

$$\lim_{t\downarrow 0}\frac{1}{t}J_1(t) = \partial_\lambda^2 H(\lambda_0, \varphi(\lambda_0); \mu_0, \nu_0) + \partial_x\partial_\lambda H(\lambda_0, \varphi(\lambda_0); \mu_0, \partial_\lambda\varphi(\lambda_0; \nu_0))$$

In a similar way,

$$
\begin{aligned}
\frac{1}{t}J_2(t) =\ &\frac{1}{t}[\partial_x H(\lambda_0, \varphi(\lambda_0 + t\nu_0); \partial_\lambda\varphi(\lambda_0 + \nu_0; \mu_0)) \\
&- \partial_x H(\lambda_0, \varphi(\lambda_0 + t\nu_0); \partial_\lambda\varphi(\lambda_0 + t\nu_0; \mu_0))] \\
&+ \frac{1}{t}[\partial_x H(\lambda_0, \varphi(\lambda_0 + t\nu_0); \partial_\lambda\varphi(\lambda_0 + t\nu_0; \mu_0)) \\
&- \partial_x H(\lambda_0, \varphi(\lambda_0); \partial_\lambda\varphi(\lambda_0 + t\nu_0; \mu_0))], \\
\lim_{t\downarrow 0}\frac{1}{t}J_2(t) =\ &\partial_\lambda\partial_x H(\lambda_0, \varphi(\lambda_0); \partial_\lambda\varphi(\lambda_0; \mu_0), \nu_0) \\
&+ \partial_x^2 H(\lambda_0, \varphi(\lambda_0); \partial_\lambda\varphi(\lambda_0; \mu_0)\partial_\lambda\varphi(\lambda_0; \nu_0)]. \quad \square
\end{aligned}
$$

Assume now that we have a sequence of mappings H_n, $n = 1, \ldots$; and a mapping H such that

$$\|H(\lambda, x) - H(\lambda, y)\|_E \;\leq\; \alpha\|x - y\|, \qquad x, y \in E, \lambda \in \Lambda, \tag{10.7}$$

$$\|H_n(\lambda, x) - H_n(\lambda, y)\|_E \;\leq\; \alpha\|x - y\|, \qquad n \in \mathbb{N}, \; x, y \in E, \; \lambda \in \Lambda, \tag{10.8}$$

If $\alpha \in [0, 1[$ than for each $\lambda \in \Lambda$, equations

$$x \;=\; H(\lambda, x) \tag{10.9}$$
$$x \;=\; H_n(\lambda, x) \tag{10.10}$$

have unique solutions denoted by $\varphi(\lambda)$ and $\varphi_n(\lambda)$ respectively. Thus

$$\varphi(\lambda) = H(\lambda, \varphi(\lambda)), \quad \varphi_n(\lambda) = H_n(\lambda, \varphi_n(\lambda)), \quad \lambda \in \Lambda, \; n \in \mathbb{N}. \tag{10.11}$$

Theorem 10.6. *Assume that functions H, H_n, $n \in \mathbb{N}$, satisfy 10.7 and 10.8 with $\alpha \in [0, 1[$.*

i) *If $H_n(\lambda, x) \to H(\lambda, x)$, for all $(\lambda, x) \in \Lambda \times E$ then $\varphi_n(\lambda) \to \varphi(\lambda)$, for all $\lambda \in \Lambda$*

ii) *If H, H_n, $n \in \mathbb{N}$ are continuous and $H_n \to H$ uniformly on compact sets then $\varphi_n \to \varphi$ uniformly on compact sets.*

Proof. i) Note that

$$
\begin{aligned}
\|\varphi_n(\lambda) - \varphi(\lambda)\|_E \;\leq\;\; & \|H_n(\lambda, \varphi_n(\lambda)) - H_n(\lambda, \varphi(\lambda))\|_E \\
& + \|H_n(\lambda, \varphi(\lambda)) - H(\lambda, \varphi(\lambda))\|_E \\
\leq\;\; & \alpha\|\varphi_n(\lambda) - \varphi(\lambda)\|_E \\
& + \|H_n(\lambda, \varphi(\lambda)) - H(\lambda, \varphi(\lambda))\|_E
\end{aligned}
$$

Therefore

$$\|\varphi_n(\lambda) - \varphi(\lambda)\|_E \;\leq\; \frac{1}{1 - \alpha}\|H_n(\lambda, \varphi(\lambda)) - H(\lambda, \varphi(\lambda))\|_E \tag{10.12}$$

and $\varphi_n(\lambda) \to \varphi(\lambda)$ for each $\lambda \in \Lambda$

ii) Let $K \subset \Lambda$ be a compact set. Since φ is a continuous function, the set $\{(\lambda, \varphi(\lambda)); \lambda \in K\}$ is compact as well. Formula 10.12 implies now that $\varphi_n \to \varphi$ uniformly on K. $\qquad\square$

The convergence result from Theorem 10.6 can be extended to directional derivatives of φ as well.

Theorem 10.7. *Assume that mappings H, H_n, $n \in \mathbb{N}$ satisfy conditions of Theorem 10.2. If*

$$H_n(\lambda, x) \to H(\lambda, x), \qquad \partial_x H_n(\lambda, x; y) \to \partial_x H(\lambda, x; y) \tag{10.13}$$

$$\partial_\lambda H_n(\lambda, x; \mu) \to \partial_\lambda H(\lambda, x; \mu) \qquad \text{as } n \to +\infty \tag{10.14}$$

uniformly in (λ, x) *form compact sets and uniformly in* y *and* μ *from bounded sets, then*

$$\partial_\lambda \varphi_n(\lambda; \mu) \to \partial_\lambda \varphi(\lambda; \mu) \quad as\ n \to +\infty \tag{10.15}$$

uniformly in λ *from compact sets and* μ *from bounded sets.*

Proof. The result follows from the explicit formula 10.4 for the partial derivatives. □

By the same reasons the following result holds true:

Theorem 10.8. *Assume that mappings* H, H_n $n \in \mathbb{N}$ *satisfy conditions of Theorem 10.4 with the same space* E_0. *Assume that conditions of Theorem 10.7 are satisfied both in* E *and* E_0. *If, in addition:*

$$\partial_\lambda \partial_x H_n(\lambda, x; y; \mu) \to \partial_\lambda \partial_x H(\lambda, x; y; \mu), \quad as\ n \to +\infty$$
$$\partial_x \partial_\lambda H_n(\lambda, x; \mu, y) \to \partial_\lambda \partial_x H(\lambda, x; \mu, y), \quad as\ n \to +\infty$$
$$\partial_\lambda^2 H_n(\lambda, x; \mu, \nu) \to \partial_\lambda^2 H(\lambda, x; \mu, \nu), \quad as\ n \to +\infty$$
$$\partial_x^2 H_n(\lambda, x; y, z) \to \partial_x^2 H(\lambda, x; y, z), \quad as\ n \to +\infty$$

uniformly in (λ, x) *from compact subsets of* $\Lambda \times E_0$ *and uniformly in* (μ, ν, y, z) *from bounded sets of* $\Lambda \times \Lambda \times E \times E$, *then*

$$\partial_\lambda^2 \varphi_n(\lambda; \mu, \nu) \to \partial_\lambda^2 \varphi(\lambda; \mu, \nu), \quad as\ n \to +\infty \tag{10.16}$$

uniformly in λ *from compact subsets of* Λ *and uniformly in* μ, ν *from bounded subsets of* $\Lambda \times \Lambda$.

References

[AhFuZa] N.U.Ahmed, M.Fuhrman and J. Zabczyk, *On filtering equations in infinite dimensions*, J. Func. Anal., 143(1997), 180-204.

[Bi] P. Billingsley *Probability and Measure* , John Wiley and Sons, 1979

[CaDaPr] P.Cannarsa and G.Da Prato,*Second-order Hamilton-Jacobi equations in infinite dimensions*, SIAM J.Control and optimization, 29(1991), 474-492 .

[CeGg] S.Cerrai and F.Gozzi,*Strong solutions of Cauchy problems associated to weakly continuous semigroups*, Differential and Integral Equations, 8(1994), 465-486.

[ChojGo1] A. Chojnowska-Michalik and B. Goldys , *On regularity properties of nonsymmetric Ornstein-Uhlenbeck semigroup in* L^p *spaces*, Stochastics Stochastics Reports, 59 (1996), 183-209.

[ChojGo2] A. Chojnowska-Michalik and B. Goldys, *Nonsymmetric Ornstein- Uhlenbeck semigroup as a second quantized operator*, submitted.

[Chow] P.L.Chow,*Infinite-dimensional Kolmogorov Equations in Gauss -Sobolev spaces*, Stochastic Analysis and Applications,(to appear).

[ChowMel] P.L.Chow and J.L.Menaldi,*Variational inequalities for the control of stochastic partial differential equations*,Proceedings of the Trento Conference 1988,LN in math.,1390(1989).

[CrLi] M.G.Crandall and P.L.Lions.*Hamilton -Jacobi equations in infinite dimen-sions,Part IV,Unbounded linear terms*,J.Functional Analysis,90(1900),237-283.

[Dal] Yu. Daleckij, *Differential equations with functional derivatives and stochas-tic equations for generalized random processes*, Dokl. Akad. Nauk SSSR, 166(1966), 1035-38 .

[DaPr1] G.Da Prato,*Some results on elliptic and parabolic equations in Hilbert spaces*,Rend.Mat.Acc.Lincei,9(1996),181-199.

[DaPr2] G. Da Prato,*Parabolic equations in Hilbert spaces*, Lecture Notes, May 1996.

[DaPr3] G. Da Prato,*Stochastic control and Hamilton-Jacobi equations*, Lecture Notes, February 1998.

[DaPrElZa] G. Da Prato, D. Elworthy and J. Zabczyk *Strong Feller property for stochastic semilinear equations*, Stochastic Analysis and Applications, 13(1995), no.1, 35-45.

[DaPrGoZa] G.Da Prato, B. Goldys and J. Zabczyk , *Ornstein- Uhlenbeck semi-groups in open sets of Hilbert spaces*, Comptes Rendus Acad. Sci. Paris, Serie I, 325(1997), 433-438.

[DaPrZa1] G.Da Prato and J.Zabczyk, *Smoothing properties of transition semi-groups in Hilbert Spaces*,Stochastic and Stochastics Reports,35 (1991),63-77.

[DaPrZa2] G. Da Prato and J. Zabczyk, *Stochastic Equations in Infinite Dimen-sions*, Cambridge University Press, 1992.

[DaPrZa3] G.Da Prato and J.Zabczyk,*Regular densities of invariant measures in Hilbert spaces* ,J.Functional Analysis,130(1995), 427 -449.

[DaPrZa4] G. Da Prato and J. Zabczyk, *Ergodicity for Infinite Dimensional Sys-tems*, Cambridge University Press, 1996.

[Da] E.B. Davis, *One-parameter Semigroups*, Academic Press, 1980.

[Do] R. Douglas *On majoration, factorization and range inclusion of operators on Hilbert spaces* , Proc. Amer. Math. Soc., 17, 413-415.

[DS] N. Dunford and J.T. Schwartz *Linear Operators. Part II: Spectral Theory*, Interscience Publishers, New York, London 1963.

[Dy] E.B. Dynkin, *Markov Processes*, Vol I, Springer Verlag, 1965.

[GoMu] B. Goldys and M. Musiela, *On partial differential differential equations related to term structure models*, Working paper, University of New South Wales, 1996.

[Go] F.Gozzi,*Regularity of solutions of a second order Hamilton- Jacobi equation and application to a control problem*, Commun. in Partial Differential Equa-tions,20(1995), 775-826 .

[Gu] P. Guiotto, *Non-differentiability of heat semigroups in infinite dimensional Hilbert space*, forthcoming in Semigroup Forum.

[Fr] A. Friedman, *Partial Differential Equations of Parabolic Type*, Englewood Cliffs, New York, Prentice Hall, 1964

[GiSk] I.I. Gikhman and A.V. Shorokhod, The Theory of Stochastic Processes, Vols. I, II and III. Springer-Verlag, 1974, 1975, 1979.

[Gr] L. Gross, *Potential theory on Hilbert spaces*, J. Funct. Anal., 1(1967), 123-181 .

[HeJaMo] D. Heath, R. Jarrow and A. Morton, *Bond pricing and term structure of interest rates: a new methodology for contingent claim valuation*, Econo-metrica, 60(1992), 77-105.

[It] K.ItÔ, *On a stochastic integral equation* , Proc.Imp.Acad., Tokyo, 22(1946), 32-35.

[Ko] A.N.Kolmogorov,*Uber die analitischen methoden in der Wahrscheinlichkeit-srechnung*,Math. Ann. 104(1931),415-458 .

[Kr]   N.V.Krylov,*Lectures on elliptic and parabolic equations in Holder spaces*, AMS, Providence, 1996.

[Ku]   H.H. Kuo, *Gaussian measures in Banach spaces*, Lect. Notes in Math., 463(1975), Springer Verlag.

[Le]   P.Levy *Problemes Concrets d'Analyse Fonctionelle*, Gauthier-Villar, Paris, 1951.

[MaRa] Z.M.Ma and M.Rockner,*Introduction to the Theory of (Non Symmetric) Dirichlet Forms* Springer Verlag, 1992.

[Mu]   M. Musiela, *Stochastic PDEs and the term structure models* , Journees Internationales de Finance, IGR-AFFI. La Baule, June 1993.

[MuRu] M. Musiela and M. Rutkowski, *Martingale Methods in Financial Modelling* , Springer, 1997.

[NeZa] J.M.A.M. Neerven and J. Zabczyk, *Norm discontinuity of Ornstein-Uhlenbeck semigroups*, forthcoming in Semigroup Forum.

[NuUs] D.Nualart and A.S. Ustunel, *Une extension du laplacian sur l'espace de Wiener et la formule d'Itô associee*, C.R.Acad.Sci. Paris, t.309, Serie I, p. 383-386, 1984.

[Pa]   K.R. Parthasarathy, *Probability Measures on Metric Spaces*, Academic Press, 1967.

[Pe]   S.Peszat, *Large deviation principle for stochastic evolution equations*, PTRF, 98 (1994), pp. 113–136.

[PeZa] S. Peszat and J. Zabczyk , *Strong Feller property and irreducibility for diffusions on Hilbert spaces*, Annals of Probability, 23(1995), 157-172.

[Ph]   R.R. Phelps *Gaussian null sets and differentiability of Lipschitz map on Banach spaces*,Pac. J. Math. 77(1978), 523-531.

[Pie]   A. Piech, *A fundamental solution of the parabolic equation on Hilbert space*, J.Func.Anal.,3 (1969), 85-114.

[Pr]   E. Priola, *π -semigroups and applications* , Preprint 9, Scuola Normale Superiore, Pisa, 1998.

[Sw]   A.Swiech,*"Unbounded" second order partial differential equations in infinite dimensional Hilbert spaces*, Commun.in Partial Differential Equations, 19(1994), 1999- 2036.

[TeZa] G.Tessitore and J. Zabczyk *Comments on transition semigroups and stochastic invariance* , Preprint, Scuola Normale Superiore, Pisa, 1998.

[Yo]   K. Yosida ,*Functional Analysis*, Springer, 1965.

[Va]   T. Vargiolu, *Invariant measures for a Langevin equation describing forward rates in an arbitrage free market*, forthcoming in Finance Stochastic.

[Za1]   J.Zabczyk, *Linear stochastic systems in Hilbert spaces: spectral properties and limit behavior*, Report No. 236, Institute of Mathematics, Polish Academy of Sciences, 1981. Also in Banach Center Publications, 41(1985), 591-609.

[Za2]   J.Zabczyk, *Mathematical Control Theory. An Introduction*, Birkhauser, 1992.

[Za3]   J.Zabczyk,*Bellman's inclusions and excessive measures*,Preprint 8 Scuola Normale Superiore, March 1998.

[Za4]   J.Zabczyk,*Infinite dimensional diffusions in Modelling and Analysis*,, Jber.d Dt.Math.- Verein., (to appear).

1954 -   1. Analisi funzionale   C.I.M.E.

         2. Quadratura delle superficie e questioni connesse   "

         3. Equazioni differenziali non lineari   "

1955 -   4. Teorema di Riemann-Roch e questioni connesse   "

         5. Teoria dei numeri   "

         6. Topologia   "

         7. Teorie non linearizzate in elasticità, idrodinamica,aerodinamica   "

         8. Geometria proiettivo-differenziale   "

1956 -   9. Equazioni alle derivate parziali a caratteristiche reali   "

        10. Propagazione delle onde elettromagnetiche   "

        11. Teoria della funzioni di più variabili complesse e delle

            funzioni automorfe   "

1957 - 12. Geometria aritmetica e algebrica (2 vol.)   "

        13. Integrali singolari e questioni connesse   "

        14. Teoria della turbolenza (2 vol.)   "

1958 - 15. Vedute e problemi attuali in relatività generale   "

        16. Problemi di geometria differenziale in grande   "

        17. Il principio di minimo e le sue applicazioni alle equazioni

            funzionali   "

1959 - 18. Induzione e statistica   "

        19. Teoria algebrica dei meccanismi automatici (2 vol.)   "

        20. Gruppi, anelli di Lie e teoria della coomologia   "

1960 - 21. Sistemi dinamici e teoremi ergodici   "

        22. Forme differenziali e loro integrali   "

1961 - 23. Geometria del calcolo delle variazioni (2 vol.)   "

        24. Teoria delle distribuzioni   "

        25. Onde superficiali   "

1962 - 26. Topologia differenziale   "

        27. Autovalori e autosoluzioni   "

        28. Magnetofluidodinamica   "

1972 - 59. Non-linear mechanics   "
       60. Finite geometric structures and their applications   "
       61. Geometric measure theory and minimal surfaces   "

1973 - 62. Complex analysis   "
       63. New variational techniques in mathematical physics   "
       64. Spectral analysis   "

1974 - 65. Stability problems   "
       66. Singularities of analytic spaces   "
       67. Eigenvalues of non linear problems   "

1975 - 68. Theoretical computer sciences   "
       69. Model theory and applications   "
       70. Differential operators and manifolds   "

1976 - 71. Statistical Mechanics   Ed Liguori, Napoli
       72. Hyperbolicity   "
       73. Differential topology   "

1977 - 74. Materials with memory   "
       75. Pseudodifferential operators with applications   "
       76. Algebraic surfaces   "

1978 - 77. Stochastic differential equations   "
       78. Dynamical systems   Ed Liguori, Napoli and Birhäuser Verlag

1979 - 79. Recursion theory and computational complexity   "
       80. Mathematics of biology   "

1980 - 81. Wave propagation   "
       82. Harmonic analysis and group representations   "
       83. Matroid theory and its applications   "

1981 - 84. Kinetic Theories and the Boltzmann Equation (LNM 1048) Springer-Verlag
       85. Algebraic Threefolds   (LNM 947)   "
       86. Nonlinear Filtering and Stochastic Control   (LNM 972)   "

1982 - 87. Invariant Theory   (LNM 996)   "
       88. Thermodynamics and Constitutive Equations   (LN Physics 228) "
       89. Fluid Dynamics   (LNM 1047)   "

1983 - 90. Complete Intersections	(LNM 1092)	Springer-Verlag
91. Bifurcation Theory and Applications	(LNM 1057)	"
92. Numerical Methods in Fluid Dynamics	(LNM 1127)	"
1984 - 93. Harmonic Mappings and Minimal Immersions	(LNM 1161)	"
94. Schrödinger Operators	(LNM 1159)	"
95. Buildings and the Geometry of Diagrams	(LNM 1181)	"
1985 - 96. Probability and Analysis	(LNM 1206)	"
97. Some Problems in Nonlinear Diffusion	(LNM 1224)	"
98. Theory of Moduli	(LNM 1337)	"
1986 - 99. Inverse Problems	(LNM 1225)	"
100. Mathematical Economics	(LNM 1330)	"
101. Combinatorial Optimization	(LNM 1403)	"
1987 - 102. Relativistic Fluid Dynamics	(LNM 1385)	"
103. Topics in Calculus of Variations	(LNM 1365)	"
1988 - 104. Logic and Computer Science	(LNM 1429)	"
105. Global Geometry and Mathematical Physics	(LNM 1451)	"
1989 - 106. Methods of nonconvex analysis	(LNM 1446)	"
107. Microlocal Analysis and Applications	(LNM 1495)	"
1990 - 108. Geometric Topology: Recent Developments	(LNM 1504)	"
109. H_∞ Control Theory	(LNM 1496)	"
110. Mathematical Modelling of Industrial Processes	(LNM 1521)	"
1991 - 111. Topological Methods for Ordinary Differential Equations	(LNM 1537)	"
112. Arithmetic Algebraic Geometry	(LNM 1553)	"
113. Transition to Chaos in Classical and Quantum Mechanics	(LNM 1589)	"
1992 - 114. Dirichlet Forms	(LNM 1563)	"
115. D-Modules, Representation Theory, and Quantum Groups	(LNM 1565)	"
116. Nonequilibrium Problems in Many-Particle Systems	(LNM 1551)	"

1993 - 117. Integrable Systems and Quantum Groups   (LNM 1620)   Springer-Verla
       118. Algebraic Cycles and Hodge Theory   (LNM 1594)
       119. Phase Transitions and Hysteresis   (LNM 1584)

1994 - 120. Recent Mathematical Methods in   (LNM 1640)
            Nonlinear Wave Propagation
       121. Dynamical Systems   (LNM 1609)
       122. Transcendental Methods in Algebraic   (LNM 1646)
            Geometry

1995 - 123. Probabilistic Models for Nonlinear PDE's (LNM 1627)
       124. Viscosity Solutions and Applications   (LNM 1660)
       125. Vector Bundles on Curves. New Directions (LNM 1649)

1996 - 126. Integral Geometry, Radon Transforms   (LNM 1684)
            and Complex Analysis
       127. Calculus of Variations and Geometric   LNM 1713
            Evolution Problems
       128. Financial Mathematics   LNM 1656

1997 - 129. Mathematics Inspired by Biology   LNM 1714
       130. Advanced Numerical Approximation of   LNM 1697
            Nonlinear Hyperbolic Equations
       131. Arithmetic Theory of Elliptic Curves   LNM 1716
       132. Quantum Cohomology   to appear

1998 - 133. Optimal Shape Design   to appear
       134. Dynamical Systems and Small Divisors   to appear
       135. Mathematical Problems in Semiconductor   to appear
            Physics
       136. Stochastic PDE's and Kolmogorov Equations LNM 1715
            in Infinite Dimension
       137. Filtration in Porous Media and Industrial to appear
            Applications

1999 - 138. Computional Mathematics driven by Industrual
            Applicationa   to appear
       139. Iwahori-Hecke Algebras and Representation
            Theory   to appear

FONDAZIONE C.I.M.E.
CENTRO INTERNAZIONALE MATEMATICO ESTIVO
INTERNATIONAL MATHEMATICAL SUMMER
CENTER

"Computational Mathematics driven by Industrial Applications"

is the subject of the first 1999 C.I.M.E. Session.

The session, sponsored by the Consiglio Nazionale delle Ricerche (C.N.R.), the Ministero dell'Università e della Ricerca Scientifica e Tecnologica (M.U.R.S.T.) and the European Community, will take place, under the scientific direction of Professors Vincenzo CAPASSO (Università di Milano), Heinz W. ENGL (Johannes Kepler Universitaet, Linz) and Doct. Jacques PERIAUX (Dassault Aviation) at the Ducal Palace of Martina Franca (Taranto), from 21 to 27 June, 1999.

Courses

a) **Paths, trees and flows: graph optimisation problems with industrial applications** (5 lectures in English) Prof. Rainer BURKARD (Technische Universität Graz)

Abstract

Graph optimisation problems play a crucial role in telecommunication, production, transportation, and many other industrial areas. This series of lectures shall give an overview about exact and heuristic solution approaches and their inherent difficulties. In particular the essential algorithmic paradigms such as greedy algorithms, shortest path computation, network flow algorithms, branch and bound as well as branch and cut, and dynamic programming will be outlined by means of examples stemming from applications.

References

1) R. K. Ahuja, T. L. Magnanti & J. B. Orlin, *Network Flows: Theory, Algorithms and Applications*, Prentice Hall, 1993

2) R. K. Ahuja, T. L. Magnanti, J.B.Orlin & M. R. Reddy, *Applications of Network Optimization*. Chapter 1 in: Network Models (Handbooks of Operations Research and Management Science, Vol. 7), ed. by M. O. Ball et al., North Holland 1995, pp. 1-83

3) R. E. Burkard & E. Cela, *Linear Assignment Problems and Extensions*, Report 127, June 1998 (to appear in Handbook of Combinatorial Optimization, Kluwer, 1999).

Can be downloaded by anonymous ftp from

ftp.tu-graz.ac.at, directory/pub/papers/math

4) R. E. Burkard, E. Cela, P. M. Pardalos & L. S. Pitsoulis, *The Quadratic Assignment Problem*, Report 126 May 1998 (to appear in Handbook of Combinatorial Optimization, Kluwer, 1999). Can be downloaded by anonymous ftp from ftp.tu-graz.ac.at, directory /pub/papers/math.

5) E. L. Lawler, J. K. Lenstra, A. H. G.Rinnooy Kan & D. B. Shmoys (Eds.), *The Travelling Salesman Problem*, Wiley, Chichester, 1985.

b) **New Computational Concepts, Adaptive Differential Equations Solvers and Virtual Labs** (5 lectures in English) Prof. Peter DEUFLHARD (Konrad Zuse Zentrum, Berlin).

Abstract

The series of lectures will address computational mathematical projects that have been tackled by the speaker and his group. In all the topics to be presented novel mathematical modelling, advanced algorithm developments. and efficient visualisation play a joint role to solve problems of practical relevance. Among the applications to be exemplified are:

1) Adaptive multilevel FEM in clinical cancer therapy planning;

2) Adaptive multilevel FEM in optical chip design;

3) Adaptive discrete Galerkin methods for countable ODEs in polymer chemistry;

4) Essential molecular dynamics in RNA drug design.

References

1) P. Deuflhard & A Hohmann, *Numerical Analysis. A first Course in Scientific Computation*, Verlag de Gruyter, Berlin, 1995

2) P. Deuflhard et al *A nonlinear multigrid eigenproblem solver for the complex Helmoltz equation*, Konrad Zuse Zentrum Berlin SC 97-55 (1997)

3) P. Deuflhard et al. *Recent developments in chemical computing*,
Computers in Chemical Engineering, **14**, (1990),pp.1249-1258.

4) P. Deuflhard et al. (eds) *Computational molecular dynamics: challenges, methods, ideas*, Lecture Notes in Computational Sciences and Engineering, vol.4 Springer Verlag, Heidelberg, 1998.

5) P.Deuflhard & M. Weiser, *Global inexact Newton multilevel FEM for nonlinear elliptic problems*, Konrad Zuse Zentrum SC 96-33, 1996.

c) Computational Methods for Aerodynamic Analysis and Design. (5 lectures in English) Prof. Antony JAMESON (Stanford University, Stanford).

Abstract

The topics to be discussed will include: - Analysis of shock capturing schemes, and fast solution algorithms for compressible flow; - Formulation of aerodynamic shape optimisation based on control theory; - Derivation of the adjoint equations for compressible flow modelled by the potential Euler and Navies-Stokes equations; - Analysis of alternative numerical search procedures; - Discussion of geometry control and mesh perturbation methods; - Discussion of numerical implementation and practical applications to aerodynamic design.

d) Mathematical Problems in Industry (5 lectures in English) Prof. Jacques-Louis LIONS (Collège de France and Dassault Aviation, France).

Abstract

1. Interfaces and scales. The industrial systems are such that for questions of reliability, safety, cost no subsystem can be underestimated. Hence the need to address problems of scales, both in space variables and in time and the crucial importance of modelling and numerical methods.

2. Examples in Aerospace Examples in Aeronautics and in Spatial Industries. Optimum design.

3. Comparison of problems in Aerospace and in Meteorology. Analogies and differences

4 Real time control. Many methods can be thought of. Universal decomposition methods will be presented.

References

1) J. L. Lions, *Parallel stabilization hyperbolic and Petrowsky systems*, WCCM4 Conference, CDROM Proceedings, Buenos Aires, June 29- July 2, 1998.

2) W. Annacchiarico & M. Cerolaza, *Structural shape optimization of 2-D finite elements models using Beta-splines and genetic algorithms*, WCCM4 Conference, CDROM Proceedings, Buenos Aires, June 29- July 2, 1998.

3) J. Periaux, M. Sefrioui & B. Mantel, *Multi-objective strategies for complex optimization problems in aerodynamics using genetic algorithms*, ICAS '98 Conference, Melbourne, September '98, ICAS paper 98-2.9.1

e) Wavelet transforms and Cosine Transform in Signal and Image Processing (5 lectures in English) Prof. Gilbert STRANG (MIT, Boston).

Abstract

In a series of lectures we will describe how a linear transform is applied to the sampled data in signal processing, and the transformed data is compressed (and quantized to a string of bits). The quantized signal is transmitted and then the inverse transform reconstructs a very good approximation to the original signal. Our analysis concentrates on the construction of the transform. There are several important constructions and we emphasise two: 1) the discrete cosine transform (DCT); 2) discrete wavelet transform (DWT). The DCT is an orthogonal transform (for which we will give a new proof). The DWT may be orthogonal, as for the Daubechies family of wavelets. In other cases it may be biorthogonal - so the reconstructing transform is the inverse but not the transpose of the analysing transform. The reason for this possibility is that orthogonal wavelets cannot also be symmetric, and symmetry is essential property in image processing (because our visual system objects to lack of symmetry). The wavelet construction is based on a "bank" of filters - often a low pass and high pass filter. By iterating the low pass filter we decompose the input space into "scales" to produce a multiresolution. An infinite iteration yields in the limit the scaling function and a wavelet: the crucial equation for the theory is the refinement equation or dilatation equation that yields the scaling function. We discuss the mathematics of the refinement equation: the existence and the smoothness of the solution, and the construction by the cascade algorithm. Throughout these lectures we will be developing the mathematical ideas, but always for a purpose. The insights of wavelets have led to new bases for function spaces and there is no doubt that other ideas are waiting to be developed. This is applied mathematics.

References

1) I. Daubechies, *Ten lectures on wavelets*, SIAM, 1992.

2) G. Strang & T. Nguyen, *Wavelets and filter banks*, Wellesley-Cambridge, 1996.

3) Y. Meyer, *Wavelets: Algorithms and Applications*, SIAM, 1993.

Seminars

Two hour seminars will be held by the Scientific Directors and Professor R. Mattheij.

1) **Mathematics of the crystallisation process of polymers.** Prof. Vincenzo CAPASSO (Un. di Milano).

2) **Inverse Problems: Regularization methods, Application in Industry.** Prof. H. W. ENGL (Johannes Kepler Un., Linz).

3) **Mathematics of Glass.** Prof. R. MATTHEIJ (TU Eindhoven).

4) **Combining game theory and genetic algorithms for solving multiobjective shape optimization problems in Aerodynamics Engineering.** Doct. J. PERIAUX (Dassault Aviation).

Applications

Those who want to attend the Session should fill in an application to C.I.M.E Foundation at the address below, **not later than April 30**, 1999. An important consideration in the acceptance of applications is the scientific relevance of the Session to the field of interest of the applicant. Applicants are requested, therefore, to submit, along with their application, a scientific curriculum and a letter of recommendation. Participation will only be allowed to persons who have applied in due time and have had their application accepted. CIME will be able to partially support some of the youngest participants. Those who plan to apply for support have to mention it explicitely in the application form.

Attendance

No registration fee is requested. Lectures will be held at Martina Franca on June 21, 22, 23, 24, 25, 26, 27. Participants are requested to register on June 20, 1999.

Site and lodging

Martina Franca is a delightful baroque town of white houses of Apulian spontaneous architecture. Martina Franca is the major and most aristocratic centre of the "Murgia dei Trulli" standing on an hill which dominates the well known Itria valley spotted with "Trulli" conical dry stone houses which go back to the 15th century. A masterpiece of baroque architecture is the Ducal palace where the workshop will be hosted. Martina Franca is part of the province of Taranto, one of the major centres of Magna Grecia, particularly devoted to mathematics. Taranto houses an outstanding museum of Magna Grecia with fabulous collections of gold manufactures.

Lecture Notes

Lecture notes will be published as soon as possible after the Session.

Arrigo CELLINA　　　　　　　　　Vincenzo VESPRI
CIME Director　　　　　　　　　　CIME Secretary

Fondazione C.I.M.E. c/o Dipartimento di Matematica ?U. Dini? Viale Morgagni, 67/A - 50134 FIRENZE (ITALY) Tel. +39-55-434975 / +39-55-4237123 FAX +39-55-434975 / +39-55-4222695 E-mail CIME@UDINI.MATH.UNIFI.IT

Information on CIME can be obtained on the system World-Wide-Web on the file HTTP: //WWW.MATH.UNIFI.IT/CIME/WELCOME.TO.CIME

FONDAZIONE C.I.M.E.
CENTRO INTERNAZIONALE MATEMATICO ESTIVO
INTERNATIONAL MATHEMATICAL SUMMER
CENTER

"Iwahori-Hecke Algebras and Representation Theory"

is the subject of the second 1999 C.I.M.E. Session.

The session, sponsored by the Consiglio Nazionale delle Ricerche (C.N.R.), the Ministero dell'Università e della Ricerca Scientifica e Tecnologica (M.U.R.S.T.) and the European Community, will take place, under the scientific direction of Professors Velleda BALDONI (Università di Roma "Tor Vergata") and Dan BARBASCH (Cornell University) at the Ducal Palace of Martina Franca (Taranto), from June 28 to July 6, 1999.

Courses

a) Double HECKE algebras and applications (6 lectures in English)
    Prof. Ivan CHEREDNIK (Un. of North Carolina at Chapel Hill, USA)
    Abstract:

The starting point of many theories in the range from arithmetic and harmonic analysis to path integrals and matrix models is the formula:

$$\Gamma(k + 1/2) = 2 \int_0^\infty e^{-x^2} x^{2k} dx.$$

Recently a q-generalization was found based on the Hecke algebra technique, which completes the 15 year old Macdonald program.

The course will be about applications of the double affine Hecke algebras (mainly one-dimensional) to the Macdonald polynomials, Verlinde algebras, Gauss integrals and sums. It will be understandable for those who are not familiar with Hecke algebras and (hopefully) interesting to the specialists.

1) *q-Gauss integrals.* We will introduce a q-analogue of the classical integral formula for the gamma-function and use it to generalize the Gaussian sums at roots of unity.

2) *Ultraspherical polynomials.* A connection of the q-ultraspherical polynomials (the Rogers polynomials) with the one-dimensional double affine Hecke algebra will be established.

3) *Duality.* The duality for these polynomials (which has no classical counterpart) will be proved via the double Hecke algebras in full details.

4) *Verlinde algebras.* We will study the polynomial representation of the 1-dim. DHA at roots of unity, which leads to a generalization and a simplification of the Verlinde algebras.

5) *$PSL_2(\mathbf{Z})$-action.* The projective action of the $PSL_2(\mathbf{Z})$ on DHA and the generalized Verlinde algebras will be considered for A_1 and arbitrary root systems.

6) *Fourier transform of the q-Gaussian.* The invariance of the q-Gaussian with respect to the q-Fourier transform and some applications will be discussed.

References:

1) *From double Hecke algebra to analysis*, Proceedings of ICM98, Documenta Mathematica (1998).

2) *Difference Macdonald–Mehta conjecture*, IMRN:10, 449–467 (1997).

3) *Lectures on Knizhnik-Zamolodchikov equations and Hecke algebras*, MSJ Memoirs (1997).

b) Representation theory of affine Hecke algebras

Prof. Gert HECKMAN (Catholic Un., Nijmegen, Netherlands)

Abstract.

1. The Gauss hypergeometric equation.
2. Algebraic aspects of the hypergeometric system for root systems.
3. The hypergeometric function for root systems.
4. The Plancherel formula in the hypergeometric context.
5. The Lauricella hypergeometric function.
6. A root system analogue of 5.

I will assume that the audience is familiar with the classical theory of ordinary differential equations in the complex plane, in particular the concept of regular singular points and monodromy (although in my first lecture I will give a brief review of the Gauss hypergeometric function). This material can be found in many text books, for example E.L. Ince, Ordinary differential equations, Dover Publ, 1956. E.T. Whittaker and G.N. Watson, A course of modern analysis, Cambridge University Press, 1927.

I will also assume that the audience is familiar with the theory of root systems and reflection groups, as can be found in N. Bourbaki, Groupes et algèbres de Lie, Ch. 4,5 et 6, Masson, 1981. J. E. Humphreys, Reflection groups and Coxeter groups, Cambridge University Press, 1990. or in one of the text books on semisimple groups.

For the material covered in my lectures references are W.J. Couwenberg, Complex reflection groups and hypergeometric functions, Thesis Nijmegen, 1994. G.J. Heckman, Dunkl operators, Sem Bourbaki no 828, 1997. E.M. Opdam, Lectures on Dunkl operators, preprint 1998.

c) Representations of affine Hecke algebras.

Prof. George LUSZTIG (MIT, Cambridge, USA)

Abstract

Affine Hecke algebras appear naturally in the representation theory of p-adic groups. In these lectures we will discuss the representation theory of affine Hecke algebras and their graded version using geometric methods such as equivariant K-theory or perverse sheaves.

References.

1. V. Ginzburg, *Lagrangian construction of representations of Hecke algebras*, Adv. in Math. 63 (1987), 100-112.

2. D. Kazhdan and G. Lusztig, *Proof of the Deligne-Langlands conjecture for Hecke algebras.*, Inv. Math. 87 (1987), 153-215.

3. G. Lusztig, *Cuspidal local systems and graded Hecke algebras*, I, IHES Publ. Math. 67 (1988),145-202; II, in "Representation of groups" (ed. B. Allison and G. Cliff), Conf. Proc. Canad. Math. Soc.. 16, Amer. Math. Soc. 1995, 217-275.

4. G. Lusztig, *Bases in equivariant K-theory*, Represent. Th., 2 (1998).

d) Affine-like Hecke Algebras and p-adic representation theory

Prof. Roger HOWE (Yale Un., New Haven, USA)

Abstract

Affine Hecke algebras first appeared in the study of a special class of representations (the spherical principal series) of reductive groups with coefficients in p-adic fields. Because of their connections with this and other topics, the structure and representation theory of affine Hecke algebras has been intensively studied by a variety of authors. In the meantime, it has gradually emerged that affine Hecke algebras, or slight generalizations of them, allow one to understand far more of the representations of p-adic groups than just the spherical principal series. Indeed, it seems possible that such algebras will allow one to understand all representations of p-adic groups. These lectures will survey progress in this approach to p-adic representation theory.

Topics:

1) Generalities on spherical function algebras on p-adic groups.

2) Iwahori Hecke algebras and generalizations.

3) - 4) Affine Hecke algebras and harmonic analysis

5) - 8) Affine-like Hecke algebras and representations of higher level.

References:

J. Adler, *Refined minimal K-types and supercuspidal representations*, Ph.D. Thesis, University of Chicago.

D. Barbasch, *The spherical dual for p-adic groups*, in Geometry and Representation Theory of Real and p-adic Groups, J. Tirao, D. Vogan, and J. Wolf, eds, Prog. In Math. 158, Birkhauser Verlag, Boston, 1998, 1 - 20.

D. Barbasch and A. Moy, *A unitarity criterion for p-adic groups*, Inv. Math. 98 (1989), 19 - 38.

D. Barbasch and A. Moy, *Reduction to real infinitesimal character in affine Hecke algebras*, J. A. M. S.6 (1993), 611- 635.

D. Barbasch, *Unitary spherical spectrum for p-adic classical groups*, Acta. Appl. Math. 44 (1996), 1 - 37.

C. Bushnell and P. Kutzko, *The admissible dual of GL(N) via open subgroups*, Ann. of Math. Stud. 129, Princeton University Press, Princeton, NJ, 1993.

C. Bushnell and P. Kutzko, *Smooth representations of reductive p-adic groups*: Structure theory via types, D. Goldstein, *Hecke algebra isomorphisms for tamely ramified characters*, R. Howe and A. Moy, *Harish-Chandra Homomorphisms for p-adic Groups*, CBMS Reg. Conf. Ser. 59, American Mathematical Society, Providence, RI, 1985.

R. Howe and A. Moy, *Hecke algebra isomorphisms for GL(N) over a p-adic field*, J. Alg. 131 (1990), 388 - 424.

J-L. Kim, *Hecke algebras of classical groups over p-adic fields and supercuspidal representations,I, II, III*, preprints, 1998.

G. Lusztig, *Classification of unipotent representations of simple p-adic groups*, IMRN 11 (1995), 517 - 589.

G. Lusztig, *Affine Hecke algebras and their graded version*, J. A. M. S. 2 (1989), 599 - 635.

L. Morris, *Tamely ramified supercuspidal representations of classical groups, I, II*, Ann. Ec. Norm. Sup 24, (1991) 705 - 738; 25 (1992), 639 - 667.

L. Morris, *Tamely ramified intertwining algebras*, Inv. Math. 114 (1994), 1 - 54.

A. Roche, *Types and Hecke algebras for principal series representations of split reductive p-adic groups*, preprint, (1996).

J-L. Waldspurger, *Algebres de Hecke et induites de representations cuspidales pour GLn*, J. reine u. angew. Math. 370 (1986), 27 - 191.

J-K. Yu, *Tame construction of supercuspidal representations*, preprint, 1998.

Applications

Those who want to attend the Session should fill in an application to the Director of C.I.M.E at the address below, not later than April 30, 1999.

An important consideration in the acceptance of applications is the scientific relevance of the Session to the field of interest of the applicant.

Applicants are requested, therefore, to submit, along with their application, a scientific curriculum and a letter of recommendation.

Participation will only be allowed to persons who have applied in due time and have had their application accepted.

CIME will be able to partially support some of the youngest participants. Those who plan to apply for support have to mention it explicitly in the application form.

Attendance

No registration fee is requested. Lectures will be held at Martina Franca on June 28, 29, 30, July 1, 2, 3, 4, 5, 6. Participants are requested to register on June 27, 1999.

Site and lodging

Martina Franca is a delightful baroque town of white houses of Apulian spontaneous architecture. Martina Franca is the major and most aristocratic centre of the Murgia dei Trulli standing on an hill which dominates the well known Itria valley spotted with Trulli conical dry stone houses which go back to the 15th century. A masterpiece of baroque architecture is the Ducal palace where the workshop will be hosted. Martina Franca is part of the province of Taranto, one of the major centres of Magna Grecia, particularly devoted to mathematics. Taranto houses an outstanding museum of Magna Grecia with fabulous collections of gold manufactures.

Lecture Notes

Lecture notes will be published as soon as possible after the Session.

Arrigo CELLINA
CIME Director

Vincenzo VESPRI
CIME Secretary

Fondazione C.I.M.E. c/o Dipartimento di Matematica U. Dini Viale Morgagni, 67/A - 50134 FIRENZE (ITALY) Tel. +39-55-434975 / +39-55-4237123 FAX +39-55-434975 / +39-55-4222695 E-mail CIME@UDINI.MATH.UNIFI.IT

Information on CIME can be obtained on the system World-Wide-Web on the file HTTP: //WWW.MATH.UNIFI.IT/CIME/WELCOME.TO.CIME.

FONDAZIONE C.I.M.E.
CENTRO INTERNAZIONALE MATEMATICO ESTIVO
INTERNATIONAL MATHEMATICAL SUMMER
CENTER
"Theory and Applications of Hamiltonian Dynamics"

is the subject of the third 1999 C.I.M.E. Session.

The session, sponsored by the Consiglio Nazionale delle Ricerche (C.N.R.), the Ministero dell'Università e della Ricerca Scientifica e Tecnologica (M.U.R.S.T.) and the European Community, will take place, under the scientific direction of Professor Antonio GIORGILLI (Un. di Milano), at Grand Hotel San Michele, Cetraro (Cosenza), from July 1 to July 10, 1999.

Courses

a) Physical applications of Nekhoroshev theorem and exponential estimates (6 lectures in English)

Prof. Giancarlo BENETTIN (Un. di Padova, Italy)

Abstract

The purpose of the lectures is to introduce exponential estimates (i.e., construction of normal forms up to an exponentially small remainder) and Nekhoroshev theorem (exponential estimates plus geometry of the action space) as the key to understand the behavior of several physical systems, from the Celestial mechanics to microphysics.

Among the applications of the exponential estimates, we shall consider problems of adiabatic invariance for systems with one or two frequencies coming from molecular dynamics. We shall compare the traditional rigorous approach via canonical transformations, the heuristic approach of Jeans and of Landau–Teller, and its possible rigorous implementation via Lindstet series. An old conjecture of Boltzmann and Jeans, concerning the possible presence of very long equilibrium times in classical gases (the classical analog of "quantum freezing") will be reconsidered. Rigorous and heuristic results will be compared with numerical results, to test their level of optimality.

Among the applications of Nekhoroshev theorem, we shall study the fast rotations of the rigid body, which is a rather complete problem, including degeneracy and singularities. Other applications include the stability of elliptic equilibria, with special emphasis on the stability of triangular Lagrangian points in the spatial restricted three body problem.

References:

For a general introduction to the subject, one can look at chapter 5 of V.I. Arnold, V.V. Kozlov and A.I. Neoshtadt, in *Dynamical Systems III*, V.I. Arnold Editor (Springer, Berlin 1988). An introduction to physical applications of Nekhorshev theorem and exponential estimates is in the proceeding of the Noto School "Non-Linear Evolution and Chaotic Phenomena", G. Gallavotti and P.W. Zweifel Editors (Plenum Press, New York, 1988), see the contributions by G. Benettin, L. Galgani and A. Giorgilli.

General references on Nekhoroshev theorem and exponential estimates: N.N. Nekhoroshev, Usp. Mat. Nauk. **32**:6, 5-66 (1977) [Russ. Math. Surv. **32**:6, 1-65

(1977)]; G. Benettin, L. Galgani, A. Giorgilli, Cel. Mech. **37**, 1 (1985); A. Giorgilli and L. Galgani, Cel. Mech. **37**, 95 (1985); G. Benettin and G. Gallavotti, Journ. Stat. Phys. **44**, 293-338 (1986); P. Lochak, Russ. Math. Surv. **47**, 57-133 (1992); J. Pöschel, Math. Z. **213**, 187-216 (1993).

Applications to statistical mechanics: G. Benettin, in: *Boltzmann's legacy 150 years afrer his birth*, Atti Accad. Nazionale dei Lincei **131**, 89-105 (1997); G. Benettin, A. Carati and P. Sempio, Journ. Stat. Phys. **73**, 175-192 (1993); G. Benettin, A. Carati and G. Gallavotti, Nonlinearity **10**, 479-505 (1997); G. Benettin, A. Carati e F. Fassò, Physica D **104**, 253-268 (1997); G. Benettin, P. Hjorth and P. Sempio, *Exponentially long equilibrium times in a one dimensional collisional model of a classical gas*, in print in Journ. Stat. Phys.

Applications to the rigid body: G. Benettin and F. Fassò, Nonlinearity **9**, 137-186 (1996); G. Benettin, F. Fassò e M. Guzzo, Nonlinearity **10**, 1695-1717 (1997).

Applications to elliptic equilibria (recent nonisochronous approach): F. Fassò, M. Guzzo e G. Benettin, Comm. Math. Phys. **197**, 347-360 (1998); L. Niederman, *Nonlinear stability around an elliptic equilibrium point in an Hamiltonian system*, preprint (1997). M. Guzzo, F. Fasso' e G. Benettin, Math. Phys. Electronic Journal, Vol. **4**, paper 1 (1998); G. Benettin, F. Fassò e M. Guzzo, *Nekhoroshev-stability of L4 and L5 in the spatial restricted three-body problem*, in print in Regular and Chaotic Dynamics.

b) KAM-theory (6 lectures in English)
Prof. Hakan ELIASSON (Royal Institute of Technology, Stockholm, Sweden)
Abstract

Quasi-periodic motions (or invariant tori) occur naturally when systems with periodic motions are coupled. The perturbation problem for these motions involves small divisors and the most natural way to handle this difficulty is by the quadratic convergence given by Newton's method. A basic problem is how to implement this method in a particular perturbative situation. We shall describe this difficulty, its relation to linear quasi-periodic systems and the way given by KAM-theory to overcome it in the most generic case. Additional difficulties occur for systems with elliptic lower dimensional tori and even more for systems with weak non-degeneracy.

We shall also discuss the difference between initial value and boundary value problems and their relation to the Lindstedt and the Poincaré-Lindstedt series.

The classical books Lectures in Celestial Mechanics by Siegel and Moser (Springer 1971) and Stable and Random Motions in Dynamical Systems by Moser (Princeton University Press 1973) are perhaps still the best introductions to KAM-theory. The development up to middle 80's is described by Bost in a Bourbaki Seminar (no. 6 1986). After middle 80's a lot of work have been devoted to elliptic lower dimensional tori, and to the study of systems with weak non-degeneracy starting with the work of Cheng and Sun (for example *"Existence of KAM-tori in Degenerate Hamiltonian systems"*, J. Diff. Eq. 114, 1994). Also on linear quasi-periodic systems there has been some progress which is described in my article *"Reducibility and point spectrum for quasi-periodic skew-products"*, Proceedings of the ICM, Berlin volume II 1998.

c) The Adiabatic Invariant in Classical Dynamics: Theory and applications (6 lectures in English).
Prof. Jacques HENRARD (Facultés Universitaires Notre Dame de la Paix, Namur, Belgique).
Abstract

The adiabatic invariant theory applies essentially to oscillating non-autonomous Hamiltonian systems when the time dependance is considerably slower than the oscillation periods. It describes "easy to compute" and "dynamicaly meaningful" quasi-invariants by which on can predict the approximate evolution of the system on very large time scales. The theory makes use and may serve as an illustration of several classical results of Hamiltonian theory.

1) Classical Adiabatic Invariant Theory (Including an introduction to angle-action variables)

2) Classical Adiabatic Invariant Theory (continued) and some applications (including an introduction to the "magnetic bottle")

3) Adiabatic Invariant and Separatrix Crossing (Neo-adiabatic theory)

4) Applications of Neo-Adiabatic Theory: Resonance Sweeping in the Solar System

5) The chaotic layer of the "Slowly Modulated Standard Map"

References:

J.R. Cary, D.F. Escande, J.L. Tennison: Phys.Rev. A, 34, 1986, 3256-4275

J. Henrard, in "*Dynamics reported*" (n=B02- newseries), Springer Verlag 1993; pp 117-235)

J. Henrard: in "*Les méthodes moderne de la mécanique céleste*" (Benest et Hroeschle eds), Edition Frontieres, 1990, 213-247

J. Henrard and A. Morbidelli: Physica D, 68, 1993, 187-200.

d) **Some aspects of qualitative theory of Hamiltonian PDEs** (6 lectures in English).

Prof. Sergei B. KUKSIN (Heriot-Watt University, Edinburgh, and Steklov Institute, Moscow)

Abstract.

I) Basic properties of Hamiltonian PDEs. Symplectic structures in scales of Hilbert spaces, the notion of a Hamiltonian PDE, properties of flow-maps, etc.

II) Around Gromov's non-squeezing property. Discussions of the finite-dimensional Gromov's theorem, its version for PDEs and its relevance for mathematical physics, infinite-dimensional symplectic capacities.

III) Damped Hamiltonian PDEs and the turbulence-limit. Here we establish some qualitative properties of PDEs of the form <non-linear Hamiltonian PDE>+<small linear damping> and discuss their relations with theory of decaying turbulence

Parts I)-II) will occupy the first three lectures, Part III - the last two.

References

[1] S.K., *Nearly Integrable Infinite-dimensional Hamiltonian Systems*. LNM 1556, Springer 1993.

[2] S.K., *Infinite-dimensional symplectic capacities and a squeezing theorem for Hamiltonian PDE's*. Comm. Math. Phys. 167 (1995), 531-552.

[3] Hofer H., Zehnder E., *Symplectic invariants and Hamiltonian dynamics*. Birkhauser, 1994.

[4] S.K. *Oscillations in space-periodic nonlinear Schroedinger equations*. Geometric and Functional Analysis 7 (1997), 338-363.

For I) see [1] (Part 1); for II) see [2,3]; for III) see [4]."

e) **An overview on some problems in Celestial Mechanics** (6 lectures in English)

Prof. Carles SIMO' (Universidad de Barcelona, Spagna)

Abstract

1. Introduction. The N-body problem. Relative equilibria. Collisions.

2. The 3D restricted three-body problem. Libration points and local stability analysis.

3. Periodic orbits and invariant tori. Numerical and symbolical computation.

4. Stability and practical stability. Central manifolds and the related stable/unstable manifolds. Practical confiners.

5. The motion of spacecrafts in the vicinity of the Earth-Moon system. Results for improved models. Results for full JPL models.

References:

C. Simò, *An overview of some problems in Celestial Mechanics*, available at http://www-ma1.upc.es/escorial .

Click of "curso completo" of Prof. Carles Simó

Applications

Deadline for application: **May 15, 1999.**

Applicants are requested to submit, along with their application, a scientific curriculum and a letter of recommendation.

CIME will be able to partially support some of the youngest participants. Those who plan to apply for support have to mention it explicitly in the application form.

Attendance

No registration fee is requested. Lectures will be held at Cetraro on July 1, 2, 3, 4, 5, 6, 7, 8, 9, 10. Participants are requested to register on June 30, 1999.

Site and lodging

The session will be held at Grand Hotel S. Michele at Cetraro (Cosenza), Italy. Prices for full board (bed and meals) are roughly 150.000 italian liras p.p. day in a single room, 130.000 italian liras in a double room. Cheaper arrangements for multiple lodging in a residence are avalaible. More detailed information may be obtained from the Direction of the hotel (tel. +39-098291012, Fax +39-098291430, email: sanmichele@antares.it.

Further information on the hotel at the web page www.sanmichele.it

Arrigo CELLINA
CIME Director

Vincenzo VESPRI
CIME Secretary

Fondazione C.I.M.E. c/o Dipartimento di Matematica U. Dini Viale Morgagni, 67/A - 50134 FIRENZE (ITALY) Tel. +39-55-434975 / +39-55-4237123 FAX +39-55-434975 / +39-55-4222695 E-mail CIME@UDINI.MATH.UNIFI.IT

Information on CIME can be obtained on the system World-Wide-Web on the file HTTP: //WWW.MATH.UNIFI.IT/CIME/WELCOME.TO.CIME.

FONDAZIONE C.I.M.E.
CENTRO INTERNAZIONALE MATEMATICO ESTIVO
INTERNATIONAL MATHEMATICAL SUMMER CENTER

"Global Theory of Minimal Surfaces in Flat Spaces"

is the subject of the fourth 1999 C.I.M.E. Session.

The session, sponsored by the Consiglio Nazionale delle Ricerche (C.N.R.), the Ministero dell'Università e della Ricerca Scientifica e Tecnologica (M.U.R.S.T.) and the European Community, will take place, under the scientific direction of Professor Gian Pietro PIROLA (Un. di Pavia), at Ducal Palace of Martina Franca (Taranto), from July 7 to July 15, 1999.

Courses

a) Asymptotic geometry of properly embedded minimal surfaces (6 lecture in English)

Prof. William H. MEEKS, III (Un. of Massachusetts, Amherst, USA).

Abstract:

In recent years great progress has been made in understanding the asymptotic geometry of properly embedded minimal surfaces. The first major result of this type was the solution of the generalized Nitsch conjecture by P. Collin, based on earlier work by Meeks and Rosenberg. It follows from the resolution of this conjecture that whenever M is a properly embedded minimal surface with more than one end and $E \subset M$ is an annular end representative, then E has finite total curvature and is asymptotic to an end of a plan or catenoid. Having finite total curvature in the case of an annular end is equivalent to proving the end has quadratic area growth with respect to the radial function r. Recently Collin, Kusner, Meeks and Rosenberg have been able to prove that any middle end of M, even one with infinite genus, has quadratic area growth. It follows from this result that middle ends are never limit ends and hence M can only have one or two limit ends which must be top or bottom ends. With more work it is shown that the middle ends of M stay a bounded distance from a plane or an end of a catenoid.

The goal of my lectures will be to introduce the audience to the concepts in the theory o f properly embedded minimal surfaces needed to understand the above results and to understand some recent classification theorems on proper minimal surfaces of genus 0 in flat three-manifolds.

References

1) H. Rosenberg, *Some recent developments in the theory of properly embedded minimal surfaces in E*, Asterisque **206**, (19929, pp. 463-535;

2) W. Meeks & H. Rosenberg, *The geometry and conformal type of properly embedded minimal surfaces in E*, Invent.Math. **114**, (1993), pp. 625-639;

3) W. Meeks, J. Perez & A. Ros, *Uniqueness of the Riemann minimal examples*, Invent. Math. **131**, (1998), pp. 107-132;

4) W. Meeks & H. Rosenberg, *The geometry of periodic minimal surfaces*, Comm. Math. Helv. **68**, (1993), pp. 255-270;

5) P. Collin, *Topologie et courbure des surfaces minimales proprement plongees dans E*, Annals of Math. **145**, (1997), pp. 1-31;

6) H. Rosenberg, *Minimal surfaces of finite type*, Bull. Soc. Math. France **123**, (1995), pp. 351-359;

7) Rodriquez & H. Rosenberg, *Minimal surfaces in E with one end and bounded curvature*, Manusc. Math. **96**, (1998), pp. 3-9.

b) Properly embedded minimal surfaces with finite total curvature (6 lectures in English)

Prof. Antonio ROS (Universidad de Granada, Spain)

Abstract:

Among properly embedded minimal surfaces in Euclidean 3-space, those that have finite total curvature form a natural and important subclass. These surfaces have finitely many ends which are all parallel and asymptotic to planes or catenoids. Although the structure of the space M of surfaces of this type which have a fixed topology is not well understood, we have a certain number of partial results and some of them will be explained in the lectures we will give.

The first nontrivial examples, other than the plane and the catenoid, were constructed only ten years ago by Costa, Hoffman and Meeks. Schoen showed that if the surface has two ends, then it must be a catenoid and López and Ros proved that the only surfaces of genus zero are the plane and the catenoid. These results give partial answers to an interesting open problem: decide which topologies are supported by this kind of surfaces. Ros obtained certain compactness properties of M. In general this space is known to be noncompact but he showed that M is compact for some fixed topologies. Pérez and Ros studied the local structure of M around a nondegenerate surface and they proved that around these points the moduli space can be naturally viewed as a Lagrangian submanifold of the complex Euclidean space.

In spite of that analytic and algebraic methods compete to solve the main problems in this theory, at this moment we do not have a satisfactory idea of the behaviour of the moduli space M. Thus the above is a good research field for young geometers interested in minimal surfaces.

References

1) C. Costa, *Example of a compete minimal immersion in $\mathbb{R}^3$ of genus one and three embedded ends*, Bull. SOc. Bras. Math. **15**, (1984), pp. 47-54;

2) D. Hoffman & H. Karcher, *Complete embedded minimal surfaces of finite total curvature*, R. Osserman ed., Encyclopedia of Math., vol. of Minimal Surfaces, **5-90**, Springer 1997;

3) D. Hoffman & W. H. Meeks III, *Embedded minimal surfaces of finite topology*, Ann. Math. **131**, (1990), pp. 1-34;

4) F. J. Lòpez & A. Ros, *On embedded minimal surfaces of genus zero*, J. Differential Geometry **33**, (1991), pp. 293-300;

5) J. P. Perez & A. Ros, *Some uniqueness and nonexistence theorems for embedded minimal surfaces*, Math. Ann. **295** (3), (1993), pp. 513-525;

6) J. P. Perez & A. Ros, *The space of properly embedded minimal surfaces with finite total curvature*, Indiana Univ. Math. J. **45** 1, (1996), pp.177-204.

c) Minimal surfaces of finite topology properly embedded in E (Euclidean 3-space).(6 lectures in English)

Prof. Harold ROSENBERG (Univ. Paris VII, Paris, France)

Abstract:

We will prove that a properly embedded minimal surface in E of finite topology and at least two ends has finite total curvature. To establish this we first prove that each annular end of such a surface M can be made transverse to the horizontal planes

(after a possible rotation in space), [Meeks-Rosenberg]. Then we will prove that such an end has finite total curvature [Pascal Collin]. We next study properly embedded minimal surfaces in E with finite topology and one end. The basic unsolved problem is to determine if such a surface is a plane or helicoid when simply connected. We will describe partial results. We will prove that a properly immersed minimal surface of finite topology that meets some plane in a finite number of connected components, with at most a finite number of singularities, is of finite conformal type. If in addition the curvature is bounded, then the surface is of finite type. This means M can be parametrized by meromorphic data on a compact Riemann surface. In particular, under the above hypothesis, M is a plane or helicoid when M is also simply connected and embedded. This is work of Rodriquez- Rosenberg, and Xavier. If time permits we will discuss the geometry and topology of constant mean curvature surfaces properly embedded in E.

References

1) H. Rosenberg, *Some recent developments in the theory of properly embedded minimal surfaces in E*, Asterique **206**, (1992), pp. 463-535;

2) W.Meeks & H. Rosenberg, *The geometry and conformal type of properly embedded minimal surfaces in E*, Invent. **114**, (1993), pp.625-639;

3) P. Collin, *Topologie et courbure des surfaces minimales proprement plongées dans E*, Annals of Math. **145**, (1997), pp. 1-31

4) H. Rosenberg, *Minimal surfaces of finite type*, Bull. Soc. Math. France **123**, (1995), pp. 351-359;

5) Rodriquez & H. Rosenberg, *Minimal surfaces in E with one end and bounded curvature*, Manusc. Math. **96**, (1998), pp. 3-9.

Applications

Those who want to attend the Session should fill in an application to the C.I.M.E Foundation at the address below, **not later than** May 15, 1999.

An important consideration in the acceptance of applications is the scientific relevance of the Session to the field of interest of the applicant.

Applicants are requested, therefore, to submit, along with their application, a scientific curriculum and a letter of recommendation.

Participation will only be allowed to persons who have applied in due time and have had their application accepted.

CIME will be able to partially support some of the youngest participants. Those who plan to apply for support have to mention it explicitly in the application form

Attendance

No registration fee is requested. Lectures will be held at Martina Franca on July 7, 8, 9, 10, 11, 12, 13, 14, 15. Participants are requested to register on July 6, 1999.

Site and lodging

Martina Franca is a delightful baroque town of white houses of Apulian spontaneous architecture. Martina Franca is the major and most aristocratic centre of the Murgia dei Trulli standing on an hill which dominates the well known Itria valley spotted with Trulli conical dry stone houses which go back to the 15th century. A masterpiece of baroque architecture is the Ducal palace where the workshop will be hosted. Martina Franca is part of the province of Taranto, one of the major centres of Magna Grecia, particularly devoted to mathematics. Taranto houses an outstanding museum of Magna Grecia with fabulous collections of gold manufactures.

Lecture Notes

Lecture notes will be published as soon as possible after the Session.

Arrigo CELLINA   Vincenzo VESPRI
CIME Director   CIME Secretary

Fondazione C.I.M.E. c/o Dipartimento di Matematica U. Dini Viale Morgagni, 67/A - 50134 FIRENZE (ITALY) Tel. +39-55-434975 / +39-55-4237123 FAX +39-55-434975 / +39-55-4222695 E-mail CIME@UDINI.MATH.UNIFI.IT

Information on CIME can be obtained on the system World-Wide-Web on the file HTTP: //WWW.MATH.UNIFI.IT/CIME/WELCOME.TO.CIME.

FONDAZIONE C.I.M.E.
CENTRO INTERNAZIONALE MATEMATICO ESTIVO
INTERNATIONAL MATHEMATICAL SUMMER
CENTER

"Direct and Inverse Methods in Solving Nonlinear Evolution Equations"

is the subject of the fifth 1999 C.I.M.E. Session.

The session, sponsored by the Consiglio Nazionale delle Ricerche (C.N.R.), the Ministero dell'Università e della Ricerca Scientifica e Tecnologica (M.U.R.S.T.) and the European Community, will take place, under the scientific direction of Professor Antonio M. Greco (Università di Palermo), at Grand Hotel San Michele,Cetraro (Cosenza), from September 8 to September 15, 1999.

a) **Exact solutions of nonlinear PDEs by singularity analysis** (6 lectures in English)

Prof. Robert CONTE (Service de physique de l'état condensé, CEA Saclay, Gif-sur-Yvette Cedex, France)

Abstract

1) Criteria of integrability : Lax pair, Darboux and Bäcklund transformations. Partial integrability, examples. Importance of involutions.

2) The Painlevé test for PDEs in its invariant version.

3) The "truncation method" as a Darboux transformation, ODE and PDE situations.

4) The one-family truncation method (WTC), integrable (Korteweg-de Vries, Boussinesq, Hirota-Satsuma, Sawada-Kotera) and partially integrable (Kuramoto-Sivashinsky) cases.

5) The two-family truncation method, integrable (sine-Gordon, mKdV, Broer-Kaup) and partially integrable (complex Ginzburg-Landau and degeneracies) cases.

6) The one-family truncation method based on the scattering problems of Gambier: BT of Kaup-Kupershmidt and Tzitzéica equations.

References

References are divided into three subsets: prerequisite (assumed known by the attendant to the school), general (not assumed known, pedagogical texts which would greatly benefit the attendant if they were read before the school), research (research papers whose content will be exposed from a synthetic point of view during the course).

Prerequisite bibliography.

The following subjects will be assumed to be known : the Painlevé property for nonlinear ordinary differential equations, and the associated Painlevé test.

Prerequisite recommended texts treating these subjects are

[P.1] E. Hille, *Ordinary differential equations in the complex domain* (J. Wiley and sons, New York, 1976).

[P.2] R. Conte, *The Painlevé approach to nonlinear ordinary differential equations, The Painlevé property, one century later*, 112 pages, ed. R. Conte, CRM series in mathematical physics (Springer, Berlin, 1999). Solv-int/9710020.

The interested reader can find many applications in the following review, which should not be read before [P.2] :

[P.3] A. Ramani, B. Grammaticos, and T. Bountis, *The Painlevé property and singularity analysis of integrable and nonintegrable systems*, Physics Reports 180 (1989) 159–245.

A text to be avoided by the beginner is Ince's book, the ideas are much clearer in Hille's book.

There exist very few pedagogical texts on the subject of this school.

A general reference, covering all the above program, is the course delivered at a Cargèse school in 1996 :

[G.1] M. Musette, *Painlevé analysis for nonlinear partial differential equations, The Painlevé property, one century later*, 65 pages, ed. R. Conte, CRM series in mathematical physics (Springer, Berlin, 1999). Solv-int/9804003.

A short subset of [G.1], with emphasis on the ideas, is the conference report

[G.2] R. Conte, *Various truncations in Painlevé analysis of partial differential equations*, 16 pages, Nonlinear dynamics : integrability and chaos, ed. M. Daniel, to appear (Springer? World Scientific?). Solv-int/9812008. Preprint S98/047.

Research papers.

[R.2] J. Weiss, M. Tabor and G. Carnevale, *The Painlevé property for partial differential equations*, J. Math. Phys. 24 (1983) 522–526.

[R.3] Numerous articles of Weiss, from 1983 to 1989, all in J. Math. Phys. [singular manifold method].

[R.4] M. Musette and R. Conte, *Algorithmic method for deriving Lax pairs from the invariant Painlevé analysis of nonlinear partial differential equations*, J. Math. Phys. 32 (1991) 1450–1457 [invariant singular manifold method].

[R.5] R. Conte and M. Musette, *Linearity inside nonlinearity: exact solutions to the complex Ginz-burg-Landau equation*, Physica D 69 (1993) 1–17 [Ginzburg-Landau].

[R.6] M. Musette and R. Conte, *The two–singular manifold method, I. Modified KdV and sine-Gordon equations*, J. Phys. A 27 (1994) 3895–3913 [Two–singular manifold method].

[R.7] R. Conte, M. Musette and A. Pickering, *The two–singular manifold method, II. Classical Boussinesq system*, J. Phys. A 28 (1995) 179–185 [Two–singular manifold method].

[R.8] A. Pickering, *The singular manifold method revisited*, J. Math. Phys. 37 (1996) 1894–1927 [Two–singular manifold method].

[R.9] M. Musette and R. Conte, *Bäcklund transformation of partial differential equations from the Painlevé-Gambier classification, I. Kaup-Kupershmidt equation*, J. Math. Phys. 39 (1998) 5617–5630. [Lecture 6].

[R.10] R. Conte, M. Musette and A. M. Grundland, *Bäcklund transformation of partial differential equations from the Painlevé-Gambier classification, II. Tzitzéica equation*, J. Math. Phys. 40 (1999) to appear. [Lecture 6].

b) Integrable Systems and Bi-Hamiltonian Manifolds (6 lectures in English)

Prof. Franco MAGRI (Università di Milano, Milano, Italy)

Abstract

1) Integrable systems and bi-hamiltonian manifolds according to Gelfand and Zakharevich.

2) Examples: KdV, KP and Sato's equations.

3) The rational solutions of KP equation.

4) Bi-hamiltonian reductions and completely algebraically integrable systems.

5) Connections with the separabilty theory.

6) The τ function and the Hirota's identities from a bi-hamiltonian point of view.

References

1) R. Abraham, J.E. Marsden, *Foundations of Mechanics*,Benjamin/Cummings, 1978

2) P. Libermann, C. M. Marle, *Symplectic Geometry and Analytical Mechanics*, Reidel Dordrecht, 1987

3) L. A. Dickey, *Soliton Equations and Hamiltonian Systems*, World Scientific, Singapore, 1991, Adv. Series in Math. Phys Vol. 12

4) I. Vaisman, *Lectures on the Geometry of Poisson Manifolds*, Progress in Math., Birkhäuser, 1994

5) P. Casati, G. Falqui, F. Magri, M. Pedroni (1996), *The KP theory revisited. I,II,III,IV*. Technical Reports, SISSA/2,3,4,5/96/FM, SISSA/ISAS, Trieste, 1995

c) Hirota Methods for non Linear Differential and Difference Equations

(6 lectures in English)

Prof. Junkichi SATSUMA (University of Tokyo, Tokyo, Japan)

Abstract

1) Introduction;

2) Nonlinear differential systems;

3) Nonlinear differential-difference systems;

4) Nonlinear difference systems;

5) Sato theory;

6) Ultra-discrete systems.

References.

1) M.J.Ablowitz and H.Segur, *Solitons and the Inverse Scattering Transform*, (SIAM, Philadelphia, 1981).

2) Y.Ohta, J.Satsuma, D.Takahashi and T.Tokihiro, " Prog. Theor. Phys. Suppl. No.94, p.210-241 (1988)

3) J.Satsuma, *Bilinear Formalism in Soliton Theory*, Lecture Notes in Physics No.495, Integrability of Nonlinear Systems, ed. by Y.Kosmann-Schwarzbach, B.Grammaticos and K.M.Tamizhmani p.297-313 (Springer, Berlin, 1997).

d) Lie Groups and Exact Solutions of non Linear Differential and Difference Equations (6 lectures in English)

Prof. Pavel WINTERNITZ (Universitè de Montreal, Montreal, Canada) 3J7

Abstract

1) Algorithms for calculating the symmetry group of a system of ordinary or partial differential equations. Examples of equations with finite and infinite Lie point symmetry groups;

2) Applications of symmetries. The method of symmetry reduction for partial differential equations. Group classification of differential equations;

3) Classification and identification of Lie algebras given by their structure constants. Classification of subalgebras of Lie algebras. Examples and applications;

4) Solutions of ordinary differential equations. Lowering the order of the equation. First integrals. Painlevè analysis and the singularity structure of solutions;

5) Conditional symmetries. Partially invariant solutions.

6) Lie symmetries of difference equations.

References.

1) P. J. Olver, *Applications of Lie Groups to Differential Equations*, Springer,1993,

2) P. Winternitz, *Group Theory and Exact Solutions of Partially Integrable Differential Systems*, in Partially Integrable Evolution Equations in Physics, Kluwer, Dordrecht, 1990, (Editors R.Conte and N.Boccara).

3) P. Winternitz, in *"Integrable Systems, Quantum Groups and Quantum Field Theories"*, Kluwer, 1993 (Editors L .A. Ibort and M. A. Rodriguez).

Applications

Those who want to attend the Session should fill in an application to the C.I.M.E Foundation at the address below, **not later than May 30**, 1999.

An important consideration in the acceptance of applications is the scientific relevance of the Session to the field of interest of the applicant.

Applicants are requested, therefore, to submit, along with their application, a scientific curriculum and a letter of recommendation.

Participation will only be allowed to persons who have applied in due time and have had their application accepted.

CIME will be able to partially support some of the youngest participants. Those who plan to apply for support have to mention it explicitly in the application form.

Attendance

No registration fee is requested. Lectures will be held at Cetraro on September 8, 9, 10, 11, 12, 13, 14, 15. Participants are requested to register on September 7, 1999.

Site and lodging

The session will be held at Grand Hotel S. Michele at Cetraro (Cosenza), Italy.
Prices for full board (bed and meals) are roughly 150.000 italian liras p.p. day in a single room, 130.000 italian liras in a double room. Cheaper arrangements for multiple lodging in a residence are avalaible. More detailed informations may be obtained from the Direction of the hotel (tel. +39-098291012, Fax +39-098291430, email: sanmichele@antares.it.

Further information on the hotel at the web page www.sanmichele.it

Lecture Notes

Lecture notes will be published as soon as possible after the Session.

Arrigo CELLINA   Vincenzo VESPRI
CIME Director   CIME Secretary

Fondazione C.I.M.E. c/o Dipartimento di Matematica U. Dini Viale Morgagni, 67/A
- 50134 FIRENZE (ITALY) Tel. +39-55-434975 / +39-55-4237123 FAX
+39-55-434975 / +39-55-4222695 E-mail CIME@UDINI.MATH.UNIFI.IT

Information on CIME can be obtained on the system World-Wide-Web on the file
HTTP: //WWW.MATH.UNIFI.IT/CIME/WELCOME.TO.CIME.

Printing: Druckhaus Beltz, Hemsbach
Binding: Buchbinderei Schäffer, Grünstadt

4. Lecture Notes are printed by photo-offset from the master-copy delivered in camera-ready form by the authors. Springer-Verlag provides technical instructions for the preparation of manuscripts. Macro packages in T_EX, L^AT_EX2e, L^AT_EX2.09 are available from Springer's web-pages at

http://www.springer.de/math/authors/b-tex.html.

Careful preparation of the manuscripts will help keep production time short and ensure satisfactory appearance of the finished book.

The actual production of a Lecture Notes volume takes approximately 12 weeks.

5. Authors receive a total of 50 free copies of their volume, but no royalties. They are entitled to a discount of 33.3% on the price of Springer books purchase for their personal use, if ordering directly from Springer-Verlag.

Commitment to publish is made by letter of intent rather than by signing a formal contract. Springer-Verlag secures the copyright for each volume. Authors are free to reuse material contained in their LNM volumes in later publications: A brief written (or e-mail) request for formal permission is sufficient.

Addresses:

Professor F. Takens, Mathematisch Instituut,
Rijksuniversiteit Groningen, Postbus 800,
9700 AV Groningen, The Netherlands
E-mail: F.Takens@math.rug.nl

Professor B. Teissier, DMI, École Normale Supérieure
45, rue d'Ulm,
F-7500 Paris, France
E-mail: Teissier@ens.fr

Springer-Verlag, Mathematics Editorial, Tiergartenstr. 17,
D-69121 Heidelberg, Germany,
Tel.: *49 (6221) 487-701
Fax: *49 (6221) 487-355
E-mail: lnm@Springer.de